STUDENT'S SOLUTIONS MANUAL

FINITE MATHEMATICS
BITTINGER/CROWN

Judith A. Penna

with the assistance of Barbara L. Johnson

ADDISON-WESLEY PUBLISHING COMPANY

Reading, Massachusetts • Menlo Park, California • New York
Don Mills, Ontario • Wokingham, England • Amsterdam • Bonn
Sydney • Singapore • Tokyo • Madrid • San Juan

ISBN 0-201-10816-X

Reproduced by Addison-Wesley from camera-ready copy supplied by the authors.

ABCDEFGHIJ-BA-898

TABLE OF CONTENTS

Special thanks are extended to Patsy Hammond for her
excellent typing. Her skill, patience, and efficiency
made the author's work much easier.

Exercise Set 1.1

1. Yes, there is a 4th of July in England, but it is not a holiday.

3. You can swim halfway into a lake. After that you are swimming out of the lake.

5. All twelve months have 28 days.

7. The pills would last one hour. For example, if the first pill is taken at 8:00, then the second is taken at 8:30, the third is taken at 9:00, and one hour has elapsed.

9.-19. The answers to familiarize questions are open-ended. They do not have absolute answers for the most part.

Exercise Set 1.2

1. $5^3 = 5 \cdot 5 \cdot 5$, or 125

   $\underbrace{}_{\text{3 factors}}$

3. $(-7)^2 = \underbrace{(-7)(-7)}_{\text{2 factors}}$, or 49

5. $(1.01)^2 = \underbrace{(1.01)(1.01)}_{\text{2 factors}}$, or 1.0201

7. $\left(\frac{1}{2}\right)^4 = \underbrace{\frac{1}{2} \cdot \frac{1}{2} \cdot \frac{1}{2} \cdot \frac{1}{2}}_{\text{4 factors}}$, or $\frac{1}{16}$

9. $(6x)^0 = 1$ $a^0 = 1$, for any nonzero real number a

11. $t^1 = t$ $a^1 = a$, for any real number a

13. $\left(\frac{1}{3}\right)^0 = 1$ $a^0 = 1$, for any nonzero real number a

15. 3^{-2}

    $= \frac{1}{3^2}$ $a^{-n} = \frac{1}{a^n}$, for any nonzero real number a

    $= \frac{1}{3 \cdot 3}$ $\underbrace{3^2 = 3 \cdot 3}_{\text{2 factors}}$

    $= \frac{1}{9}$

    Note: 3^{-2}, or $\frac{1}{9}$, is a positive number.

17. $\left(\frac{1}{2}\right)^{-3}$

    $= \frac{1}{\left(\frac{1}{2}\right)^3}$ $a^{-n} = \frac{1}{a^n}$ for any nonzero real number a

    $= \frac{1}{\frac{1}{2} \cdot \frac{1}{2} \cdot \frac{1}{2}}$

    $= \frac{1}{\frac{1}{8}}$

    $= 1 \cdot \frac{8}{1}$ Multiplying by the reciprocal

    $= 8$

19. $10^{-1} = \frac{1}{10^1} = \frac{1}{10}$, or 0.1

21. $e^{-b} = \frac{1}{e^b}$

23. $b^{-1} = \frac{1}{b^1} = \frac{1}{b}$

25. To multiply when bases are the same, add the exponents.
    $x^2 \cdot x^3 = x^{2+3} = x^5$ $a^n \cdot a^m = a^{n+m}$

    $\begin{bmatrix} \text{Think:} \\ x^2 \cdot x^3 = \underbrace{x \cdot x}_{\substack{2 \\ \text{factors}}} \cdot \underbrace{x \cdot x \cdot x}_{\substack{3 \\ }} \\ = \underbrace{x \cdot x \cdot x \cdot x \cdot x}_{\text{5 factors}} \\ = x^5 \end{bmatrix}$

27. $x^{-7} \cdot x$
    $= x^{-7} \cdot x^1$ $x = x^1$
    $= x^{-7+1}$ $a^n \cdot a^m = a^{n+m}$
    $= x^{-6}$, or $\frac{1}{x^6}$

29. $5x^2 \cdot 7x^3$
    $= 5 \cdot 7 \cdot x^2 \cdot x^3$
    $= 35x^{2+3}$ $a^n \cdot a^m = a^{n+m}$
    $= 35x^5$

31. $x^{-4} \cdot x^7 \cdot x$
    $= x^{-4} \cdot x^7 \cdot x^1$ $x = x^1$
    $= x^{-4+7+1}$ $a^n \cdot a^m = a^{n+m}$
    $= x^4$

33. $e^{-t} \cdot e^t$
    $= e^{-t+t}$ $a^n \cdot a^m = a^{n+m}$
    $= e^0$
    $= 1$ $a^0 = 1$, for any nonzero real number a

35. To divide when the bases are the same subtract the exponent in the denominator from the exponent in the numerator.

$\frac{x^5}{x^2} = x^{5-2} = x^3$ $\frac{a^n}{a^m} = a^{n-m}$

$$
\begin{array}{ll}
\text{Think:} & \\
\frac{x^5}{x^2} = \frac{x \cdot x \cdot x \cdot x \cdot x}{x \cdot x} & \text{5 factors} \\
& \text{2 factors} \\
= \frac{x \cdot x}{x \cdot x} \cdot \frac{x \cdot x \cdot x}{1} & \text{Factoring the fraction} \\
= 1 \cdot \frac{x \cdot x \cdot x}{1} & \frac{x \cdot x}{x \cdot x} = 1 \\
= \frac{x \cdot x \cdot x}{1} & \\
= x^3 &
\end{array}
$$

37. $\frac{x^2}{x^5}$

$= x^{2-5}$ $\frac{a^n}{a^m} = a^{n-m}$

$= x^{-3}$, or $\frac{1}{x^3}$

39. $\frac{e^k}{e^k}$

$= e^{k-k}$ $\frac{a^n}{a^m} = a^{n-m}$

$= e^0$

$= 1$ $a^0 = 1$, for any nonzero real number a

41. $\frac{e^t}{e^4} = e^{t-4}$ $\frac{a^n}{a^m} = a^{n-m}$

43. $\frac{t^6}{t^{-8}}$

$= t^{6-(-8)}$ $\frac{a^n}{a^m} = a^{n-m}$

$= t^{14}$ $6 - (-8) = 6 + 8 = 14$

45. $\frac{t^{-9}}{t^{-11}}$

$= t^{-9-(-11)}$ $\frac{a^n}{a^m} = a^{n-m}$

$= t^2$ $-9 - (-11) = -9 + 11 = 2$

47. To raise a power to a power, multiply the exponents.

$(t^{-2})^3 = t^{-2 \cdot 3} = t^{-6}$, or $\frac{1}{t^6}$ $(a^n)^m = a^{n \cdot m}$

$$
\begin{array}{l}
\text{Think:} \\
(t^{-2})^3 = (t^{-2})(t^{-2})(t^{-2}) \\
\quad\quad\quad = t^{-2+(-2)+(-2)} \\
\quad\quad\quad = t^{-6}
\end{array}
$$

49. $(e^x)^4$

$= e^{x \cdot 4}$ $(a^n)^m = a^{n \cdot m}$

$= e^{4x}$

51. $(2x^2y^4)^3$

$= 2^3 \cdot (x^2)^3 \cdot (y^4)^3$ Each factor is raised to the third power

$= 2^3 x^6 y^{12}$ $(a^n)^m = a^{n \cdot m}$

$= 8x^6 y^{12}$ $2^3 = 8$

53. $(3x^{-2}y^{-5}z^4)^{-4}$

$= 3^{-4} \cdot (x^{-2})^{-4} \cdot (y^{-5})^{-4} \cdot (z^4)^{-4}$

   Each factor is raised to the −4 power

$= 3^{-4} x^8 y^{20} z^{-16}$ $(a^n)^m = a^{n \cdot m}$

$= \frac{1}{81} x^8 y^{20} z^{-16}$ $3^{-4} = \frac{1}{3^4} = \frac{1}{81}$

or $\frac{x^8 y^{20}}{81 z^{16}}$

55. $(-3x^{-8}y^7z^2)^2$

$= (-3)^2 \cdot (x^{-8})^2 \cdot (y^7)^2 \cdot (z^2)^2$

   Each factor is raised to the second power

$= (-3)^2 x^{-16} y^{14} z^4$ $(a^n)^m = a^{n \cdot m}$

$= 9x^{-16} y^{14} z^4$ $(-3)^2 = (-3) \cdot (-3) = 9$

or $\frac{9y^{14}z^4}{x^{16}}$

57. $5(x - 7)$

$= 5 \cdot x - 5 \cdot 7$ Using a distributive law

$= 5x - 35$

59. $x(1 - t)$

$= x \cdot 1 - x \cdot t$ Using a distributive law

$= x - xt$

61. $(x - 5)(x - 2)$

$= (x - 5)x - (x - 5)2$ Using a distributive law

$= x \cdot x - 5 \cdot x - x \cdot 2 + 5 \cdot 2$ Using a distributive law

$= x^2 - 5x - 2x + 10$

$= x^2 - 7x + 10$ Collecting like terms

Also consider the following method.

61. (continued)

To multiply two binomials, we multiply each term of one by every term of the other.

$$(A + B)(C + D) = AC + AD + BC + BD$$

1. Multiply First terms: AC
2. Multiply Outside terms: AD
3. Multiply Inside terms: BC
4. Multiply Last terms: BD

FOIL

$(x - 5)(x - 2)$

$\quad$ F $\quad$ O $\quad$ I $\quad$ L
$= x \cdot x + x(-2) + (-5)x + (-5)(-2)$
$= x^2 - 2x - 5x + 10$
$= x^2 - 7x + 10 \qquad$ Collecting like terms

63.

$(a - b)(a^2 + ab + b^2)$
$= (a - b)a^2 + (a - b)ab + (a - b)b^2$
$\qquad\qquad$ Using a distributive law
$= a \cdot a^2 - b \cdot a^2 + a \cdot ab - b \cdot ab + a \cdot b^2 - b \cdot b^2$
$\qquad\qquad$ Using a distributive law
$= a^3 - a^2b + a^2b - ab^2 + ab^2 - b^3$
$= a^3 - b^3 \qquad$ Collecting like terms

65.

$(2x + 5)(x - 1)$
$= (2x + 5)x - (2x + 5)1 \qquad$ Using a distributive law
$= 2x \cdot x + 5 \cdot x - 2x \cdot 1 - 5 \cdot 1 \qquad$ Using a distributive law
$= 2x^2 + 5x - 2x - 5$
$= 2x^2 + 3x - 5 \qquad$ Collecting like terms

Also consider the FOIL method explained in Exercise 61.

$(2x + 5)(x - 1)$

$\quad$ F $\quad$ O $\quad$ I $\quad$ L
$= 2x \cdot x + 2x(-1) + 5 \cdot x + 5(-1)$
$= 2x^2 - 2x + 5x - 5$
$= 2x^2 + 3x - 5 \qquad$ Collecting like terms

67. $(a - 2)(a + 2)$
$= a^2 - 2^2 \qquad (A - B)(A + B) = A^2 - B^2$
$= a^2 - 4$

69. $(5x + 2)(5x - 2)$
$= (5x)^2 - 2^2 \qquad (A + B)(A - B) = A^2 - B^2$
$= 25x^2 - 4$

71. $(a - h)^2$
$= a^2 - 2 \cdot a \cdot h + h^2 \qquad (A - B)^2 = A^2 - 2AB + B^2$
$= a^2 - 2ah + h^2$

73. $(5x + t)^2$
$= (5x)^2 + 2(5x)(t) + t^2$
$\qquad\qquad (A + B)^2 = A^2 + 2AB + B^2$
$= 25x^2 + 10xt + t^2$

75. $5x(x^2 + 3)^2$
$= 5x[(x^2)^2 + 2(x^2)(3) + 3^2]$
$\qquad\qquad (A + B)^2 = A^2 + 2AB + B^2$
$= 5x(x^4 + 6x^2 + 9)$
$= 5x \cdot x^4 + 5x \cdot 6x^2 + 5x \cdot 9 \qquad$ Using a distributive law
$= 5x^5 + 30x^3 + 45x$

77. $(a + b)^3 = a^3 + 3a^2b + 3ab^2 + b^3$
$\qquad\qquad$ Using Equation 1

79. $(x - 5)^3$
$= [x + (-5)]^3$
$= x^3 + 3x^2(-5) + 3x(-5)^2 + (-5)^3$
$\qquad\qquad [(x + h)^3 = x^3 + 3x^2h + 3xh^2 + h^3]$
$= x^3 - 15x^2 + 75x - 125$

81. $x - xt$
$= x \cdot 1 - x \cdot t$
$= x(1 - t) \qquad$ Factoring out the common factor, x

83. $x^2 + 6xy + 9y^2$
$= x^2 + 2 \cdot x \cdot 3y + (3y)^2$
$= (x + 3y)^2 \qquad A^2 + 2AB + B^2 = (A + B)^2$

85. $x^2 - 2x - 15$

We look for two numbers whose product is -15 and whose sum is -2.
They are -5 and 3. $\qquad -5 \cdot 3 = -15, \; -5 + 3 = -2$

$x^2 - 2x - 15 = (x - 5)(x + 3) \qquad$ Factoring

87. $x^2 - x - 20$

We look for two numbers whose product is -20 and whose sum is -1.
They are -5 and 4. $\qquad -5 \cdot 4 = -20, \; -5 + 4 = -1$

$x^2 - x - 20 = (x - 5)(x + 4) \qquad$ Factoring

89. $49x^2 - t^2$
$= (7x)^2 - t^2$
$= (7x - t)(7x + t) \qquad A^2 - B^2 = (A - B)(A + B)$

91. $36t^2 - 16m^2$

= $4(9t^2 - 4m^2)$ Factoring out the common factor, 4

= $4[(3t)^2 - (2m)^2]$

= $4(3t - 2m)(3t + 2m)$ $A^2 - B^2 = (A - B)(A + B)$

93. $a^3b - 16ab^3$

= $ab(a^2 - 16b^2)$ Factoring out the common factor, ab

= $ab[a^2 - (4b)^2]$

= $ab(a - 4b)(a + 4b)$ $A^2 - B^2 = (A - B)(A + B)$

95. $a^8 - b^8$

= $(a^4)^2 - (b^4)^2$

= $(a^4 + b^4)(a^4 - b^4)$ $A^2 - B^2 = (A + B)(A - B)$

= $(a^4 + b^4)[(a^2)^2 - (b^2)^2]$

= $(a^4 + b^4)(a^2 + b^2)(a^2 - b^2)$

= $(a^4 + b^4)(a^2 + b^2)(a + b)(a - b)$

97. $10a^2x - 40b^2x$

= $10x(a^2 - 4b^2)$ Factoring out the common factor, 10x

= $10x[a^2 - (2b)^2]$

= $10x(a - 2b)(a + 2b)$ $A^2 - B^2 = (A - B)(A + B)$

99. $2 - 32x^4$

= $2(1 - 16x^4)$ Factoring out the common factor, 2

= $2[1^2 - (4x^2)^2]$

= $2(1 - 4x^2)(1 + 4x^2)$ $A^2 - B^2 = (A - B)(A + B)$

= $2[1^2 - (2x)^2](1 + 4x^2)$

= $2(1 - 2x)(1 + 2x)(1 + 4x^2)$

$A^2 - B^2 = (A - B)(A + B)$

101. $9x^2 + 17x - 2$

First we look for a common factor. There is none other than 1.

Next we look for pairs of numbers whose product is 9. Since it is common practice that both factors be positive, we only consider 1, 9 and 3, 3.

We have these possibilities:

(x   )(9x   )   (3x   )(3x   )

Next we look for pairs of numbers whose product is -2. These are 1, -2 and -1, 2.

Using these pairs of numbers we have six possible factorizations:

(x + 1)(9x - 2)   (3x + 1)(3x - 2)
(x - 2)(9x + 1)   (3x - 1)(3x + 2)
(x - 1)(9x + 2)
(x + 2)(9x - 1)

We multiply and find that the desired factorization is (x + 2)(9x - 1).

$9x^2 + 17x - 2 = (x + 2)(9x - 1)$

103. $x^3 + 8$

= $x^3 + 2^3$

= $(x + 2)(x^2 - x \cdot 2 + 2^2)$

$A^3 + B^3 = (A + B)(A^2 - AB + B^2)$

= $(x + 2)(x^2 - 2x + 4)$

105. $y^3 - 64t^3$

= $y^3 - (4t)^3$

= $(y - 4t)[y^2 + y \cdot 4t + (4t)^2]$

$A^3 - B^3 = (A - B)(A^2 + AB + B^2)$

= $(y - 4t)(y^2 + 4yt + 16t^2)$

107.   a) Substituting 4 for x and 0.1 for h in

$(x + h)^2 - x^2 = h(2x + h)$

we get

$(4.1)^2 - 4^2 = 0.1(2 \cdot 4 + 0.1)$

$= 0.1(8.1)$

$= 0.81$

So $(4.1)^2$ differs from 4^2 by 0.81.

b) Substituting 4 for x and 0.01 for h in

$(x + h)^2 - x^2 = h(2x + h)$

we get

$(4.01)^2 - 4^2 = 0.01(2 \cdot 4 + 0.01)$

$= 0.01(8.01)$

$= 0.0801$

So $(4.01)^2$ differs from 4^2 by 0.0801.

c) Substituting 4 for x and 0.001 for h in

$(x + h) - x^2 = h(2x + h)$

we get

$(4.001)^2 - 4^2 = 0.001(2 \cdot 4 + 0.001)$

$= 0.001(8.001)$

$= 0.008001$

So $(4.001)^2$ differs from 4^2 by 0.008001.

109.   a) $A = P(1 + i)^t$

$A = 1000(1 + 0.16)^1$ Substituting

$= 1000(1.16)$

$= \$1160$

b) $A = P(1 + \frac{i}{2})^{2t}$

$= 1000(1 + \frac{0.16}{2})^{2 \cdot 1}$ Substituting

$= 1000(1.08)^2$

$= 1000(1.1664)$

$= \$1166.40$

109. (continued)

c) $A = P(1 + \frac{i}{4})^{4t}$

$= 1000(1 + \frac{0.16}{4})^{4 \cdot 1}$ Substituting

$= 1000(1.04)^4$

$= 1000(1.16985856)$

$= \$1169.86$ Rounding to the nearest cent

d) $A = P(1 + \frac{i}{365})^{365t}$

$= 1000(1 + \frac{0.16}{365})^{365 \cdot 1}$ Substituting

$= 1000(1.000438356)^{365}$

$= 1000(1.17346975)$

$= \$1173.47$

e) $A = P(1 + \frac{i}{8760})^{8760t}$

$= 1000(1 + \frac{0.16}{8760})^{8760 \cdot 1}$ Substituting

$= 1000(1.000018265)^{8760}$

$= 1000(1.17350874)$

$= \$1173.51$

111. $A = P(1 + \frac{i}{4})^{4t}$

$= 25,000(1 + \frac{0.084}{4})^{4 \cdot 1}$ Substituting

$= 25,000(1.021)^4$

$= 25,000(1.086683238)$

$= \$27,167.08$ Rounding to the nearest cent

113. a) $M = P\left[\dfrac{\frac{i}{12}(1 + \frac{i}{12})^n}{(1 + \frac{i}{12})^n - 1}\right]$

We substitute \$43,000 for P, $8\frac{3}{4}$ % (or 0.0875) for i, and 300 for n (25 × 12 = 300).

$M = 43,000\left[\dfrac{\frac{0.0875}{12}(1 + \frac{0.0875}{12})^{300}}{(1 + \frac{0.0875}{12})^{300} - 1}\right]$

$= 43,000\left[\dfrac{0.00729167(1.00729167)^{300}}{(1.00729167)^{300} - 1}\right]$

$= 43,000\left[\dfrac{0.00729167(8.84245268)}{8.84245268 - 1}\right]$

$= 43,000\left[\dfrac{0.06447625}{7.84245268}\right]$

$= \$353.52$

113. (continued)

b) The total amount paid back is the sum of the 300 monthly payments of \$353.52 each.

$T = 300 \times 353.52$

$= \$106,056$

c) The interest paid is the difference between the amount paid back on the loan and the amount of the mortgage.

$I = 106,056 - 43,000$

$= \$63,056$

Exercise Set 1.3

1. $-8x + 9 = 4x - 70$

$-8x + 9 - 4x - 9 = 4x - 70 - 4x - 9$

Adding $-4x$ and also -9 to get all terms with variables on one side and all other terms on the other side

$-12x = -79$ Collecting like terms

$-\frac{1}{12}(-12x) = -\frac{1}{12} \cdot (-79)$ Multiplying by $-\frac{1}{12}$

$x = \frac{79}{12}$ Simplifying

The solution set is $\left\{\frac{79}{12}\right\}$.

3. $5x - 2 + 3x = 2x + 6 - 4x$

$8x - 2 = -2x + 6$ Collecting like terms on each side

$8x - 2 + 2x + 2 = -2x + 6 + 2x + 2$ Adding 2x and also 2

$10x = 8$ Collecting like terms

$\frac{1}{10} \cdot 10x = \frac{1}{10} \cdot 8$ Multiplying by $\frac{1}{10}$

$x = \frac{8}{10}$, or $\frac{4}{5}$ Simplifying

The solution set is $\left\{\frac{4}{5}\right\}$.

5. $x + 0.5x = 210$

$1.5x = 210$ Collecting like terms

$\frac{1}{1.5}(1.5x) = \frac{1}{1.5}(210)$ Multiplying by $\frac{1}{1.5}$

$x = 140$ Simplifying

The solution set is {140}.

7. $x + 0.05x = 210$

$1.05x = 210$

$x = \frac{210}{1.05}$, or 200

The solution set is {200}.

9. $4 + \frac{1}{2}x = 1$

$\qquad \frac{1}{2}x = -3$

$\qquad 2 \cdot \frac{1}{2}x = 2(-3)$

$\qquad x = -6$

The solution set is {-6}.

11. $7(3x + 6) = 11 - (x + 2)$

$\quad 21x + 42 = 11 - x - 2 \qquad$ Removing parentheses

$\quad 21x + 42 = 9 - x \qquad$ Collecting like terms on the right side

$\quad 22x = -33 \qquad$ Adding x and also -42

$\quad x = -\frac{33}{22},\ \text{or}\ -\frac{3}{2}$

The solution set is $\left\{-\frac{3}{2}\right\}$.

13. $2x - (5 + 7x) = 4 - [x - (2x + 3)]$

$2x - (5 + 7x) = 4 - [x - 2x - 3] \quad$ Removing parentheses inside the brackets

$2x - (5 + 7x) = 4 - (-x - 3) \quad$ Collecting like terms inside the brackets

$2x - 5 - 7x = 4 + x + 3 \qquad$ Removing parentheses

$-5x - 5 = x + 7 \qquad$ Collecting like terms on each side

$-6x = 12 \qquad$ Adding -x and also 5

$x = -2$

The solution set is {-2}.

15. $2x - 7 = 9 + 2x$

$2x - 7 - 2x + 7 = 9 + 2x - 2x + 7$

$0 = 16$

We get a false equation. No replacement for x will make the equation true. There are no solutions, so the solution set is the empty set denoted ∅.

17. $11 + y = y + 11$

$11 + y - y = y + 11 - y$

$11 = 11$

We get a true equation. Any replacement for y will make the equation true. The solution set is the set of all real numbers.

19. $2x(x + 3)(5x - 4) = 0$

$2x = 0 \ \text{or}\ x + 3 = 0 \ \text{or}\ 5x - 4 = 0 \quad$ Principle of Zero Products

$x = 0 \ \text{or} \qquad x = -3 \ \text{or} \qquad 5x = 4 \quad$ Solving each equation separately

$x = 0 \ \text{or} \qquad x = -3 \ \text{or} \qquad x = \frac{4}{5}$

The solution set is $\left\{0, -3, \frac{4}{5}\right\}$.

21. $x^2 + 1 = 2x + 1$

$\quad x^2 = 2x \qquad$ Adding -1

$\quad x^2 - 2x = 0 \qquad$ Adding -2x

$\quad x(x - 2) = 0 \qquad$ Factoring

$\quad x = 0 \ \text{or}\ x - 2 = 0 \qquad$ Principle of Zero Products

$\quad x = 0 \ \text{or} \qquad x = 2 \qquad$ Solving each equation separately

The solution set is {0,2}.

23. $t^2 - 2t = t$

$\quad t^2 - 3t = 0 \qquad$ Adding -t

$\quad t(t - 3) = 0 \qquad$ Factoring

$\quad t = 0 \ \text{or}\ t - 3 = 0 \qquad$ Principle of Zero Products

$\quad t = 0 \ \text{or} \qquad t = 3 \qquad$ Solving each equation separately

The solution set is {0,3}.

25. $6x - x^2 = -x$

$\quad 0 = x^2 - 7x \qquad$ Adding x^2 and also -6x

$\quad 0 = x(x - 7) \qquad$ Factoring

$\quad x = 0 \ \text{or}\ x - 7 = 0 \qquad$ Principle of Zero Products

$\quad x = 0 \ \text{or} \qquad x = 7$

The solution set is {0,7}.

27. $9x^3 = x$

$\quad 9x^3 - x = 0$

$\quad x(9x^2 - 1) = 0 \qquad$ Factoring

$\quad x(3x + 1)(3x - 1) = 0 \qquad$ Factoring

$\quad x = 0 \ \text{or}\ 3x + 1 = 0 \ \text{or}\ 3x - 1 = 0 \quad$ Principle of Zero Products

$\quad x = 0 \ \text{or} \qquad 3x = -1 \ \text{or} \qquad 3x = 1$

$\quad x = 0 \ \text{or} \qquad x = -\frac{1}{3} \ \text{or} \qquad x = \frac{1}{3}$

The solution set is $\left\{0, -\frac{1}{3}, \frac{1}{3}\right\}$.

29. $(x - 3)^2 = x^2 + 2x + 1$

$\quad x^2 - 6x + 9 = x^2 + 2x + 1$ Squaring on the left side

$\quad\quad -6x + 9 = 2x + 1$ Adding $-x^2$

$\quad\quad\quad -8x = -8$ Adding $-2x$ and also -9

$\quad\quad\quad\quad x = 1$

The solution set is {1}.

31. Method 1: $x + 6 \geqslant 5x - 6$

$\quad\quad\quad\quad x \geqslant 5x - 12$ Adding -6

$\quad\quad\quad -4x \geqslant -12$ Adding $-5x$

$\quad -\frac{1}{4}(-4x) \leqslant -\frac{1}{4}(-12)$ Multiplying by $-\frac{1}{4}$ and reversing the inequality sign

$\quad\quad\quad\quad x \leqslant 3$

The solution set is $\{x | x \leqslant 3\}$.

Method 2: In order to avoid multiplying by a negative number and, consequently reversing the inequality sign, we can use the addition principle to make the coefficient of the variable positive.

$\quad x + 6 \geqslant 5x - 6$

$\quad\quad 6 \geqslant 4x - 6$ Adding $-x$ to make the coefficient of x positive

$\quad\quad 12 \geqslant 4x$ Adding 6

$\quad \frac{1}{4} \cdot 12 \geqslant \frac{1}{4} \cdot 4x$ Multiplying by $\frac{1}{4}$

$\quad\quad 3 \geqslant x$

The solution set is $\{x | 3 \geqslant x\}$. This is equivalent to the solution set in Method 1 above.

33. $3x - 3 + 3x < 1 - 7x - 9$

$\quad\quad 6x - 3 < -7x - 8$ Collecting like terms on each side

$\quad\quad 6x < -7x - 5$ Adding 3

$\quad\quad 13x < -5$ Adding $7x$

$\quad \frac{1}{13} \cdot 13x < \frac{1}{13}(-5)$ Multiplying by $\frac{1}{13}$

$\quad\quad\quad x < -\frac{5}{13}$

The solution set is $\left\{x \middle| x < -\frac{5}{13}\right\}$.

35. $\quad -5x \leqslant 6$

$\quad -\frac{1}{5}(-5x) \geqslant -\frac{1}{5} \cdot 6$ Multiplying by $-\frac{1}{5}$ and reversing the inequality sign

$\quad\quad\quad x \geqslant -\frac{6}{5}$

The solution set is $\left\{x \middle| x \geqslant -\frac{6}{5}\right\}$.

37. $\quad -\frac{3}{4}x \geqslant -\frac{5}{8} + \frac{2}{3}x$

$\quad 24\left[-\frac{3}{4}x\right] \geqslant 24\left[-\frac{5}{8} + \frac{2}{3}x\right]$ Multiplying by 24 to clear of fractions

$\quad\quad -18x \geqslant -15 + 16x$ Simplifying

$\quad\quad -34x \geqslant -15$ Adding $-16x$

$\quad -\frac{1}{34}(-34x) \leqslant -\frac{1}{34}(-15)$ Multiplying by $-\frac{1}{34}$ and reversing the inequality sign

$\quad\quad\quad x \leqslant \frac{15}{34}$

The solution set is $\left\{x \middle| x \leqslant \frac{15}{34}\right\}$.

39. $\quad 14 - 5y \leqslant 8y - 8$

$\quad\quad -13y \leqslant -22$ Adding $-8y$ and also -14

$\quad -\frac{1}{13}(-13y) \geqslant -\frac{1}{13}(-22)$ Multiplying by $-\frac{1}{13}$ and reversing the inequality sign

$\quad\quad\quad y \geqslant \frac{22}{13}$

The solution set is $\left\{y \middle| y \geqslant \frac{22}{13}\right\}$.

41. $5(y - 12) > -3(4 - 2y)$

$\quad 5y - 60 > -12 + 6y$ Multiplying on each side

$\quad\quad -y > 48$ Adding $-6y$ and also 60

$\quad\quad y < -48$ Multiplying by -1 and reversing the inequality sign

The solution set is $\{y | y < -48\}$.

43. $400 + v \geqslant 0$

$\quad\quad v \geqslant -400$ Adding -400

The solution set is $\{v | v \geqslant -400\}$.

45. $2400 - 6v \geqslant 0$

$\quad\quad -6v \geqslant -2400$ Adding -2400

$\quad\quad v \leqslant 400$ Multiplying by $-\frac{1}{6}$ and reversing the inequality sign

The solution set is $\{v | v \leqslant 400\}$.

47. $0x \geqslant -240$

$\quad 0 \geqslant -240$ $0 \cdot x = 0$

We get a true inequality. Any replacement for x will make the inequality true. The solution set is the set of all real numbers.

49. $8 < 3x + 2 < 14$

$\quad 6 < 3x < 12$ Adding -2 to each part of the inequality

$\quad \frac{1}{3} \cdot 6 < \frac{1}{3} \cdot 3x < \frac{1}{3} \cdot 12$ Multiplying each part by $\frac{1}{3}$

$\quad\quad 2 < x < 4$

The solution set is $\{x | 2 < x < 4\}$.

51. $3 \leqslant 4x - 3 \leqslant 19$

$ 6 \leqslant 4x \leqslant 22 \qquad$ Adding 3

$\frac{1}{4} \cdot 6 \leqslant \frac{1}{4} \cdot 4x \leqslant \frac{1}{4} \cdot 22 \qquad$ Multiplying by $\frac{1}{4}$

$\phantom{\frac{1}{4}} \frac{6}{4} \leqslant x \leqslant \frac{22}{4}$

$\phantom{\frac{1}{4}} \frac{3}{2} \leqslant x \leqslant \frac{11}{2} \qquad$ Simplifying

The solution set is $\left\{ x \mid \frac{3}{2} \leqslant x \leqslant \frac{11}{2} \right\}$.

53. $-7 \leqslant 5x - 2 \leqslant 12$

$ -5 \leqslant 5x \leqslant 14 \qquad$ Adding 2

$\frac{1}{5}(-5) \leqslant \frac{1}{5} \cdot 5x \leqslant \frac{1}{5} \cdot 14 \qquad$ Multiplying by $\frac{1}{5}$

$\phantom{\frac{1}{5}(-5)} -1 \leqslant x \leqslant \frac{14}{5}$

The solution set is $\left\{ x \mid -1 \leqslant x \leqslant \frac{14}{5} \right\}$.

55.

```
  +---o===========================⊕---+  .
  -1  0   1   2   3   4   5   6
```

(0,5) = the set of all numbers x
    such that 0 < x < 5.

The open circles and the parentheses indicate that 0 and 5 are not included.

57.

```
  +---●=======================o---+
 -10 -9  -8  -7  -6  -5  -4  -3
```

[-9,-4) = the set of all numbers x
    such that $-9 \leqslant x < -4$.

The solid circle and the bracket indicate that -9 is included.

The open circle and the parenthesis indicate that -4 is not included.

59.

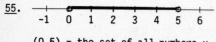

[x,x + h] = the set of all numbers greater than or equal to x and less than or equal to x + h.

The solid circles and the brackets indicate that x and x + h are included.

61.

```
  ⊕==============================>
  p
```

(p,∞) = the set of all numbers x such that x > p.

The open circle and the parenthesis on the left indicate that p is not included.

The symbol ∞ indicates that the interval is of unlimited extent in the positive direction. Since ∞ is not a number and therefore could not be an included endpoint, a parentheses is used on the right.

63. [-3,3] = the set of all numbers x
    such that $-3 \leqslant x \leqslant 3$.

The brackets indicate that both -3 and 3 are included.

65. [-14,-11) = the set of all numbers x
    such that $-14 \leqslant x < -11$.

The bracket indicates that -14 is included.
The parenthesis indicates that -11 is not included.

67. (-∞,-4] = the set of all numbers x
    such that $x \leqslant -4$.

The symbol -∞ indicates that the interval is of unlimited extent in the negative direction. Since -∞ is not a number and therefore could not be an included endpoint, a parenthesis is used.

The bracket indicates that -4 is included.

69. { x ∣ x > 13 }

```
  ┌──┬──┬──┐──────────>   The set of
  │  │  │  └──────────>   all x
  │  │  └─────────────>   such that
  │  └────────────────>   x is greater than 13
  └───────────────────>
```

71. The solution of the equation m + 9 = -23 is m = -32. We may write

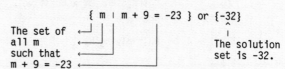

$$\{ m \mid m + 9 = -23 \} \text{ or } \{-32\}$$

The set of
all m
such that
m + 9 = -23

The solution set is -32.

73. Use a calculator to add and multiply in this exercise.

$2.905x - 3.214 + 6.789x = 3.012 + 1.805x$

$ 9.694x - 3.214 = 3.012 + 1.805x$

 Collecting like terms on the left

$ 7.889x = 6.226 \qquad$ Adding -1.805x and also 3.214

$ x = 0.7892 \qquad$ Multiplying by $\frac{1}{7.889}$ and rounding to 4 decimal places

The solution set is {0.7892}.

75. Use a calculator to add and multiply in this exercise.

$$1.52(6.51x + 7.3) < 11.2 - (7.2x + 13.52)$$

$9.8952x + 11.096 < 11.2 - 7.2x - 13.52$

Removing parentheses

$9.8952x + 11.096 < -2.32 - 7.2x$ Collecting like terms on the right side

$17.0952x < -13.416$ Adding $7.2x$ and also -11.096

$x < -0.7848$ Multiplying by $\frac{1}{17.0952}$ and rounding to 4 decimal places

The solution set is $\{x \mid x < -0.7848\}$.

77. $x^2 + 1 > 0$

$x^2 > -1$ Adding -1

We get a true inequality since $x^2 \geqslant 0$ (and hence $x^2 \geqslant -1$) for all real numbers x. Any replacement for x will make the inequality true. The solution set is the set of all real numbers.

Exercise Set 1.4

1. a) <u>Familiarize</u> and <u>translate</u> to an equation.

$79.20 is what percent of $180?

$$\begin{array}{cccccc} \downarrow & \downarrow & \downarrow & \downarrow & \downarrow & \downarrow \\ 79.20 & = & p & \% & \times & 180 \end{array}$$

<u>Carry out</u>. We solve the equation.

$79.20 = p \times 0.01 \times 180$ Writing % as "× 0.01"

$79.20 = 1.8p$ Multiplying on the right side

$44 = p$ Multiplying by $\frac{1}{1.8}$

The solution of the equation is 44.

<u>Check</u>. $44\% \times \$180 = 0.44(\$180) = \$79.2$

<u>State</u>. $79.20 is 44% of $180.

b) <u>Familiarize</u> and <u>translate</u> to an equation.

(What percent) of $25,000 is $10,750?

$$\begin{array}{ccccccc} \downarrow & & \downarrow & & \downarrow & & \downarrow \\ p & & \times & 25,000 & = & 10,750 \end{array}$$

<u>Carry out</u>. We solve the equation.

$25,000p = 10,750$

$p = \dfrac{10,750}{25,000}$ Multiplying by $\frac{1}{25,000}$

$p = 0.43$, or 43%

The solution of the equation is 43%.

<u>Check</u>. $0.43(\$25,000) = \$10,750$

<u>State</u>. $10,750 is 43% of $25,000.

3. a) <u>Familiarize</u> and <u>translate</u> to an equation.

6% of (what number) is $480?

$$\begin{array}{ccccc} \downarrow & \downarrow & & \downarrow & \downarrow \downarrow \\ 6\% & \times & n & = & 480 \end{array}$$

<u>Carry out</u>. We solve the equation.

$6\%n = 480$

$.06n = 480$

$n = \dfrac{480}{0.06}$ Multiplying by $\frac{1}{0.06}$

$n = 8000$

The solution of the equation is 8000.

<u>Check</u>. $6\%(\$8000) = 0.06(\$8000) = \$480$

<u>State</u>. 6% of $8000 is $480.

b) <u>Familiarize</u> and <u>translate</u> to an equation.

$1805.60 is 24.4% of (what number)?

$$\begin{array}{cccccc} \downarrow & \downarrow & \downarrow & \downarrow & & \downarrow \\ 1805.60 & = & 24.4\% & \times & & n \end{array}$$

<u>Carry out</u>. We solve the equation.

$1805.60 = 24.4\%n$

$1805.60 = 0.244n$

$\dfrac{1805.60}{0.244} = n$ Multiplying by $\frac{1}{0.244}$

$7400 = n$

The solution of the equation is 7400.

<u>Check</u>. $24.4\%(\$7400) = 0.244(\$7400) = \$1805.60$

<u>State</u>. $1805.60 is 24.4% of $7400.

5. <u>Familiarize</u> and <u>translate</u> to an equation.

$$\begin{array}{ccccc} \begin{bmatrix} \text{Original} \\ \text{weight} \end{bmatrix} & + & \begin{bmatrix} \text{Weight} \\ \text{gain} \end{bmatrix} & \text{is} & (363.8 \text{ lb}) \\ \downarrow & & \downarrow & \downarrow & \downarrow \\ w & + & 7\%w & = & 363.8 \end{array}$$

<u>Carry out</u>. We solve the equation.

$w + 7\%w = 363.8$

$1 \cdot w + 0.07w = 363.8$

$(1 + 0.07)w = 363.8$ Factoring, using a distributive law in reverse

$1.07w = 363.8$

$w = 340$ Multiplying by $\frac{1}{1.07}$

The solution of the equation is 340.

<u>Check</u>. $340 \text{ lb} + 7\%(340 \text{ lb}) = 340 \text{ lb} + 23.8 \text{ lb} = 363.8 \text{ lb}$

<u>State</u>. The original weight is 340 lb.

7. <u>Familiarize</u> and <u>translate</u> to an equation.
Remember that the interest formula is I = Prt.

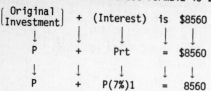

$$\begin{bmatrix}\text{Original} \\ \text{Investment}\end{bmatrix} + (\text{Interest})\ \ \text{is}\ \ \$8560$$

$$P \quad + \quad Prt \quad = \quad \$8560$$

$$P \quad + \quad P(7\%)1 \quad = \quad 8560$$

<u>Carry out</u>. We solve the equation.

$$P + 7\%P = 8560$$
$$1 \cdot P + 0.07P = 8560$$
$$(1 + 0.07)P = 8560 \qquad \text{Factoring, using a distributive law in reverse}$$
$$1.07P = 8560$$
$$P = 8000 \qquad \text{Multiplying by } \frac{1}{1.07}$$

The solution of the equation is 8000.

<u>Check</u>. $8000 + 7\%(\$8000) = \$8000 + 0.07(\$8000) = \$8000 + \$560 = \$8560$

<u>State</u>. The original investment was $8000.

9. First find the old salary.

<u>Familiarize</u> and <u>translate</u> to an equation.

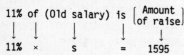

$$11\% \text{ of (Old salary) is } \begin{bmatrix}\text{Amount} \\ \text{of raise}\end{bmatrix}$$

$$11\% \quad \times \quad s \quad = \quad 1595$$

<u>Carry out</u>. We solve the equation.

$$(11\%)s = 1595$$
$$0.11s = 1595$$
$$s = 14,500 \qquad \text{Multiplying by } \frac{1}{0.11}$$

The solution of the equation is 14,500.

<u>Check</u>. $11\%(\$14,500) = 0.11(\$14,500) = \$1595$

<u>State</u>. The old salary was $14,500.

Next find the new salary.

<u>Familiarize</u> and <u>translate</u> to an equation.

$$\begin{bmatrix}\text{Old} \\ \text{salary}\end{bmatrix} + (\text{Raise}) \text{ is } \begin{bmatrix}\text{New} \\ \text{salary}\end{bmatrix}$$

$$14,500 \quad + \quad 1595 \quad = \quad n$$

<u>Carry out</u>. We solve the equation.

$$14,500 + 1595 = n$$
$$16,095 = n$$

The solution of the equation is 16,095.

<u>Check</u>. $14,500 + 1595 = 16,095$

<u>State</u>. The new salary is $16,095.

11. <u>Familiarize</u> and <u>translate</u> to an equation.

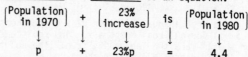

$$\begin{bmatrix}\text{Former} \\ \text{population}\end{bmatrix} + \begin{bmatrix}2\% \\ \text{increase}\end{bmatrix} \text{ is } 826,200$$

$$p \quad + \quad 2\%p \quad = \quad 826,200$$

<u>Carry out</u>. We solve the equation.

$$p + 2\%p = 826,200$$
$$1 \cdot p + 0.02p = 826,200$$
$$1.02p = 826,200$$
$$p = 810,000 \qquad \text{Multiplying by } \frac{1}{1.02}$$

The solution of the equation is 810,000.

<u>Check</u>. $810,000 + 2\%(810,000) = 810,000 + 16,200 = 826,200$

<u>State</u>. The former population was 810,000.

13. <u>Familiarize</u> and <u>translate</u> to an equation.

$$\begin{bmatrix}\text{Population} \\ \text{in 1970}\end{bmatrix} + \begin{bmatrix}23\% \\ \text{increase}\end{bmatrix} \text{ is } \begin{bmatrix}\text{Population} \\ \text{in 1980}\end{bmatrix}$$

$$p \quad + \quad 23\%p \quad = \quad 4.4$$

<u>Carry out</u>. We solve the equation.

$$p + 23\% = 4.4$$
$$1 \cdot p + 0.23p = 4.4$$
$$1.23p = 4.4$$
$$p = 3.6 \qquad \text{Multiplying by } \frac{1}{1.23} \text{ and rounding to the nearest tenth}$$

The solution of the equation is 3.6.

<u>Check</u>. $3.6 + 23\%(3.6) = 3.6 + 0.8 = 4.4$

<u>State</u>. In 1970 the population was 3.6 billion.

15. <u>Familiarize</u> and <u>translate</u> to an inequality.
Remember that "x is at least y" translates to
$x \geqslant y$.

$$\begin{bmatrix}\text{Total} \\ \text{cost}\end{bmatrix} \text{ is at least } (\$22,000)$$

$$5x + 100 \quad \geqslant \quad 22,000$$

<u>Carry out</u>. We solve the inequality.

$$5x + 100 \geqslant 22,000$$
$$5x \geqslant 21,900 \qquad \text{Adding } -100$$
$$x \geqslant 4380 \qquad \text{Multiplying by } \frac{1}{5}$$

<u>Check</u>. We cannot check all the solutions since there are infinitely many. Let's check one number, say 4400. If the firm manufactures 4400 units of the product, its cost will be 5($4400) + $100, or $22,100. This is more than $22,000.

<u>State</u>. The total cost will be at least $22,000 when the firm manufactures 4380 or more units of the product. We express the answer in set builder notation as $\{x \mid x \geqslant 4380\}$.

17. <u>Familiarize</u> and <u>translate</u> to an inequality.

$$\underset{\underset{\displaystyle \frac{65 + 83 + 82 + s}{4}}{\downarrow}}{\text{(Average)}} \quad \underset{\underset{\displaystyle \geqslant}{\downarrow}}{\overset{\text{is at}}{\text{least}}} \quad \underset{\underset{\displaystyle 70}{\downarrow}}{(70\%)}$$

<u>Carry out</u>. We solve the inequality.

$$\frac{65 + 83 + 82 + s}{4} \geqslant 70$$

$$\frac{230 + s}{4} \geqslant 70 \qquad \begin{array}{l}\text{Adding in the} \\ \text{numerator on the} \\ \text{left side}\end{array}$$

$$230 + s \geqslant 280 \qquad \text{Multiplying by 4}$$

$$s \geqslant 50 \qquad \text{Adding -230}$$

<u>Check</u>. We cannot check all the possible solutions since there are infinitely many. Let's check one number, say 52. If a student scores 52% on the fourth test the average will be

$$\frac{65 + 83 + 82 + 52}{4} = \frac{282}{4}, \text{ or } 70.5\%, \text{ which is more}$$

than 70%.

<u>State</u>. Scores of 50% or more will yield a C or better. We express the answer in set builder notation as $\{s \mid s \geqslant 50\%\}$.

19. <u>Familiarize</u> and <u>translate</u> to an inequality.

$$\underset{\underset{\displaystyle \frac{9}{5}C + 32}{\downarrow}}{\begin{bmatrix}\text{Celsius} \\ \text{temperature}\end{bmatrix}} \quad \underset{\underset{\displaystyle \geqslant}{\downarrow}}{\overset{\text{is at}}{\text{least}}} \quad \underset{\underset{\displaystyle 98.6}{\downarrow}}{\begin{pmatrix}\text{the equivalent} \\ \text{of } 98.6^\circ \text{ F}\end{pmatrix}}$$

<u>Carry out</u>. We solve the inequality.

$$\frac{9}{5}C + 32 \geqslant 98.6$$

$$\frac{9}{5}C \geqslant 66.6 \qquad \text{Adding -32}$$

$$C \geqslant 37 \qquad \text{Multiplying by } \frac{5}{9}$$

<u>Check</u>. We cannot check all the possible solutions since there are infinitely many. Let's check one number, say 40. Using the conversion formula we see that

$$F = \frac{9}{5} \cdot 40 + 32 = 72 + 32, \text{ or } 104^\circ, \text{ which is more}$$

than 98.6°.

<u>State</u>. The body is fevered for Celsius temperatures of 37° or more. We express the answer in set builder notation as $\{C \mid C \geqslant 37^\circ\}$.

21. <u>Familiarize</u> and <u>translate</u> to an equation. Let s represent a score.

$$\underset{\underset{\displaystyle 1776}{\downarrow}}{(1776)} \quad \underset{\underset{\displaystyle +}{\downarrow}}{\text{plus}} \quad \underset{\underset{\displaystyle 4s + 7}{\downarrow}}{\begin{bmatrix}\text{Four score and} \\ \text{seven years}\end{bmatrix}} \quad \underset{\underset{\displaystyle =}{\downarrow}}{\text{is}} \quad \underset{\underset{\displaystyle 1863}{\downarrow}}{(1863)}$$

<u>Carry out</u>. We solve the equation.

$$1776 + 4s + 7 = 1863$$

$$1783 + 4s = 1863 \qquad \begin{array}{l}\text{Collecting like terms} \\ \text{on the left side}\end{array}$$

$$4s = 80 \qquad \text{Adding -1783}$$

$$s = 20 \qquad \text{Multiplying by } \frac{1}{4}$$

21. (continued)

The solution of the equation is 20.

<u>Check</u>. $1776 + 4 \cdot 20 + 7 = 1776 + 80 + 7 = 1863$

<u>State</u>. A score is 20 years.

23. <u>Familiarize</u> and <u>translate</u> to an inequality. Let S represent gross sales.

$$\underset{\underset{\displaystyle 600 + 4\%S}{\downarrow}}{\begin{bmatrix}\text{Amount earned} \\ \text{under Plan A}\end{bmatrix}} \quad \underset{\underset{\displaystyle >}{\downarrow}}{\overset{\text{is more}}{\text{than}}} \quad \underset{\underset{\displaystyle 800 + 6\%(S - 10{,}000)}{\downarrow}}{\begin{bmatrix}\text{Amount earned} \\ \text{under Plan B}\end{bmatrix}}$$

<u>Carry out</u>. We solve the inequality.

$$600 + 4\%S > 800 + 6\%(S - 10{,}000)$$

$$600 + 0.04S > 800 + 0.06(S - 10{,}000)$$

$$600 + 0.04S > 800 + 0.06S - 600$$

$$600 + 0.04S > 200 + 0.06S$$

$$-0.02S > -400$$

$$S < 20{,}000 \qquad \begin{array}{l}\text{Multiplying by } -\dfrac{1}{0.02} \\ \text{and reversing the} \\ \text{inequality sign}\end{array}$$

<u>Check</u>. We cannot check all the possible solutions since there are infinitely many. Let's check one number, say 19,500. Under Plan A the salesperson would earn $600 + 4\%(\$19{,}500) = \$600 + 0.04(19{,}500) = \$600 + \$780$, or $1380. Under Plan B the salesperson would earn $800 + 6\%(\$19{,}500 - \$10{,}000) = \$800 + 0.06(\$9{,}500) = \$800 + \$570$, or $1370 which is less than the amount earned under Plan A. For gross sales of $19,500, Plan A is better than Plan B.

<u>State</u>. Plan A is better than Plan B for gross sales less than $20,000. We express the answer in set builder notation as $\{S \mid S < \$20{,}000\}$.

Exercise Set 1.5

1. a) $f(x) = 2x + 3$

   $f(4.1) = 2(4.1) + 3 = 8.2 + 3 = 11.2$

   $f(4.01) = 2(4.01) + 3 = 8.02 + 3 = 11.02$

   $f(4.001) = 2(4.001) + 3 = 8.002 + 3 = 11.002$

   $f(4) = 2(4) + 3 = 8 + 3 = 11$

Input	Output
4.1	11.2
4.01	11.02
4.001	11.002
4	11

   b) $f(x) = 2x + 3$

   $f(5) = 2(5) + 3 = 10 + 3 = 13$

   $f(-1) = 2(-1) + 3 = -2 + 3 = 1$

   $f(k) = 2(k) + 3 = 2k + 3$

   $f(1 + t) = 2(1 + t) + 3 = 2 + 2t + 3 = 2t + 5$

   $f(x + h) = 2(x + h) + 3 = 2x + 2h + 3$

3. $g(x) = x^2 - 3$

$g(-1) = (-1)^2 - 3 = 1 - 3 = -2$

$g(0) = 0^2 - 3 = 0 - 3 = -3$

$g(1) = 1^2 - 3 = 1 - 3 = -2$

$g(5) = 5^2 - 3 = 25 - 3 = 22$

$g(u) = u^2 - 3$

$g(a + h) = (a + h)^2 - 3 = a^2 + 2ah + h^2 - 3$

$g(1 - h) = (1 - h)^2 - 3 = 1 - 2h + h^2 - 3$
$\qquad\qquad = h^2 - 2h - 2$

5. a) $f(x) = (x - 3)^2$

$f(4) = (4 - 3)^2 = 1^2 = 1$

$f(-2) = (-2 - 3)^2 = (-5)^2 = 25$

$f(0) = (0 - 3)^2 = (-3)^2 = 9$

$f(a) = (a - 3)^2 = a^2 - 6a + 9$

$f(t + 1) = (t + 1 - 3)^2 = (t - 2)^2 = t^2 - 4t + 4$

$f(t + 3) = (t + 3 - 3)^2 = t^2$

$f(x + h) = (x + h - 3)^2$
$\qquad\quad = [(x + h) - 3]^2$
$\qquad\quad = (x + h)^2 - 6(x + h) + 9$
$\qquad\quad = x^2 + 2xh + h^2 - 6x - 6h + 9$

b) $f(x) = x^2 - 6x + 9$

This function subtracts six times the input from the square of the input and then adds 9.

7. Graph $f(x) = 2x + 3$.

We first choose any number for x and then determine $f(x)$, or y.

$f(-2) = 2(-2) + 3 = -4 + 3 = -1$

$f(-1) = 2(-1) + 3 = -2 + 3 = 1$

$f(0) = 2 \cdot 0 + 3 = 0 + 3 = 3$

$f(1) = 2 \cdot 1 + 3 = 2 + 3 = 5$

x	y	(x, y)
-2	-1	(-2,-1)
-1	1	(-1, 1)
0	3	(0, 3)
1	5	(1, 5)

Next we plot the input-output pairs from the table and draw the graph.

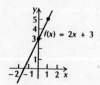

9. Graph $g(x) = -4x$.

We first choose any number for x and then determine $g(x)$, or y.

$g(-1) = -4(-1) = 4$

$g(0) = -4 \cdot 0 = 0$

$g(1) = -4 \cdot 1 = -4$

x	y	(x, y)
-1	4	(-1, 4)
0	0	(0, 0)
1	-4	(1,-4)

Next we plot the input-output pairs from the table and draw the graph.

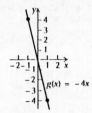

11. Graph $f(x) = x^2 - 5$.

We first choose any number for x and then determine $f(x)$, or y.

$f(-3) = (-3)^2 - 5 = 9 - 5 = 4$

$f(-1) = (-1)^2 - 5 = 1 - 5 = -4$

$f(0) = 0^2 - 5 = 0 - 5 = -5$

$f(1) = 1^2 - 5 = 1 - 5 = -4$

$f(3) = 3^2 - 5 = 9 - 5 = 4$

x	y	(x, y)
-3	4	(-3, 4)
-1	-4	(-1,-4)
0	-5	(0,-5)
1	-4	(1,-4)
3	4	(3, 4)

Next we plot the input-output pairs from the table and draw the graph.

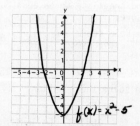

13. Graph g(x) = x³.

We first choose any number for x and then determine g(x), or y.

g(-2) = (-2)³ = -8

g(-1) = (-1)³ = -1

g(0) = 0³ = 0

g(1) = 1³ = 1

g(2) = 2³ = 8

x	y	(x, y)
-2	-8	(-2,-8)
-1	-1	(-1,-1)
0	0	(0, 0)
1	1	(1, 1)
2	8	(2, 8)

Next we plot the input-output pairs from the table and draw the graph.

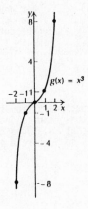

15. Vertical line test:

If it is possible for a vertical line to meet a graph more than once, the graph is not the graph of a function.

Vertical
line

Visualize moving this vertical line across the graph. Ask yourself the question:

Will this line ever meet the graph more than once?

If the answer is yes, the graph is not a graph of a function.

If the answer is no, the graph is a graph of a function.

In this problem no vertical line meets the graph more than once. Thus the graph is a graph of a function.

17. Vertical line test:

If it is possible for a vertical line to meet a graph more than once, the graph is not the graph of a function.

Vertical
line

Visualize moving this vertical line across the graph. Ask yourself the question:

Will this line ever meet the graph more than once?

If the answer is yes, the graph is not a graph of a function.

If the answer is no, the graph is a graph of a function.

In this problem no vertical line meets the graph more than once. Thus the graph is a graph of a function.

19. Vertical line test:

If it is possible for a vertical line to meet a graph more than once, the graph is not the graph of a function.

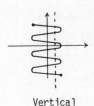

Vertical
line

In this problem a vertical line (in fact many) meets the graph more than once. Therefore the graph is not the graph of a function.

21.

Vertical
line

Visualize moving the vertical line on the left across the graph. Ask yourself the question:

Will this line ever meet the graph more than once?

The answer is yes. Thus the graph is not the graph of a function.

23. a) Graph x = y² - 1.

We first choose any number for y (since x is expressed in terms of y) and then determine x.

For y = -2, x = (-2)² - 1 = 4 - 1 = 3.
For y = -1, x = (-1)² - 1 = 1 - 1 = 0.
For y = 0, x = 0² - 1 = 0 - 1 = -1.
For y = 1, x = 1² - 1 = 1 - 1 = 0.
For y = 2, x = 2² - 1 = 4 - 1 = 3.

x	y	(x, y)
3	-2	(3,-2)
0	-1	(0,-1)
-1	0	(-1, 0)
0	1	(0, 1)
3	2	(3, 2)

Next we plot the ordered pairs and draw the graph.

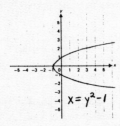

b) Vertical line test:

If it is possible for a vertical line to meet a graph more than once, the graph is not the graph of a function.

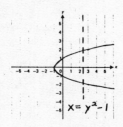

Vertical
line

A vertical line (in fact many) meets the graph more than once. Therefore the graph of x = y² - 1 is not the graph of a function.

25. f(x) = x² - 3x

f(x + h) = (x + h)² - 3(x + h)   Substituting
        = x² + 2xh + h² - 3x - 3h

27. C(x) = 2x + 50

C(10) = 2·10 + 50 = 20 + 50 = $70
C(100) = 2·100 + 50 = 200 + 50 = $250

29. Graph: f(x) = { 1 for x < 0,
                     -1 for x ≥ 0

First graph f(x) = 1 for inputs less than 0.

x	y	(x, y)
-½	1	(-½, 1)
-1	1	(-1, 1)
-2	1	(-2, 1)
-3.21	1	(-3.21,1)
-4	1	(-4, 1)

For any input less than 0, the output is 1.

Plot the input-output pairs from the table and draw this part of the graph. Since the number 0 is not an input, the point (0,1) is not part of the graph. Therefore an open circle is used at the point (0,1).

Next graph f(x) = -1 for inputs greater than or equal to 0.

x	y	(x, y)
0	-1	(0,-1)
1	-1	(1,-1)
1¾	-1	(1¾,-1)
2.7	-1	(2.7,-1)
3	-1	(3,-1)

For any input greater than or equal to 0, the output is -1.

Plot the input-output pairs from the table and draw this part of the graph. Since the number 0 is an input, the point (0,-1) is part of the graph. Therefore a solid circle is used at the point (0,-1).

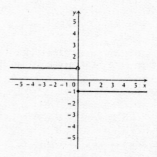

31. Graph: $f(x) = \begin{cases} -3 & \text{for } x = -2 \\ x^2 & \text{for } x \neq -2 \end{cases}$

First graph f(x) = -3 for x = -2. This graph consists of only one point, (-2,-3).

Next graph f(x) = x² for all inputs except -2.

x	y	(x, y)
-3	9	(-3, 9)
-1	1	(-1, 1)
0	0	(0, 0)
1	1	(1, 1)
2	4	(2, 4)

Plot the input-output pairs from the table and draw this part of the graph. Since the number -2 <u>is not</u> an input, the point (-2,4) <u>is not</u> part of the graph. Therefore an open circle is used at the point (-2,4).

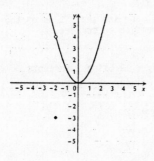

33. 2x + y - 16 = 4 - 3y + 2x

y - 16 = 4 - 3y	Adding -2x
4y - 16 = 4	Adding 3y
4y = 20	Adding 16
y = 5	Multiplying by $\frac{1}{5}$

Graph y = 5.

x	y	(x, y)	
-2	5	(-2, 5)	The input, x, can be any number.
0	5	(0, 5)	
3	5	(3, 5)	The output, y, must always be 5.

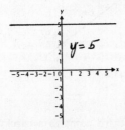

No vertical line meets the graph more than once. Thus the equation represents a function.

35. $(4y^{2/3})^3 = 64x$

$4^3 \cdot (y^{2/3})^3 = 4^3 x$

$y^2 = x$

$y = \pm \sqrt{x}$

This equation does not yield a function that maps a number x to a unique number y.

Exercise Set 1.6

1. Graph y = -4.

Any ordered pair (x,-4) is a solution. The variable y must be -4, but the x variable can be any number we choose. A few solutions are listed below. Plot these and draw the line.

x	y	
-3	-4	y must be -4;
0	-4	x can be any number
2	-4	

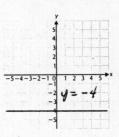

3. Graph x = 4.5.

Any ordered pair (4.5,y) is a solution. The variable x must be 4.5, but the y variable can be any number we choose. A few solutions are listed below. Plot these and draw the line.

x	y	
4.5	-2	x must be 4.5;
4.5	0	y can be any number
4.5	4	

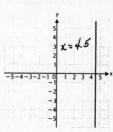

5. Graph 4x + 5y = 20 using intercepts.

To find the y-intercept, set x = 0 and solve for y:

4·0 + 5y = 20

0 + 5y = 20

5y = 20

y = 4

The y-intercept is (0,4).

5. (continued)

To find the x-intercept, set y = 0 and solve for x:

$$4x + 5·0 = 20$$
$$4x + 0 = 20$$
$$4x = 20$$
$$x = 5$$

The x-intercept is (5,0).

Plot the points (0,4) and (5,0) and draw a line through them.

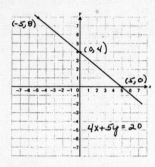

We will use a third point as a check against possible computational errors. We choose -5 for x and compute y:

$$4(-5) + 5y = 20$$
$$-20 + 5y = 20$$
$$5y = 40$$
$$y = 8$$

We plot (-5,8) and see that it is on the graph, so our graph seems to be correct.

7. Graph 2x - y = 4 using intercepts.

To find the y-intercept, set x = 0 and solve for y:

$$2·0 - y = 4$$
$$0 - y = 4$$
$$-y = 4$$
$$y = -4$$

The y-intercept is (0,-4).

To find the x-intercept, set y = 0 and solve for x:

$$2x - 0 = 4$$
$$2x = 4$$
$$x = 2$$

The x-intercept is (2,0).

Plot the points (0,-4) and (2,0) and draw a line through them.

7. (continued)

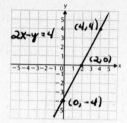

We will use a third point as a check against possible computational errors. We choose 4 for x and compute y:

$$2·4 - y = 4$$
$$8 - y = 4$$
$$-y = -4$$
$$y = 4$$

We plot (4,4) and see that it is on the graph, so our graph seems to be correct.

9. Graph 3x + 4y = 12 using intercepts.

To find the y-intercept, set x = 0 and solve for y:

$$3·0 + 4y = 12$$
$$0 + 4y = 12$$
$$4y = 12$$
$$y = 3$$

The y-intercept is (0,3).

To find the x-intercept, set y = 0 and solve for x:

$$3x + 4·0 = 12$$
$$3x + 0 = 12$$
$$3x = 12$$
$$x = 4$$

The x-intercept is (4,0).

We plot the points (0,3) and (4,0) and draw a line through them.

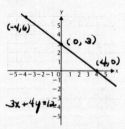

We will use a third point as a check against possible computational errors. We choose -4 for x and compute y:

$$3(-4) + 4y = 12$$
$$-12 + 4y = 12$$
$$4y = 24$$
$$y = 6$$

<u>9</u>. (continued)

We plot (-4,6) and see that it is on the graph, so our graph seems to be correct.

<u>11</u>. Graph y - x = 0 using intercepts.

To find the y-intercept, set x = 0 and solve for y:

$$y - 0 = 0$$
$$y = 0$$

The y-intercept is (0,0). Then the x-intercept is also (0,0). In this case we must find a second point on the graph. Let's choose 4 for x and solve for y:

$$y - 4 = 0$$
$$y = 4$$

The point (4,4) is on the graph.

Plot the points (0,0) and (4,4) and draw a line through them.

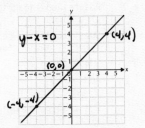

We will use a third point as a check against possible computational errors. We choose -4 for x and compute y:

$$y - (-4) = 0$$
$$y + 4 = 0$$
$$y = -4$$

We plot (-4,-4) and see that it is on the graph, so our graph seems to be correct.

<u>13</u>. Graph 2x + 3y = 5 using intercepts.

To find the y-intercept, set x = 0 and solve for y:

$$2 \cdot 0 + 3y = 5$$
$$0 + 3y = 5$$
$$3y = 5$$
$$y = \frac{5}{3}$$

The y-intercept is $\left[0, \frac{5}{3}\right]$.

To find the x-intercept, set y = 0 and solve for x:

$$2x + 3 \cdot 0 = 5$$
$$2x + 0 = 5$$
$$2x = 5$$
$$x = \frac{5}{2}$$

The x-intercept is $\left[\frac{5}{2}, 0\right]$.

<u>13</u>. (continued)

We plot the points $\left[0, \frac{5}{3}\right]$ and $\left[\frac{5}{2}, 0\right]$ and draw a line through them.

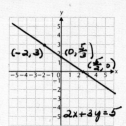

We will use a third point as a check against possible computational errors. We choose -2 for x and compute y:

$$2(-2) + 3y = 5$$
$$-4 + 3y = 5$$
$$3y = 9$$
$$y = 3$$

We plot (-2,3) and see that it is on the graph, so our graph seems to be correct.

<u>15</u>. Graph y = -3x.

We first make a table of values. We choose <u>any</u> number for x and then determine y by substitution.

When x = -2, y = -3(-2) = 6.

When x = 0, y = -3·0 = 0.

When x = 1, y = -3·1 = -3.

Plot these ordered pairs and draw the graph.

x	y
-2	6
0	0
1	-3

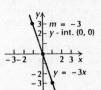

[The graph of the function given by y = mx is the straight line through the origin (0,0) and the point (1,m). The constant m is called the slope of the line.]

The graph of the function given by y = -3x is the straight line through the origin (0,0) and the point (1,-3). The constant -3 is called the slope of the line. The y-intercept is the point (0,0).

<u>17</u>. Graph y = 0.5x.

We first make a table of values. We choose <u>any</u> number for x and then determine y by substitution. It can be helpful to think of y = 0.5x as $y = \frac{1}{2}x$.

When x = -4, y = 0.5(-4) = $\frac{1}{2}$(-4) = -2.

When x = 0, y = 0.5(0) = 0.

When x = 2, y = 0.5(2) = $\frac{1}{2}$(2) = 1.

17. (continued)

Plot these ordered pairs and draw the graph.

x	y
-4	-2
0	0
2	1

The graph of the function given by $y = 0.5x$ is the straight line through the origin $(0,0)$ and the point $(1, 0.5)$. The constant 0.5 is called the slope of the line. The y-intercept is the point $(0,0)$.

19. Graph $y = -2x + 3$.

We first make a table of values. We choose any number for x and then determine y by substitution.

When $x = -1$, $y = -2(-1) + 3 = 2 + 3 = 5$.

When $x = 0$, $y = -2 \cdot 0 + 3 = 0 + 3 = 3$.

When $x = 3$, $y = -2 \cdot 3 + 3 = -6 + 3 = -3$.

Plot these ordered pairs and draw the graph.

x	y
-1	5
0	3
3	-3

[A linear function is given by $y = mx + b$ and has a graph which is the straight line parallel to $y = mx$ with y-intercept $(0,b)$. The constant m is called the slope.]

The graph of the linear function $y = -2x + 3$ is the straight line parallel to $y = -2x$ with y-intercept $(0,3)$. The constant -2 is called the slope.

21. Graph $y = -x - 2$.

$$y = -1 \cdot x - 2 \qquad -x = -1 \cdot x$$

We first make a table of values. We choose any number for x and then determine y by substitution.

When $x = -5$, $y = -1(-5) - 2 = 5 - 2 = 3$.

When $x = 0$, $y = -1 \cdot 0 - 2 = 0 - 2 = -2$.

When $x = 2$, $y = -1 \cdot 2 - 2 = -2 - 2 = -4$.

Plot these ordered pairs and draw the graph.

x	y
-5	3
0	-2
2	-4

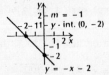

The graph of the linear function $y = -x - 2$ is the straight line parallel to $y = -x$ with y-intercept $(0,-2)$. The constant -1 is called the slope.

23. $y = mx + b$ is called the slope intercept equation of a line. The constant m is the slope of the line. The y-intercept is the point $(0,b)$.

Solve the equation for y.

$2x + y - 2 = 0$

$\qquad y = -2x + 2 \qquad$ Adding $-2x + 2$
$\qquad\qquad\qquad\qquad\qquad m = -2, b = 2$

The slope is -2.
The y-intercept is $(0,2)$.

25. Solve the equation for y.

$2x + 2y + 5 = 0$

$\qquad 2y = -2x - 5 \qquad$ Adding $-2x - 5$

$\qquad y = -x - \frac{5}{2} \qquad$ Multiplying by $\frac{1}{2}$

$\qquad y = -1 \cdot x + (-\frac{5}{2}) \quad m = -1, b = -\frac{5}{2}$

The slope is -1.
The y-intercept is $(0, -\frac{5}{2})$.

27. $\quad y - y_1 = m(x - x_1) \qquad$ Point-slope equation of a line

$\quad y - (-5) = -5(x - 1) \qquad$ Substituting -5 for m, 1 for x_1 and -5 for y_1

$\quad y + 5 = -5x + 5 \qquad$ Simplifying

$\quad y = -5x \qquad$ Adding -5

29. $\quad y - y_1 = m(x - x_1) \qquad$ Point-slope equation of a line

$\quad y - 3 = -2(x - 2) \qquad$ Substituting -2 for m, 2 for x_1 and 3 for y_1

$\quad y - 3 = -2x + 4 \qquad$ Simplifying

$\quad y = -2x + 7 \qquad$ Adding 3

31. $\quad y = mx + b \qquad$ Slope-intercept equation of a line

$\quad y = \frac{1}{2}x + (-6) \qquad$ Substituting $\frac{1}{2}$ for m and -6 for b

$\quad y = \frac{1}{2}x - 6$

33. $\quad y - y_1 = m(x - x_1) \qquad$ Point-slope equation of a line

$\quad y - 3 = 0(x - 2) \qquad$ Substituting 0 for m, 2 for x_1, and 3 for y_1

$\quad y - 3 = 0 \qquad$ Simplifying

$\quad y = 3 \qquad$ Adding 3

35. Find the slope of the line containing $(-4,-2)$ and $(-2,1)$.

$m = \frac{y_2 - y_1}{x_2 - x_1} \qquad$ Slope of line containing points (x_1,y_1) and (x_2,y_2)

$m = \frac{1 - (-2)}{-2 - (-4)} \qquad$ Substituting 1 for y_2, -2 for y_1, -2 for x_2, and -4 for x_1

$\quad = \frac{1 + 2}{-2 + 4}$

$\quad = \frac{3}{2}$

<u>35.</u> (continued)

It does not matter which point is taken first, as long as we subtract coordinates in the same order. We could also find m as follows.

$m = \dfrac{-2 - 1}{-4 - (-2)}$ Substituting -2 for y_2, 1 for y_1, -4 for x_2, and -2 for x_1

$= \dfrac{-2 - 1}{-4 + 2}$

$= \dfrac{-3}{-2}$

$= \dfrac{3}{2}$

<u>37.</u> Find the slope of the line containing (2,-4) and (4,-3).

$m = \dfrac{y_2 - y_1}{x_2 - x_1}$ Slope of line containing points (x_1,y_1) and (x_2,y_2)

$m = \dfrac{-3 - (-4)}{4 - 2}$ Substituting -3 for y_2, -4 for y_1, 4 for x_2, and 2 for x_1

$= \dfrac{-3 + 4}{4 - 2}$

$= \dfrac{1}{2}$

<u>39.</u> Find the slope of the line containing (3,-7) and (3,-9).

$m = \dfrac{y_2 - y_1}{x_2 - x_1}$ Slope of line containing points (x_1,y_1) and (x_2,y_2)

$= \dfrac{-9 - (-7)}{3 - 3}$ Substituting -9 for y_2, -7 for y_1, 3 for x_2, and 3 for x_1

$= \dfrac{-9 + 7}{3 - 3}$

$= \dfrac{-2}{0}$

Since we cannot divide by 0, the slope of the line through (3,-7) and (3,-9) is <u>undefined</u>.

It helps to visualize the graph of the line containing (3,-7) and (3,-9). The first coordinates are the same.

The line is vertical. A vertical line has no slope.

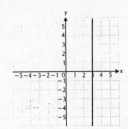

<u>41.</u> Find the slope of the line containing (2,3) and (-1,3).

$m = \dfrac{y_2 - y_1}{x_2 - x_1}$ Slope of line containing points (x_1,y_1) and (x_2,y_2)

$m = \dfrac{3 - 3}{-1 - 2}$ Substituting 3 for y_2, 3 for y_1, -1 for x_2, and 2 for x_1

$= \dfrac{0}{-3}$

$= 0$

It helps to visualize the graph of the line containing (2,3) and (-1,3). The second coordinates are the same.

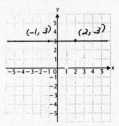

The line is horizontal. The slope of a horizontal line is 0.

<u>43.</u> Find the slope of the line containing (x,3x) and (x + h,3(x + h)).

$m = \dfrac{y_2 - y_1}{x_2 - x_1}$ Slope of line containing points (x_1,y_1) and (x_2,y_2)

$m = \dfrac{3(x + h) - 3x}{x + h - x}$ Substituting 3(x + h) for y_2, 3x for y_1, x + h for x_2, and x for x_1

$= \dfrac{3x + 3h - 3x}{x + h - x}$ Removing parentheses

$= \dfrac{3h}{h}$ Collecting like terms

$= 3$ Simplifying

<u>45.</u> Find the slope of the line containing (x, 2x + 3) and (x + h, 2(x + h) + 3).

$m = \dfrac{y_2 - y_1}{x_2 - x_1}$ Slope of line containing points (x_1,y_1) and (x_2,y_2)

$m = \dfrac{[2(x + h) + 3] - (2x + 3)}{(x + h) - x}$

Substituting 2(x + h) + 3 for y_2, 2x + 3 for y_1, x + h for x_2, and x for x_1

$= \dfrac{2x + 2h + 3 - 2x - 3}{x + h - x}$ Removing parentheses

$= \dfrac{2h}{h}$ Collecting like terms

$= 2$ Simplifying

47. $y - y_1 = \dfrac{y_2 - y_1}{x_2 - x_1}(x - x_1)$ Two-point equation

$y - (-2) = \dfrac{1 - (-2)}{-2 - (-4)}[x - (-4)]$

   Substituting; letting $(x_1,y_1) =$
   $(-4,-2)$ and $(x_2,y_2) = (-2,1)$

$y + 2 = \dfrac{1 + 2}{-2 + 4}(x + 4)$

$y + 2 = \dfrac{3}{2}(x + 4)$

$y + 2 = \dfrac{3}{2}x + 6$

$y = \dfrac{3}{2}x + 4$ Adding -2
   Slope-intercept form

49. $y - y_1 = \dfrac{y_2 - y_1}{x_2 - x_1}(x - x_1)$ Two-point equation

$y - (-4) = \dfrac{-3 - (-4)}{4 - 2}(x - 2)$

   Substituting; letting $(x_1,y_1) =$
   $(2,-4)$ and $(x_2,y_2) = (4,-3)$

$y + 4 = \dfrac{1}{2}(x - 2)$

$y + 4 = \dfrac{1}{2}x - 1$

$y = \dfrac{1}{2}x - 5$ Adding -4
   Slope-intercept form

51. The slope of the line containing $(3,-7)$ and $(3,-9)$ is not defined because we cannot divide by 0. The first coordinates are the same. The graph consists of all ordered pairs whose first coordinate is 3. The second coordinate can be any number. The line is vertical. The equation is $x = 3$.

53. The slope of the line containing $(2,3)$ and $(-1,3)$ is 0. The line is horizontal. The second coordinates are the same. The graph consists of all ordered pairs whose second coordinate is 3. The first coordinate can be any number. The line is horizontal. The equation is $y = 3$.

55. $y - y_1 = \dfrac{y_2 - y_1}{x_2 - x_1}(x - x_1)$ Two-point equation

$y - 3x = \dfrac{3(x + h) - 3x}{(x + h) - x} \cdot (x - x)$

   Substituting; letting $(x_1,y_1) = (x,3x)$
   and $(x_2,y_2) = (x + h, 3(x + h))$

$y - 3x = \dfrac{3h}{h} \cdot 0$

$y - 3x = 0$

$y = 3x$ Adding 3x

57. From Exercise 45 we know that the slope of the line containing $(x, 2x + 3)$ and $(x + h, 2(x + h) + 3)$ is 2. To find the equation of the line containing these two points we can use the point-slope equation and substitute 2 for m, x for x_1, and $2x + 3$ for y_1.

$y - y_1 = m(x - x_1)$ Point-slope equation of a line

$y - (2x + 3) = 2(x - x)$ Substituting

$y - (2x + 3) = 0$

$y = 2x + 3$

59. a) If R is directly proportional to T, then there is some positive constant m such that $R = mT$.

   To find m substitute 12.51 for R and 3 for T in the equation $R = mT$ and solve for m.

   $R = mT$

   $12.51 = m \cdot 3$ Substituting

   $\dfrac{12.51}{3} = m$ Multiplying by $\dfrac{1}{3}$

   $4.17 = m$

   The variation constant is 4.17.
   The equation of variation is $R = 4.17T$.

   b) $R = 4.17T$ Equation of variation

   $R = 4.17(6)$ Substituting 6 for T

   $= 25.02$

   The R-factor for insulation that is 6 inches thick is 25.02.

61. a) If B is directly proportional to W, then there is some positive constant m such that

   $B = mW.$

   To find m substitute 200 for W and 5 for B in the equation $B = mW$ and solve for m.

   $B = mW$

   $5 = m \cdot 200$ Substituting

   $\dfrac{5}{200} = m$ Multiplying by $\dfrac{1}{200}$

   $\dfrac{1}{40} = m$

   The variation constant is $\dfrac{1}{40}$.
   The equation of variation is $B = \dfrac{1}{40} W$.

   b) The variation constant, as a fraction, is $\dfrac{1}{40}$.

   First change $\dfrac{1}{40}$ to decimal notation by dividing 1 by 40.

$$
\begin{array}{r}
0.0\,2\,5 \\
40\,\overline{)\,1.0\,0\,0} \\
\underline{8\,0} \\
2\,0\,0 \\
\underline{2\,0\,0} \\
0
\end{array}
$$

$\dfrac{1}{40} = 0.025$

61. (continued)

Next change 0.025 to percent notation.

$$0.025 = 0.025 \cdot \frac{100}{100} \qquad \text{Multiplying by 1}$$

$$= 2.5 \cdot \frac{1}{100}$$

$$= 2.5\% \qquad \text{Replacing } \frac{1}{100} \text{ by \%}$$

The resulting equation is

B = 2.5% W

The weight, B, of a human's brain is 2.5% of the body weight, W.

c)   B = 2.5% W   Equation of variation

B = 2.5% · 120   Substituting 120 for W

= 0.025·120   2.5% = 0.025

= 3

The weight of the brain of a person who weighs 120 lb is 3 lb.

63. a) A is directly proportional to P if there is some positive constant m such that A = mP.

We let P represent the investment and 14% P represent the interest earned on the investment in one year. Thus,

A = P + 14% P

= 1·P + 0.14P

= (1 + 0.14)P

= 1.14P

Since A can be expressed as a positive constant times P, we say A is directly proportional to P.

b) A = 1.14P

A = 1.14(100)   Substituting 100 for P

A = $114

c)   A = 1.14P

273.60 = 1.14P   Substituting 273.60 for A

$\frac{273.60}{1.14}$ = P   Multiplying by $\frac{1}{1.14}$

$240 = P

65. a) D(F) = 2F + 115

D(0°) = 2·0 + 115 = 0 + 115 = 115 ft

D(-20°) = 2(-20) + 115 = -40 + 115 = 75 ft

D(10°) = 2·10 + 115 = 20 + 115 = 135 ft

D(32°) = 2·32 + 115 = 64 + 115 = 179 ft

b)

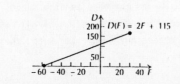

65. (continued)

c) First solve D(F) = 2F + 115 for F when D(F) = 0.

0 = 2F + 115

-115 = 2F

-57.5 = F

When the temperature is -57.5° F, the stopping distance is 0 feet. Temperatures below -57.5° F would yield negative stopping distances, which have no meaning here. For temperatures above 32° F there would be no ice.

67. a) A(t) = 1.1t + 2

A(0) = 1.1(0) + 2 = 0 + 2 = 2 sq mi

A(1) = 1.1(1) + 2 = 1.1 + 2 = 3.1 sq mi

A(4) = 1.1(4) + 2 = 4.4 + 2 = 6.4 sq mi

A(10) = 1.1(10) + 2 = 11 + 2 = 13 sq mi

b) The equation A(t) = 1.1t + 2 is linear. The graph is a straight line through (0,2) and (10,13).

c) The area is measured only from the time the organism is released. Thus only nonnegative values of t would be used as inputs.

69. a) Total costs = $\frac{\text{Variable}}{\text{costs}}$ + $\frac{\text{Fixed}}{\text{costs}}$

To produce x calculators it costs $20 per calculator in addition to the fixed costs of $100,000. That is, the variable costs are 20x dollars. The total cost is

C(x) = 20x + 100,000

b) The total revenue from the sale of x calculators is $45 per calculator. That is, the total revenue is given by the function

R(x) = 45x

c) Total profit = $\frac{\text{Total}}{\text{revenue}}$ - $\frac{\text{Total}}{\text{costs}}$

P(x) = R(x) - C(x)

P(x) = 45x - (20x + 100,000)

        Substituting 45x for R(x) and 20x + 100,000 for C(x)

P(x) = 45x - 20x - 100,000

P(x) = 25x - 100,000

d)   P(x) = 25x - 100,000   Profit function

P(150,000) = 25(150,000) - 100,000

        Substituting 150,000 for x

= 3,750,000 - 100,000

= 3,650,000

A profit of $3,650,000 will be realized if the expected sales of 150,000 calculators occur.

69. (continued)

   e) The profit will be $0 if the firm breaks even. We set the profit function equal to 0 and solve for x.

   $P(x) = 25x - 100,000$ Profit function

   $0 = 25x - 100,000$

   Setting the profit function equal to 0

   $100,000 = 25x$ Adding 100,000

   $4000 = x$ Multiplying by $\frac{1}{25}$

   Thus, to break even the firm must sell 4000 calculators.

71. a) $M(x) = 2.89x + 70.64$

   $M(45) = 2.89(45) + 70.64$ Substituting
   $= 130.05 + 70.64$
   $= 200.69$

   The male was 200.69 cm tall.

   b) $F(x) = 2.75x + 71.48$

   $F(45) = 2.75(45) + 71.48$ Substituting
   $= 123.75 + 71.48$
   $= 195.23$

   The female was 195.23 cm tall.

73. a) $S = 0.73Y - 1430.9$

   To predict the sales in 1986, substitute 1986 for Y.

   $S = 0.73(1986) - 1430.9$
   $= 1449.78 - 1430.9$
   $= 18.88$

   The predicted sales in 1986 were $18.88 billion.

   To predict the sales in 1988, substitute 1988 for Y.

   $S = 0.73(1988) - 1430.9$
   $= 1451.24 - 1430.9$
   $= 20.34$

   The predicted sales in 1988 were $20.34 billion.

   To predict the sales in 1990, substitute 1990 for Y.

   $S = 0.73(1990) - 1430.9$
   $= 1452.7 - 1430.9$
   $= 21.8$

   The predicted sales in 1990 are $21.8 billion.

   To predict the sales in 2000, substitute 2000 for Y.

   $S = 0.73(2000) - 1430.9$
   $= 1460 - 1430.9$
   $= 29.1$

   The predicted sales in 2000 are $29.1 billion.

73. (continued)

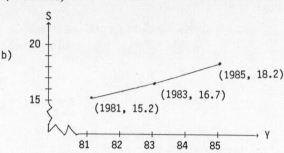

b)

Note that the equation $S = 0.73Y - 1430.9$ <u>approximates</u> the actual sales values shown by the bar graph. Therefore, the sales values shown on the graph above, which were determined using the equation, will not be <u>exactly</u> the same as those on the bar graph.

Exercise Set 1.7

---

1. Graph $y = \frac{1}{2} x^2$ and $y = -\frac{1}{2} x^2$ on the same set of axes.

   Find some ordered pairs that are solutions of $y = \frac{1}{2} x^2$, keeping the results in a table. Since the domain of $y = \frac{1}{2} x^2$ consists of <u>all</u> real numbers, we choose <u>any</u> number for x and then find y.

   When $x = 0$, $y = \frac{1}{2} \cdot 0^2 = \frac{1}{2} \cdot 0 = 0$.

   When $x = 2$, $y = \frac{1}{2} \cdot 2^2 = \frac{1}{2} \cdot 4 = 2$.

   When $x = -2$, $y = \frac{1}{2} \cdot (-2)^2 = \frac{1}{2} \cdot 4 = 2$.

   When $x = 3$, $y = \frac{1}{2} \cdot 3^2 = \frac{1}{2} \cdot 9 = \frac{9}{2}$.

   When $x = -3$, $y = \frac{1}{2} \cdot (-3)^2 = \frac{1}{2} \cdot 9 = \frac{9}{2}$.

x	y
0	0
2	2
-2	2
3	$\frac{9}{2}$
-3	$\frac{9}{2}$

   Table of values for $y = \frac{1}{2} x^2$

   Find some ordered pairs that are solutions of $y = -\frac{1}{2} x^2$, keeping the results in a table. Since the domain of $y = -\frac{1}{2} x^2$ consists of <u>all</u> real numbers, we choose any number for x and then find y.

1. (continued)

When x = 0, y = $-\frac{1}{2} \cdot 0^2 = -\frac{1}{2} \cdot 0 = 0$.

When x = 2, y = $-\frac{1}{2} \cdot 2^2 = -\frac{1}{2} \cdot 4 = -2$.

When x = -2, y = $-\frac{1}{2} \cdot (-2)^2 = -\frac{1}{2} \cdot 4 = -2$.

When x = 3, y = $-\frac{1}{2} \cdot 3^2 = -\frac{1}{2} \cdot 9 = -\frac{9}{2}$.

When x = -3, y = $-\frac{1}{2} \cdot (-3)^2 = -\frac{1}{2} \cdot 9 = -\frac{9}{2}$.

x	y
0	0
2	-2
-2	-2
3	$-\frac{9}{2}$
-3	$-\frac{9}{2}$

Table of values for $y = -\frac{1}{2} x^2$

Plot the ordered pairs for $y = \frac{1}{2} x^2$ and draw the graph with a dashed line. Plot the ordered pairs for $y = -\frac{1}{2} x^2$ and draw the graph with a solid line. For $y = \frac{1}{2} x^2$ the coefficient of x^2 is positive, thus the graph opens upward. For $y = -\frac{1}{2} x^2$ the coefficient of x^2 is negative, thus the graph opens downward.

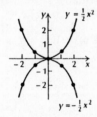

3. Graph $y = x^2$ and $y = (x - 1)^2$ on the same set of axes.

Find some ordered pairs that are solutions of $y = x^2$, keeping the results in a table. Since the domain of $y = x^2$ consists of all real numbers, we choose any number for x and then find y.

When x = 0, y = $0^2 = 0$.

When x = 1, y = $1^2 = 1$.

When x = -1, y = $(-1)^2 = 1$.

When x = 2, y = $2^2 = 4$.

When x = -2, y = $(-2)^2 = 4$.

x	y
0	0
1	1
-1	1
2	4
-2	4

Table of values for $y = x^2$

3. (continued)

Find some ordered pairs that are solutions of $y = (x - 1)^2$, keeping the results in a table. Since the domain of $y = (x - 1)^2$ consists of all real numbers, we choose any number for x and then find y.

When x = 1, y = $(1 - 1)^2 = 0^2 = 0$.

When x = 0, y = $(0 - 1)^2 = (-1)^2 = 1$.

When x = 2, y = $(2 - 1)^2 = 1^2 = 1$.

When x = -1, y = $(-1 - 1)^2 = (-2)^2 = 4$.

When x = 3, y = $(3 - 1)^2 = 2^2 = 4$.

x	y
1	0
0	1
2	1
-1	4
3	4

Table of values for $y = (x - 1)^2$

Plot the ordered pairs for $y = x^2$ and draw the graph with a dashed line. Plot the ordered pairs for $y = (x - 1)^2$, or $y = x^2 - 2x + 1$, and draw the graph with a solid line.

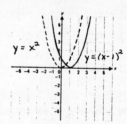

5. Graph $y = x^2$ and $y = (x + 1)^2$ on the same set of axes.

Find some ordered pairs that are solutions of $y = x^2$, keeping the results in a table. Since the domain of $y = x^2$ consists of all real numbers, we choose any number for x and then find y.

When x = 0, y = $0^2 = 0$.

When x = 1, y = $1^2 = 1$.

When x = -1, y = $(-1)^2 = 1$.

When x = 2, y = $2^2 = 4$.

When x = -2, y = $(-2)^2 = 4$.

x	y
0	0
1	1
-1	1
2	4
-2	4

Table of values for $y = x^2$

Find some ordered pairs that are solutions of $y = (x + 1)^2$, keeping the results in a table. Since the domain of $y = (x + 1)^2$ consists of all real numbers, we choose any number for x and then find y.

5. (continued)

When x = -1, y = (-1 + 1)² = 0² = 0.
When x = -2, y = (-2 + 1)² = (-1)² = 1.
When x = 0, y = (0 + 1)² = 1² = 1.
When x = -3, y = (-3 + 1)² = (-2)² = 4.
When x = 1, y = (1 + 1)² = 2² = 4.

x	y
-1	0
-2	1
0	1
-3	4
1	4

Table of values for $y = (x + 1)^2$

Plot the ordered pairs for $y = x^2$ and draw the graph with a dashed line. Plot the ordered pairs for $y = (x + 1)^2$, or $y = x^2 + 2x + 1$, and draw the graph with a solid line.

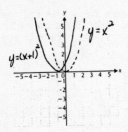

7. Graph y = |x| and y = |x + 3| on the same set of axes.

Find some ordered pairs that are solutions of y = |x|. The domain of y = |x| is the set of <u>all</u> real numbers. We choose any number for x and then find y.

When x = 0, y = |0| = 0.
When x = 2, y = |2| = 2.
When x = -2, y = |-2| = 2.
When x = 5, y = |5| = 5.
When x = -5, y = |-5| = 5.

x	y
0	0
2	2
-2	2
5	5
-5	5

Table of values for y = |x|

Find some ordered pairs that are solutions of y = |x + 3|. The domain of y = |x + 3| is the set of <u>all</u> real numbers. We choose any number for x and then find y.

When x = -3, y = |-3 + 3| = |0| = 0.
When x = -4, y = |-4 + 3| = |-1| = 1.
When x = -2, y = |-2 + 3| = |1| = 1.
When x = -6, y = |-6 + 3| = |-3| = 3.
When x = 0, y = |0 + 3| = |3| = 3.

7. (continued)

x	y
-3	0
-4	1
-2	1
-6	3
0	3

Table of values for y = |x + 3|

Plot the ordered pairs for y = |x| and draw the graph with a dashed line. Plot the ordered pairs for y = |x + 3| and draw the graph with a solid line.

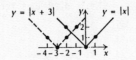

9. Graph $y = x^3$ and $y = x^3 + 1$ on the same set of axes.

Find some ordered pairs that are solutions of $y = x^3$. The domain of $y = x^3$ is the set of <u>all</u> real numbers. We choose any number for x and then find y.

When x = 0, y = 0³ = 0.
When x = -1, y = (-1)³ = -1.
When x = 1, y = 1³ = 1.
When x = -2, y = (-2)³ = -8.
When x = 2, y = 2³ = 8.

x	y
0	0
-1	-1
1	1
-2	-8
2	8

Table of values for $y = x^3$

Find some ordered pairs that are solutions of $y = x^3 + 1$. The domain of $y = x^3 + 1$ is the set of <u>all</u> real numbers. We choose any number for x and then find y.

When x = 0, y = 0³ + 1 = 1.
When x = -1, y = (-1)³ + 1 = -1 + 1 = 0.
When x = 1, y = 1³ + 1 = 1 + 1 = 2.
When x = -2, y = (-2)³ + 1 = -8 + 1 = -7.
When x = 2, y = 2³ + 1 = 8 + 1 = 9.

x	y
0	1
-1	0
1	2
-2	-7
2	9

Table of values for $y = x^3 + 1$

Plot the ordered pairs for $y = x^3$ and draw the graph with a dashed line. Plot the ordered pairs for $y = x^3 + 1$ and draw the graph with a solid line.

9. (continued)

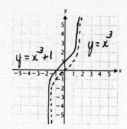

11. (continued)

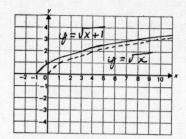

11. Graph $y = \sqrt{x}$ and $y = \sqrt{x + 1}$ on the same set of axes.

Find some ordered pairs that are solutions of $y = \sqrt{x}$. The domain consists of only the non-negative real numbers. We choose for x any number in the interval $[0, \infty)$ and then find y.

When $x = 0$, $y = \sqrt{0} = 0$.

When $x = 1$, $y = \sqrt{1} = 1$.

When $x = 4$, $y = \sqrt{4} = 2$.

When $x = 9$, $y = \sqrt{9} = 3$.

x	y
0	0
1	1
4	2
9	3

Table of values for $y = \sqrt{x}$

Find some ordered pairs that are solutions of $y = \sqrt{x + 1}$. The domain of this function is restricted to those input values that result in the value of the radicand, $x + 1$, being greater than or equal to 0. To determine the domain, we solve the inequality $x + 1 \geq 0$.

$x + 1 \geq 0$

$x \geq -1$ Adding -1

The domain consists of all real numbers greater than or equal to -1. We choose any real number greater than or equal to -1 and then find y.

When $x = -1$, $y = \sqrt{-1 + 1} = \sqrt{0} = 0$.

When $x = 0$, $y = \sqrt{0 + 1} = \sqrt{1} = 1$.

When $x = 3$, $y = \sqrt{3 + 1} = \sqrt{4} = 2$.

When $x = 8$, $y = \sqrt{8 + 1} = \sqrt{9} = 3$.

x	y
-1	0
0	1
3	2
8	3

Table of values for $y = \sqrt{x + 1}$

Plot the ordered pairs for $y = \sqrt{x}$ and draw the graph with a dashed line. Plot the ordered pairs for $y = \sqrt{x + 1}$ and draw the graph with a solid line.

13. Graph $y = x^2 - 4x + 3$.

Before making a table of values we should recognize that $y = x^2 - 4x + 3$ is a quadratic function whose cup-shaped graph opens upward (the coefficient of x^2, 1, is positive). We can find the first coordinates of points where the graph of the function intersects the x-axis, if they exist, by solving the quadratic equation

$x^2 - 4x + 3 = 0$ Quadratic equation

$(x - 3)(x - 1) = 0$ Factoring

$x - 3 = 0$ or $x - 1 = 0$ Principle of zero products

$x = 3$ or $x = 1$

The graph of $y = x^2 - 4x + 3$ has $(3,0)$ and $(1,0)$ as x-intercepts.

Find some ordered pairs that are solutions of $y = x^2 - 4x + 3$. We choose any number for x and then find y.

When $x = 2$, $y = 2^2 - 4 \cdot 2 + 3 = -1$.

When $x = 0$, $y = 0^2 - 4 \cdot 0 + 3 = 3$.

When $x = 4$, $4^2 - 4 \cdot 4 + 3 = 3$.

x	y	
3	0	x-intercept
1	0	x-intercept
2	-1	
0	3	
4	3	

Table of values for $y = x^2 - 4x + 3$

Plot these ordered pairs and draw the graph.

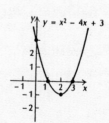

15. Graph $y = -x^2 + 2x - 1$.

We should recognize that $y = -x^2 + 2x - 1$ is a quadratic function whose graph opens downward (the coefficient of x^2, -1, is negative). We find the first coordinates of points where the graph of the function intersects the x-axis, if they exist, by solving the quadratic equation

$-x^2 + 2x - 1 = 0$ Quadratic equation

$x^2 - 2x + 1 = 0$ Multiplying by -1

$(x - 1)(x - 1) = 0$ Factoring

$x - 1 = 0$ or $x - 1 = 0$ Principle of zero products

    $x = 1$ or $x = 1$

The graph of $y = -x^2 + 2x - 1$ has one x-intercept, (1,0).

Find some ordered pairs that are solutions of $y = -x^2 + 2x - 1$.

When $x = 0$, $y = -0^2 + 2 \cdot 0 - 1 = -1$.

When $x = 2$, $y = -2^2 + 2 \cdot 2 - 1 = -1$.

When $x = -1$, $y = -(-1)^2 + 2(-1) - 1 = -4$.

When $x = 3$, $y = -3^2 + 2 \cdot 3 - 1 = -4$.

x	y		
1	0	x-intercept	Table of values for
0	-1		$y = -x^2 + 2x - 1$
2	-1		
-1	-4		
3	-4		

Plot these ordered pairs and draw the graph.

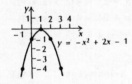

17. Graph $y = \dfrac{2}{x}$.

Note that 0 is not in the domain of this function since it would yield a 0 denominator.

We find some ordered pairs that are solutions of $y = \dfrac{2}{x}$ by choosing any real number except 0 for x and finding y.

Here we list only a few substitutions.

When $x = \dfrac{1}{2}$, $y = \dfrac{2}{\frac{1}{2}} = 2 \cdot \dfrac{2}{1} = 4$.

When $x = -2$, $y = \dfrac{2}{-2} = -1$.

When $x = 4$, $y = \dfrac{2}{4} = \dfrac{1}{2}$.

17. (continued)

x	y		x	y	
$\frac{1}{2}$	4		$-\frac{1}{2}$	-4	Table of values
1	2		-1	-2	for $y = \frac{2}{x}$
2	1		-2	-1	
4	$\frac{1}{2}$		-4	$-\frac{1}{2}$	
6	$\frac{1}{3}$		-6	$-\frac{1}{3}$	

Plot these ordered pairs and draw the graph.

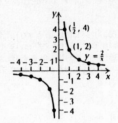

19. Graph $y = \dfrac{-2}{x}$.

Note that 0 is not in the domain of this function since it would yield a 0 denominator.

We find some ordered pairs that are solutions of $y = \dfrac{-2}{x}$ by choosing any real number except 0 for x and finding y.

Here we list only a few substitutions.

When $x = \dfrac{1}{2}$, $y = \dfrac{-2}{\frac{1}{2}} = -2 \cdot \dfrac{2}{1} = -4$.

When $x = 4$, $y = \dfrac{-2}{4} = -\dfrac{1}{2}$.

When $x = -\dfrac{1}{3}$, $y = \dfrac{-2}{-\frac{1}{3}} = -2 \cdot \dfrac{-3}{1} = 6$.

When $x = -2$, $y = \dfrac{-2}{-2} = 1$.

x	y		x	y	
$\frac{1}{2}$	-4		$-\frac{1}{2}$	4	Table of values
1	-2		-1	2	for $y = \frac{-2}{x}$
2	-1		-2	1	
4	$-\frac{1}{2}$		-4	$\frac{1}{2}$	
6	$-\frac{1}{3}$		-6	$\frac{1}{3}$	

Plot these ordered pairs and draw the graph.

19. (continued)

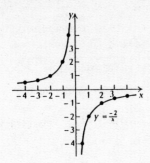

21. Graph $y = \frac{1}{x^2}$.

Note that 0 is not in the domain of this function since it would yield a 0 denominator.

We find some ordered pairs that are solutions of $y = \frac{1}{x^2}$ by choosing any real number except 0 for x and finding y.

When x = -2, $y = \frac{1}{(-2)^2} = \frac{1}{4}$.

When x = -1, $y = \frac{1}{(-1)^2} = \frac{1}{1} = 1$.

When x = $-\frac{1}{2}$, $y = \frac{1}{(-\frac{1}{2})^2} = \frac{1}{\frac{1}{4}} = 1 \cdot \frac{4}{1} = 4$.

When x = $\frac{1}{2}$, $y = \frac{1}{(\frac{1}{2})^2} = \frac{1}{\frac{1}{4}} = 4$.

When x = 1, $y = \frac{1}{1^2} = \frac{1}{1} = 1$.

When x = 2, $y = \frac{1}{2^2} = \frac{1}{4}$.

x	y
-2	$\frac{1}{4}$
-1	1
$-\frac{1}{2}$	4
$\frac{1}{2}$	4
1	1
2	$\frac{1}{4}$

Table of values for $y = \frac{1}{x^2}$

Plot these ordered pairs and draw the graph.

23. Graph $y = \sqrt[3]{x}$.

Find some ordered pairs that are solutions of $y = \sqrt[3]{x}$ keeping the results in a table. Since the domain of $y = \sqrt[3]{x}$ consists of all real numbers, we choose any number for x and then find y.

When x = -6, $y = \sqrt[3]{-6} \approx -1.817$.
When x = -4, $y = \sqrt[3]{-4} \approx -1.587$.
When x = -1, $y = \sqrt[3]{-1} \approx -1$.
When x = -0.5, $y = \sqrt[3]{-0.5} \approx -0.794$.
When x = 0, $y = \sqrt[3]{0} = 0$.
When x = 0.5, $y = \sqrt[3]{0.5} \approx 0.794$.
When x = 1, $y = \sqrt[3]{1} = 1$.
When x = 4, $y = \sqrt[3]{4} \approx 1.587$.
When x = 6, $y = \sqrt[3]{6} \approx 1.817$.

x	y
-6	-1.8
-4	-1.6
-1	-1
-0.5	-0.8
0	0
0.5	0.8
1	1
4	1.6
6	1.8

Table of values for $y = \sqrt[3]{x}$

Plot these ordered pairs and draw the graph.

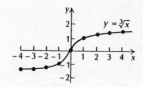

25. $x^2 - 2x = 2$
$x^2 - 2x - 2 = 0$ Adding -2
Standard form

$a = 1$, $b = -2$, and $c = -2$

Then use the quadratic formula.

$x = \dfrac{-b \pm \sqrt{b^2 - 4ac}}{2a}$ Quadratic formula

$x = \dfrac{-(-2) \pm \sqrt{(-2)^2 - 4(1)(-2)}}{2 \cdot 1}$

Substituting 1 for a, -2 for b, and -2 for c

$= \dfrac{2 \pm \sqrt{4 + 8}}{2}$

$= \dfrac{2 \pm \sqrt{12}}{2}$

$= \dfrac{2 \pm 2\sqrt{3}}{2}$ $\sqrt{12} = \sqrt{4 \cdot 3} = 2\sqrt{3}$

$= \dfrac{2(1 \pm \sqrt{3})}{2}$ Factoring in the numerator

$= \dfrac{2}{2} \cdot (1 \pm \sqrt{3})$ Factoring out a form of 1

$= 1 \pm \sqrt{3}$ Simplifying

The solutions are $1 + \sqrt{3}$ and $1 - \sqrt{3}$.

27. $x^2 + 6x = 1$
$x^2 + 6x - 1 = 0$ Adding -1
Standard notation

$a = 1$, $b = 6$, and $c = -1$

Then use the quadratic formula.

$x = \dfrac{-b \pm \sqrt{b^2 - 4ac}}{2a}$ Quadratic formula

$x = \dfrac{-6 \pm \sqrt{6^2 - 4(1)(-1)}}{2 \cdot 1}$ Substituting

$= \dfrac{-6 \pm \sqrt{36 + 4}}{2}$

$= \dfrac{-6 \pm \sqrt{40}}{2}$

$= \dfrac{-6 \pm 2\sqrt{10}}{2}$ $\sqrt{40} = \sqrt{4 \cdot 10} = 2\sqrt{10}$

$= \dfrac{2(-3 \pm \sqrt{10})}{2}$ Factoring in the numerator

$= \dfrac{2}{2} \cdot (-3 \pm \sqrt{10})$ Factoring out a form of 1

$= -3 \pm \sqrt{10}$ Simplifying

The solutions are $-3 + \sqrt{10}$ and $-3 - \sqrt{10}$.

29. $4x^2 = 4x + 1$
$4x^2 - 4x - 1 = 0$ Adding -4x - 1
Standard notation

$a = 4$, $b = -4$, and $c = -1$

Then use the quadratic formula.

$x = \dfrac{-b \pm \sqrt{b^2 - 4ac}}{2a}$ Quadratic formula

$x = \dfrac{-(-4) \pm \sqrt{(-4)^2 - 4(4)(-1)}}{2 \cdot 4}$ Substituting

$= \dfrac{4 \pm \sqrt{16 + 16}}{8}$

$= \dfrac{4 \pm \sqrt{32}}{8}$

$= \dfrac{4 \pm 4\sqrt{2}}{8}$ $\sqrt{32} = \sqrt{16 \cdot 2} = 4\sqrt{2}$

$= \dfrac{4(1 \pm \sqrt{2})}{4 \cdot 2}$ Factoring

$= \dfrac{1 \pm \sqrt{2}}{2}$ Simplifying

The solutions are $\dfrac{1 + \sqrt{2}}{2}$ and $\dfrac{1 - \sqrt{2}}{2}$.

31. $3y^2 + 8y + 2 = 0$ Standard form
$a = 3$, $b = 8$, and $c = 2$

Use the quadratic formula.

$y = \dfrac{-b \pm \sqrt{b^2 - 4ac}}{2a}$ Quadratic formula

$y = \dfrac{-8 \pm \sqrt{8^2 - 4 \cdot 3 \cdot 2}}{2 \cdot 3}$ Substituting

$= \dfrac{-8 \pm \sqrt{64 - 24}}{6}$

$= \dfrac{-8 \pm \sqrt{40}}{6}$

$= \dfrac{-8 \pm 2\sqrt{10}}{6}$ $\sqrt{40} = \sqrt{4 \cdot 10} = 2\sqrt{10}$

$= \dfrac{2(-4 \pm \sqrt{10})}{2 \cdot 3}$ Factoring

$= \dfrac{-4 \pm \sqrt{10}}{3}$ Simplifying

The solutions are $\dfrac{-4 + \sqrt{10}}{3}$ and $\dfrac{-4 - \sqrt{10}}{3}$.

33. $\sqrt{x^3} = \sqrt[2]{x^3}$ The root index is 2.
$= x^{3/2}$ $\sqrt[n]{a^m} = a^{m/n}$

35. $\sqrt[5]{a^3} = a^{3/5}$ $\sqrt[n]{a^m} = a^{m/n}$

37. $\sqrt[7]{t} = \sqrt[7]{t^1}$ $t = t^1$
$= t^{1/7}$ $\sqrt[n]{a^m} = a^{m/n}$

39. $\dfrac{1}{\sqrt[3]{t^4}} = \dfrac{1}{t^{4/3}}$ $\sqrt[n]{a^m} = a^{m/n}$
$= t^{-4/3}$ $\dfrac{1}{a^n} = a^{-n}$

41. $\dfrac{1}{\sqrt{t}} = \dfrac{1}{\sqrt[2]{t^1}}$

$\quad = \dfrac{1}{t^{1/2}}$ $\qquad \sqrt[n]{a^m} = a^{m/n}$

$\quad = t^{-1/2}$ $\qquad \dfrac{1}{a^n} = a^{-n}$

43. $\dfrac{1}{\sqrt{x^2 + 7}} = \dfrac{1}{\sqrt[2]{(x^2 + 7)^1}}$

$\qquad = \dfrac{1}{(x^2 + 7)^{1/2}}$ $\qquad \sqrt[n]{a^m} = a^{m/n}$

$\qquad = (x^2 + 7)^{-1/2}$ $\qquad \dfrac{1}{a^n} = a^{-n}$

45. $x^{1/5} = \sqrt[5]{x^1}$ $\qquad a^{m/n} = \sqrt[n]{a^m}$

$\qquad = \sqrt[5]{x}$ $\qquad x^1 = x$

47. $y^{2/3} = \sqrt[3]{y^2}$ $\qquad a^{m/n} = \sqrt[n]{a^m}$

49. $t^{-2/5} = \dfrac{1}{t^{2/5}}$ $\qquad a^{-n} = \dfrac{1}{a^n}$

$\qquad = \dfrac{1}{\sqrt[5]{t^2}}$ $\qquad a^{m/n} = \sqrt[n]{a^m}$

51. $b^{-1/3} = \dfrac{1}{b^{1/3}}$ $\qquad a^{-n} = \dfrac{1}{a^n}$

$\qquad = \dfrac{1}{\sqrt[3]{b^1}}$ $\qquad a^{m/n} = \sqrt[n]{a^m}$

$\qquad = \dfrac{1}{\sqrt[3]{b}}$ $\qquad b^1 = b$

53. $e^{-17/6} = \dfrac{1}{e^{17/6}}$ $\qquad a^{-n} = \dfrac{1}{a^n}$

$\qquad = \dfrac{1}{\sqrt[6]{e^{17}}}$ $\qquad a^{m/n} = \sqrt[n]{a^m}$

55. $(x^2 - 3)^{-1/2} = \dfrac{1}{(x^2 - 3)^{1/2}}$ $\qquad a^{-n} = \dfrac{1}{a^n}$

$\qquad = \dfrac{1}{\sqrt[2]{(x^2 - 3)^1}}$ $\qquad a^{m/n} = \sqrt[n]{a^m}$

$\qquad = \dfrac{1}{\sqrt{x^2 - 3}}$

57. $9^{3/2}$

$= (9^{1/2})^3$ $\qquad \dfrac{3}{2} = \dfrac{1}{2} \cdot 3$

$\qquad\qquad\qquad a^{m \cdot n} = (a^m)^n$

$= (\sqrt{9})^3$ $\qquad a^{1/n} = \sqrt[n]{a}$

$= 3^3$ $\qquad \sqrt{9} = 3$

$= 27$

59. $64^{2/3}$

$= (64^{1/3})^2$ $\qquad \dfrac{2}{3} = \dfrac{1}{3} \cdot 2$

$\qquad\qquad\qquad a^{m \cdot n} = (a^m)^n$

$= (\sqrt[3]{64})^2$ $\qquad a^{1/n} = \sqrt[n]{a}$

$= 4^2$ $\qquad \sqrt[3]{64} = 4$

$= 16$

61. $16^{3/4}$

$= (16^{1/4})^3$ $\qquad \dfrac{3}{4} = \dfrac{1}{4} \cdot 3$

$\qquad\qquad\qquad a^{m \cdot n} = (a^m)^n$

$= (\sqrt[4]{16})^3$ $\qquad a^{1/n} = \sqrt[n]{a}$

$= 2^3$ $\qquad \sqrt[4]{16} = 2$

$= 8$

63. The domain of a rational function is restricted to those input values that do not result in division by 0.

To determine the domain of

$f(x) = \dfrac{x^2 - 25}{x - 5}$

we set the denominator equal to 0 and solve.

$x - 5 = 0$

$\quad x = 5$ $\qquad$ Adding 5

Thus 5 is not in the domain. The domain consists of all real numbers except 5.

65. The domain of a rational function is restricted to those input values that do not result in division by 0.

To determine the domain of

$f(x) = \dfrac{x^3}{x^2 - 5x + 6}$

we set the denominator equal to 0 and solve.

$\quad x^2 - 5x + 6 = 0$

$(x - 3)(x - 2) = 0$ $\qquad$ Factoring

$x - 3 = 0$ or $x - 2 = 0$ $\quad$ Principle of zero products

$\quad x = 3$ or $\quad\quad x = 2$

Thus 3 and 2 are not in the domain. The domain consists of all real numbers except 3 and 2.

67. The domain of $f(x) = \sqrt{5x + 4}$ is restricted to those input values that result in the value of the radicand, $5x + 4$, being greater than or equal to 0. To determine the domain we solve the inequality $5x + 4 \geqslant 0$.

$5x + 4 \geqslant 0$

$\quad 5x \geqslant -4$ $\qquad$ Adding -4

$\quad\quad x \geqslant -\dfrac{4}{5}$ $\qquad$ Multiplying by $\dfrac{1}{5}$

Thus the domain consists of all real numbers greater than or equal to $-\dfrac{4}{5}$.

69. $D(x) = -2x + 8$ $S(x) = x + 2$

To find the equilibrium point we set $D(x) = S(x)$ and solve.

$-2x + 8 = x + 2$

$8 = 3x + 2$	Adding 2x
$6 = 3x$	Adding -2
$2 = x$	Multiplying by $\frac{1}{3}$

Thus $x_E = 2$ (units). To find p_E we substitute x_E into either $D(x)$ or $S(x)$.

Here we use $S(x)$.

$p_E = S(x_E) = x_E + 2$

$p_E = S(2) = 2 + 2$ Substituting 2 for x_E

$\quad = \$4$

Thus the equilibrium price is \$4 per unit and the equilibrium point is (2,\$4).

71. $D(x) = (x - 3)^2$ $S(x) = x^2 + 2x + 1$

To find the equilibrium point we set $D(x) = S(x)$ and solve.

$(x - 3)^2 = x^2 + 2x + 1$

$x^2 - 6x + 9 = x^2 + 2x + 1$	Squaring on the left
$-6x + 9 = 2x + 1$	Adding $-x^2$
$9 = 8x + 1$	Adding 6x
$8 = 8x$	Adding -1
$1 = x$	Multiplying by $\frac{1}{8}$

Thus $x_E = 1$ (unit). To find p_E we substitute x_E into either $D(x)$ or $S(x)$.

Here we use $S(x)$.

$p_E = S(x_E) = x_E^2 + 2x_E + 1$

$p_E = S(1) = 1^2 + 2 \cdot 1 + 1$ Substituting 1 for x_E

$\quad = 1 + 2 + 1$

$\quad = \$4$

Thus the equilibrium price is \$4 per unit and the equilibrium point is (1,\$4).

73. $D(x) = (x - 4)^2$ $S(x) = x^2$

To find the equilibrium point we set $D(x) = S(x)$ and solve.

$(x - 4)^2 = x^2$

$x^2 - 8x + 16 = x^2$	Squaring on the left
$-8x + 16 = 0$	Adding $-x^2$
$16 = 8x$	Adding 8x
$2 = x$	Multiplying by $\frac{1}{8}$

Thus $x_E = 2$ (units). To find p_E we substitute x_E into either $D(x)$ or $S(x)$.

73. (continued)

Here we use $S(x)$.

$p_E = S(x_E) = (x_E)^2$

$p_E = S(2) = 2^2$ Substituting

$\quad = \$4$

Thus the equilibrium price is \$4 per unit and the equilibrium point is (2,\$4).

75. Dividends (D) are inversely proportional to the prime rate (R).

$D = \frac{x}{R}$ Inverse variation

$2.09 = \frac{x}{19\%}$ Substituting

$2.09 = \frac{x}{0.19}$

$0.3971 = x$ Multiplying by 0.19

The variation constant is 0.3971.

The equation of variation is $D = \frac{0.3971}{R}$.

$D = \frac{0.3971}{R}$

$D = \frac{0.3971}{17.5\%}$ Substituting

$\quad = \frac{0.3971}{0.175}$

$\approx \$2.27$ per share

77. $f(x,y) = 3x - 4y$

$f(-2,5)$ is defined to be the value of the function found by substituting -2 for x and 5 for y.

$f(-2,5) = 3(-2) - 4 \cdot 5 = -6 - 20 = -26$

$f(4,0)$ is defined to be the value of the function found by substituting 4 for x and 0 for y.

$f(4,0) = 3 \cdot 4 - 4 \cdot 0 = 12 - 0 = 12$

$f(10,-6)$ is defined to be the value of the function found by substituting 10 for x and -6 for y.

$f(10,-6) = 3 \cdot 10 - 4(-6) = 30 + 24 = 54$

79. $f(x,y) = 2x + 4y + 7$

To find $f(3,4)$ substitute 3 for x and 4 for y.

$f(3,4) = 2 \cdot 3 + 4 \cdot 4 + 7 = 6 + 16 + 7 = 29$

To find $f(2,5)$ substitute 2 for x and 5 for y.

$f(2,5) = 2 \cdot 2 + 4 \cdot 5 + 7 = 4 + 20 + 7 = 31$

To find $f(0,1)$ substitute 0 for x and 1 for y.

$f(0,1) = 2 \cdot 0 + 4 \cdot 1 + 7 = 0 + 4 + 7 = 11$

81. $f(x,y) = y^2 + 3xy$

To find $f(-2,0)$ substitute -2 for x and 0 for y.

$f(-2,0) = 0^2 + 3(-2)(0) = 0 + 0 = 0$

To find $f(3,2)$ substitute 3 for x and 2 for y.

$f(3,2) = 2^2 + 3 \cdot 3 \cdot 2 = 4 + 18 = 22$

81. (continued)

To find f(-5,10) substitute -5 for x and 10 for y.
$$f(-5,10) = 10^2 + 3(-5)(10) = 100 - 150 = -50$$

83. $f(x,y) = x^2 - y^2$
$$f(-2,-3) = (-2)^2 - (-3)^2 = 4 - 9 = -5$$
$$f(5,0) = 5^2 - 0^2 = 25 - 0 = 25$$
$$f(0,5) = 0^2 - 5^2 = 0 - 25 = -25$$

85. $f(x,y) = (3x + 4y)^2$
$$f(-1,0) = [3(-1) + 4 \cdot 0]^2 = (-3 + 0)^2 = (-3)^2 = 9$$
$$f(2,2) = (3 \cdot 2 + 4 \cdot 2)^2 = (6 + 8)^2 = 14^2 = 196$$
$$f(-4,5) = [3(-4) + 4 \cdot 5]^2 = (-12 + 20)^2 = 8^2 = 64$$

87. $T = W^{1.31}$

We can approximate function values using a power key [y^x] on a calculator.

W	0	10	20	30	40	50	100	150
T	0	20	51	86	126	168	417	709

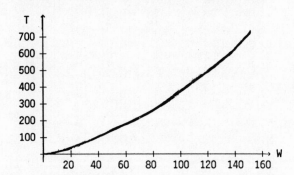

Exercise Set 2.1

<u>1</u>. We substitute 3 for x and -2 for y in each
    equation.

$$\frac{x + y = 1}{\begin{array}{c|c} 3 + (-2) & 1 \\ 1 & \end{array}} \qquad \frac{x - y = 6}{\begin{array}{c|c} 3 - (-2) & 6 \\ 3 + 2 & \\ 5 & \end{array}}$$

We see that (3, -2) is a solution of x + y = 1,
but it is not a solution of x - y = 6. Therefore,
(3, -2) is not a solution of the system of
equations.

<u>3</u>. We substitute $\frac{1}{2}$ for x and 1 for y in each
    equation.

$$\frac{3x + y = \frac{5}{2}}{\begin{array}{c|c} 3\left(\frac{1}{2}\right) + 1 & \frac{5}{2} \\ \frac{3}{2} + 1 & \\ \frac{5}{2} & \end{array}} \qquad \frac{2x - y = \frac{1}{4}}{\begin{array}{c|c} 2\left(\frac{1}{2}\right) - 1 & \frac{1}{4} \\ 1 - 1 & \\ 0 & \end{array}}$$

We see that $\left[\frac{1}{2},\ 1\right]$ is a solution of $3x + y = \frac{5}{2}$,
but it is not a solution of $2x - y = \frac{1}{4}$.
Therefore, $\left[\frac{1}{2},\ 1\right]$ is not a solution of the system
of equations.

<u>5</u>. We graph the two equations.

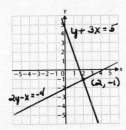

The point of intersection appears to be (2, -1).
We check.

$$\frac{y + 3x = 5}{\begin{array}{c|c} -1 + 3(2) & 5 \\ -1 + 6 & \\ 5 & \end{array}} \qquad \frac{2y - x = -4}{\begin{array}{c|c} 2(-1) - 2 & -4 \\ -2 - 2 & \\ -4 & \end{array}}$$

The solution is (2, -1).

<u>7</u>. We graph the two equations.

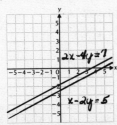

The lines are parallel, so the system has no
solution.

<u>9</u>. We graph the two equations.

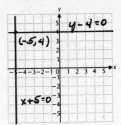

The point of intersection appears to be (-5, 4).
We check.

$$\frac{y - 4 = 0}{\begin{array}{c|c} 4 - 4 & 0 \\ 0 & \end{array}} \qquad \frac{x + 5 = 0}{\begin{array}{c|c} -5 + 5 & 0 \\ 0 & \end{array}}$$

The solution is (-5, 4).

<u>11</u>. We graph the two equations.

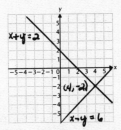

The lines intersect at exactly one point. This
indicates that the system has a solution, so it is
<u>consistent</u>. There is no way to obtain one
equation from the other by multiplying by a
constant, so the system is <u>independent</u>. The point
of intersection appears to be (4, -2). We check.

$$\frac{x + y = 2}{\begin{array}{c|c} 4 + (-2) & 2 \\ 2 & \end{array}} \qquad \frac{x - y = 6}{\begin{array}{c|c} 4 - (-2) & 6 \\ 4 + 2 & \\ 6 & \end{array}}$$

The solution is (4, -2).

<u>13</u>. We graph the two equations.

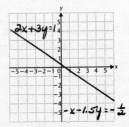

The lines are the same. This indicates that the
system has infinitely many solutions, so it is
<u>consistent</u>. We obtain the second equation from
the first by multiplying by $-\frac{1}{2}$, so the system is
<u>dependent</u>.

15. We graph the two equations.

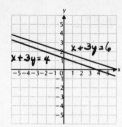

The lines are parallel. This indicates that the system has no solution, so it is <u>inconsistent</u>. There is no way to obtain one equation from the other by multiplying by a constant, so the system is <u>independent</u>.

17. We graph the two equations.

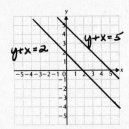

The lines are parallel. This indicates that the system has no solution, so it is <u>inconsistent</u>. There is no way to obtain one equation from the other by multiplying by a constant, so the system is <u>independent</u>.

19. We graph the two equations.

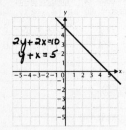

The lines are the same. This indicates that the system has infinitely many solutions, so it is <u>consistent</u>. We obtain the second equation from the first by multiplying by $\frac{1}{2}$, so the system is <u>dependent</u>.

21. Use a computer or a calculator which graphs equations to graph the two equations. Look at the graphs over smaller and smaller intervals, if possible, to get a good estimate of the solution.

21. (continued)

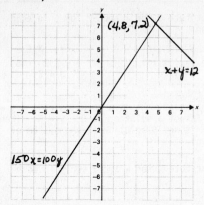

A good estimate of the solution is (4.8, 7.2). In fact, this is the exact solution.

23. Use a computer or a calculator which graphs equations to graph the two equations. Look at the graphs over smaller and smaller intervals, if possible, to get a good estimate of the solution.

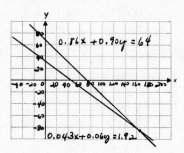

A good estimate of the solution is (163.7, -85.3).

25. We must write p as a function of x for both demand and supply in order to find the equilibrium point using the technique in Section 1.7.

For demand: $x - 5p = 6000$

$$-5p = -x + 6000$$

$$p = \frac{1}{5}x - 1200$$

For supply: $x + 25p = 7500$

$$25p = -x + 7500$$

$$p = -\frac{1}{25}x + 300$$

Now equate demand and supply.

$$\frac{1}{5}x - 1200 = -\frac{1}{25}x + 300$$

$$\frac{6}{25}x = 1500$$

$$x = 6250$$

Thus $x_E = 6250$ (units). To find p_E we substitute x_E into either the demand or the supply equation. We use the demand equation.

25. (continued)

$$p_E = \frac{1}{5}(6250) - 1200 = 1250 - 1200 = \$50$$

The equilibrium point is (6250, $50).

Now solve the system by graphing.

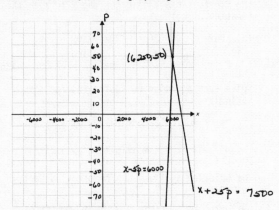

The point of intersection appears to be (6250, $50). We check.

x - 5p = 6000		x + 25p = 7500	
6250 - 5(50) 6250 - 250 6000	6000	6250 + 25(50) 6250 + 1250 7500	7500

The solution is (6250, $50). We observe that the solution of the system of equations is the equilibrium point.

Exercise Set 2.2

1.   y + 4x =   5   (1)
    -3y + 2x = 13   (2)

Since Eq. (1) has y with a coefficient of 1, we solve that equation for y.

     y = -4x + 5   (3)

Now we substitute -4x + 5 for y in Eq. (2).

     -3(-4x + 5) + 2x = 13

We now have an equation in one variable, x. We solve for x using the addition and multiplication principles.

     12x - 15 + 2x = 13

            14x = 28

              x = 2

We substitute 2 for x in Eq. (3) to find y.

     y = -4(2) + 5 = -8 + 5 = -3

The solution is (2, -3). The check is left to the reader.

---

3.   5x + y =   8   (1)
    3x - 4y = 14   (2)

Since Eq. (1) has y with coefficient 1, we solve that equation for y.

     y = -5x + 8   (3)

Now we substitute -5x + 8 for y in Eq. (2).

     3x - 4(-5x + 8) = 14

We have an equation in one variable, x. We solve for x using the addition and multiplication principles.

     3x + 20x - 32 = 14

             23x = 46

               x = 2

We substitute 2 for x in Eq. (3) to find y.

     y = -5(2) + 8 = -10 + 8 = -2

The solution is (2, -2). The check is left to the reader.

5. Our goal is to transform the system to an equivalent one of the form

     Ax   = C
       By = D.

(1)   3x + 5y = 28
      5x - 3y = 24

We first multiply the second equation by 3 to make the x-coefficient a multiple of 3:

(2)   3x + 5y = 28
      15x - 9y = 72   Multiplying by 3

Now we multiply the first equation of system (2) by -5 and add it to the second equation. This eliminates the x-term in the second equation:

(3)   3x + 5y =   28   ⌐ -15x - 25y = -140
                                Multiplying by -5
      -34y = -68 ⌐   15x - 9y =   72
                            -34y = -68   Adding

Next we multiply the first equation of system (3) by 34 to make the y-coefficient a multiple of 34:

(4)   102x + 170y = 952   Multiplying by 34
            -34y = -68

Now we multiply the second equation of system (4) by 5 and add it to the first. This eliminates the y-term in the second equation:

(5)   102x   = 612 ⟵ ⌐102x + 170y =   952
       -34y = -68   -170y = -340
                              Multiplying by 5
                  102x   = 612
                                Adding

Now we solve each equation in system (5) for its variable:

(6)   x   = 6   Multiplying by $\frac{1}{102}$

        y = 2   Multiplying by $-\frac{1}{34}$

The solution is (6, 2). The check is left to the reader.

7. Our goal is to transform the system to an equivalent one of the form

$$Ax \qquad = C$$
$$\qquad By = D.$$

(1) $5x - 4y = -3$

$\qquad 7x + 2y = 6$

We first multiply the second equation by 5 to make the x-coefficient a multiple of 5:

(2) $\quad 5x - 4y = -3$

$\qquad 35x + 10y = 30 \qquad$ Multiplying by 5

Now we multiply the first equation by –7 and add it to the second equation of system (2). This eliminates the x-term in the second equation:

(3) $5x - 4y = -3$ $\quad\left[\begin{array}{l} -35x + 28y = 21 \\ \qquad\qquad\text{Multiplying by } -7 \\ \underline{35x + 10y = 30} \\ \qquad\quad 38y = 51 \quad \text{Adding} \end{array}\right.$

$\qquad 38y = 51$

We multiply the first equation of system (3) by 38 to make the y-coefficient a multiple of 38:

(4) $190x - 152y = -114$

$\qquad\qquad 38y = 51$

Now we multiply the second equation of system (4) by 4 and add it to the first. This eliminates the y-term in the first equation:

(5) $190x \qquad = 90$ $\quad\left[\begin{array}{l} 190x - 152y = -114 \\ \\ \underline{\qquad 152y = 204} \\ \qquad\text{Multiplying by 4} \\ 190x \qquad = 90 \quad \text{Adding} \end{array}\right.$

$\qquad 38y = 51$

Solve each equation of system (5) for its variable:

$x \qquad = \dfrac{9}{19} \qquad$ Multiplying by $\dfrac{1}{190}$

$y = \dfrac{51}{38} \qquad$ Multiplying by $\dfrac{1}{38}$

The solution is $\left[\dfrac{9}{19}, \dfrac{51}{38}\right]$. The check is left to the reader.

9. Our goal is to transform this system to an equivalent one of the form

$$Ax \qquad = C$$
$$\qquad By = D.$$

(1) $4x + 2y = 11$

$\qquad 3x - y = 2$

We first multiply the second equation by 4 to make the x-coefficient a multiple of 4:

(2) $4x + 2y = 11$

$\qquad 12x - 4y = 8 \qquad$ Multiplying by 4

Now we multiply the first equation of system (2) by –3 and add it to the second. This eliminates the x-term in the second equation:

(3) $4x + 2y = 11$ $\quad\left[\begin{array}{l} -12x - 6y = -33 \\ \qquad\text{Multiplying by } -3 \\ \underline{12x - 4y = 8} \\ \quad -10y = -25 \quad \text{Adding} \end{array}\right.$

$\qquad -10y = -25$

9. (continued)

Next we multiply the first equation in system (3) by 5 to make the y-coefficient a multiple of 10:

(4) $20x + 10y = 55 \qquad$ Multiplying by 5

$\qquad\quad -10y = -25$

Now we add the second equation of system (4) to the first. This eliminates the y-term in the first equation:

(5) $20x \qquad = 30$ $\longleftarrow\left[\begin{array}{l} 20x + 10y = 55 \\ \underline{\qquad -10y = -25} \\ 20x \qquad = 30 \quad \text{Adding} \end{array}\right.$

$\qquad -10y = -25$

Solve each equation of system (5) for its variable:

$x \qquad = \dfrac{3}{2} \qquad$ Multiplying by $\dfrac{1}{20}$

$y = \dfrac{5}{2} \qquad$ Multiplying by $-\dfrac{1}{10}$

The solution is $\left[\dfrac{3}{2}, \dfrac{5}{2}\right]$. The check is left to the reader.

11. Our goal is to transform the system to an equivalent one of the form

$$Aa \qquad = C$$
$$\qquad Bb = D.$$

(1) $9a - 2b = 5$

$\qquad 3a - 3b = 11$

We first multiply the second equation by –3 to make the a-coefficient a multiple of 9:

(2) $9a - 2b = 5$

$\qquad -9a + 9b = -33 \qquad$ Multiplying by –3

Now we add the first equation of system (2) to the second. This eliminates the a-term in the second equation:

(3) $9a - 2b = 5$ $\quad\left[\begin{array}{l} 9a - 2b = 5 \\ \underline{-9a + 9b = -33} \\ \qquad 7b = -28 \quad \text{Adding} \end{array}\right.$

$\qquad 7b = -28$

Multiply the first equation of system (3) by 7 to make the b-coefficient a multiple of 7:

(4) $63a - 14b = 35 \qquad$ Multiplying by 7

$\qquad\qquad 7b = -28$

Multiply the second equation of system (4) by 2 and add it to the first equation. This eliminates the b-term in the first equation:

(5) $63a \qquad = -21$ $\longleftarrow\left[\begin{array}{l} 63a - 14b = 35 \\ \underline{\qquad 14b = -56} \\ \qquad\text{Multiplying by 2} \\ 63a \qquad = -21 \quad \text{Adding} \end{array}\right.$

$\qquad 7b = -28$

Now solve each equation in system (5) for its variable:

$a \qquad = -\dfrac{1}{3} \qquad$ Multiplying by $\dfrac{1}{63}$

$b = -4 \qquad$ Multiplying by $\dfrac{1}{7}$

The solution is $\left[-\dfrac{1}{3}, -4.\right]$ The check is left to the reader.

13. Our goal is to transform the system to an equivalent one of the form

$$At = C$$
$$Bs = D.$$

(1) $3t - 6s = 15$

$\quad\ \ 4t - 8s = 20$

We multiply the second equation by 3 to make the t-coefficient a multiple of 3:

(2) $3t - 6s = 15$

$\quad 12t - 24s = 60$

Next we multiply the first equation of system (2) by -4 and add it to the second. This eliminates the t-term in the second equation:

(3) $3t - 6s = 15$

$\qquad\qquad\qquad\quad -12t + 24s = -60$

$\qquad\qquad\qquad\qquad\qquad\qquad$ Multiplying by -4

$\qquad 0 = 0 \leftarrow \quad \underline{12t - 24s = \ 60}$

$\qquad\qquad\qquad\qquad\qquad 0 = \ \ 0 \quad$ Adding

The second equation of system (3), 0 = 0 is true for all values of s and t; hence, any solutions of the first equation are solutions of the system. The equation 3t - 6s = 15 has infinitely many solutions. Therefore, the original system of equations has infinitely many solutions.

[Observe that we can obtain the second equation of the original system from the first by multiplying by $\frac{4}{3}$.]

15. Our goal is to transform this system to an equivalent one of the form

$$Ax = C$$
$$By = D.$$

(1) $5x = 20 - 10y$

$\quad\ \ 4y = 9 - 2x$

First write the equations in the standard form ax + by = c.

(2) $5x + 10y = 20$

$\quad\ \ 2x + 4y = \ \ 9$

Multiply the second equation of system (2) by 5 to make the x-coefficient a multiple of 5:

(3) $5x + 10y = 20$

$\quad 10x + 20y = 45 \qquad$ Multiplying by 5

Now multiply the first equation of system (3) by -2 and add it to the second. This eliminates the x-term in the second equation:

(4) $5x + 10y = 20$

$\qquad\qquad\qquad\quad -10x - 20y = -40$

$\qquad\qquad\qquad\qquad\qquad\qquad$ Multiplying by -2

$\qquad 0 = 5 \leftarrow \quad \underline{10x + 20y = \ 45}$

$\qquad\qquad\qquad\qquad\qquad 0 = \ 5 \quad$ Adding

The second equation of system (4) is false; hence, the system has no solution.

17. Our goal is to transform the system to an equivalent one of the form

$$Ax = C$$
$$By = D.$$

(1) $0.3x + 0.2y = -0.9$

$\quad\ \ 0.02x - 0.03y = -0.06$

Multiply the first equation by 10 and the second by 100 to clear of decimals:

(2) $3x + 2y = -9 \qquad$ Multiplying by 10

$\quad\ \ 2x - 3y = -6 \qquad$ Multiplying by 100

Multiply the second equation of system (2) by 3 to make the x-coefficient a multiple of 3:

(3) $3x + 2y = \ -9$

$\quad\ \ 6x - 9y = -18 \qquad$ Multiplying by 3

Now multiply the first equation of system (3) by -2 and add it to the second. This eliminates the x-term in the second equation:

(4) $3x + 2y = -9$

$\qquad\qquad\qquad\quad -6x - 4y = \ \ 18$

$\qquad\qquad\qquad\qquad\qquad\qquad$ Multiplying by -2

$\quad -13y = \ 0 \leftarrow \quad \underline{6x - 9y = -18}$

$\qquad\qquad\qquad\qquad\quad -13y = \ \ 0 \quad$ Adding

Multiply the first equation of system (4) by 13 to make the y-coefficient a multiple of 13:

(5) $39x + 26y = -117$

$\qquad\quad -13y = \quad 0$

Now multiply the second equation of system (5) by 2 and add it to the first. This eliminates the y-term in the first equation:

(6) $39x \qquad\ = -117$

$\qquad -13y = \quad 0$

Solve each equation of system (6) for its variable:

$\quad x \quad\ = -3 \qquad$ Multiplying by $\frac{1}{39}$

$\qquad\ \ y = \ 0 \qquad$ Multiplying by $-\frac{1}{13}$

The solution is (-3, 0). The check is left to the reader.

19. Our goal is to transform the system to an equivalent one of the form

$$Ax = C$$
$$By = D.$$

(1) $\frac{1}{5}x + \frac{1}{2}y = 6$

$\quad\ \ \frac{3}{5}x - \frac{1}{2}y = 2$

First we multiply each equation by 10 to clear of fractions:

(2) $2x + 5y = 60$

$\quad\ \ 6x - 5y = 20$

19. (continued)

Since the x-coefficient in the second equation of system (2) is a multiple of 2, we multiply the first equation by -3 and add it to the second. This eliminates the x-term in the second equation:

(3) $2x + 5y = 60$ $\begin{array}{l} -6x - 15y = -180 \\ \qquad\qquad\qquad \text{Multiplying by } -3 \\ \underline{6x - 5y = 20} \\ -20y = -160 \quad \text{Adding} \end{array}$

$-20y = -160 \longleftarrow$

Multiply the first equation of system (3) by 4 to make the y-coefficient a multiple of 20:

(4) $8x + 20y = 240$ Multiplying by 4

$-20y = -160$

Add the second equation of system (4) to the first. This eliminates the y-term in the first equation:

(5) $8x = 80 \longleftarrow$ $\begin{array}{l} 8x + 20y = 240 \\ \underline{-20y = -160} \\ 8x = 80 \quad \text{Adding} \end{array}$

$-20y = -160$

Solve each equation of system (5) for its variable:

$x = 10$ Multiplying by $\frac{1}{8}$

$y = 8$ Multiplying by $-\frac{1}{20}$

The solution is (10, 8). The check is left to the reader.

21. <u>Familiarize and translate</u>. Let x = the first number and y = the second number. The information in the exercise gives us a system of equations:

$8x + 5y = 184$

$x - y = -3$

<u>Carry out</u>. We solve the system. First we interchange the equations to make the x-coefficient in the second a multiple of the first:

$x - y = -3$

$8x + 5y = 184$

Next we multiply the first equation by -8 and add it to the second:

$x - y = -3$

$13y = 208$

Multiply the first equation by 13:

$13x - 13y = -39$

$13y = 208$

Add the second equation to the first:

$13x = 169$

$13y = 208$

Solve each equation for its variable:

$x = 13$ Multiplying by $\frac{1}{13}$

$y = 16$ Multiplying by $\frac{1}{13}$

21. (continued)

<u>Check</u>. The solution of the system is (13, 16). See if it checks in the original problem.

$8 \cdot 13 + 5 \cdot 16 = 104 + 80 = 184$ and

$13 - 16 = -3$

The solution checks in the original problem.

<u>State</u>. The first number is 13, and the second is 16.

23. <u>Familiarize</u>. Organize the information in a table.

Let c = the number of cloth gloves sold and p = the number of pigskin gloves sold.

Kind of glove	Cloth	Pigskin	Total
Price	$4.95	$7.50	
Number sold	c	p	20	$\rightarrow c + p = 20$
Money taken in	4.95c	7.50p	137.25

$4.95c + 7.50p = 137.25$

<u>Translate</u>. Adding c and p on the third row, we get 20. This gives us one equation:

$c + p = 20$

The total amount taken in for cloth gloves was 4.95c, and the total for pigskin gloves was 7.50p. The total amount taken in was $137.25, so we have:

$4.95c + 7.50p = 137.25$

Now we have a system of equations:

$c + p = 20$

$4.95c + 7.50p = 137.25$

<u>Carry out</u>. We solve the system. Multiply the second equation by 100 to clear of decimals:

$c + p = 20$

$495c + 750p = 13,725$

Multiply the first equation by -495 and add it to the second:

$c + p = 20$

$255p = 3825$

Multiply the first equation by -255:

$-255c - 255p = -5100$

$255p = 3825$

Add the second equation to the first:

$-255c = -1275$

$255p = 3825$

Solve each equation for its variable:

$c = 5$ Multiplying by $-\frac{1}{255}$

$p = 15$ Multiplying by $\frac{1}{255}$

23. (continued)

Check. The solution of the system is (5, 15).
See if these numbers check in the original
problem.

Number of gloves sold: 5 + 15 = 20

Money from cloth gloves: \$4.95(5) = \$24.75

Money from pigskin gloves: \$7.50(15) = \$112.50

Since \$24.75 + \$112.50 = \$137.25, the total amount
taken in checks.

State. The business sold 5 pairs of cloth gloves
and 15 pairs of pigskin gloves.

25. Familiarize. Organize the information in a table.

Let x = the amount of solution A to be used and
y = the amount of solution B.

Solution	A	B	Mixture
Amount of solution	x	y	60 L
Percent of alcohol	2%	6%	3.2%
Amount of alcohol in solution	2%x	6%y	3.2%x60 L, or 1.92 L

$\rightarrow$ x + y = 60

2%x + 6%y = 1.92

Translate. Adding x and y in the second row, we
get 60. This gives us one equation:

x + y = 60

Adding the amounts of alcohol in the fourth row,
we get 1.92. This gives us another equation:

2%x + 6%y = 1.92, or

0.02x + 0.06y = 1.92   Expressing percents
as decimals

Now we have a system of equations:

x + y = 60

0.02x + 0.06y = 1.92

Carry out. We solve the system. First multiply
the second equation by 100 to clear of decimals:

x + y = 60

2x + 6y = 192

Multiply the first equation by -2 and add it to
the second:

x + y = 60

4y = 72

Multiply the first equation by -4:

-4x - 4y = -240

4y =   72

Add the second equation to the first:

-4x   = -168

4y =   72

25. (continued)

Solve each equation for its variable:

x   = 42   Multiplying by $-\frac{1}{4}$

y = 18   Multiplying by $\frac{1}{4}$

Check. The solution of the system is (42, 18).
See if 42 L of solution A and 18 L of solution B
check in the original problem. Add the amounts
of solution: 42 L + 18 L = 60 L. The amount of
solution checks. Now check the amount of
alcohol:

2%(42) + 6%(18) = 0.02(42) + 0.06(18) =

0.84 + 1.08 = 1.92 L

The amount of alcohol checks.

State. The technician should use 42 L of
solution A and 18 L of solution B.

27. Familiarize. Organize the information in a table.

Let x = the amount invested at 12% and y = the
amount invested at 13%.

Since the time we are interested in is 1 year, the
interest formula I = Prt becomes I = Pr(1) or
I = Pr in this application. We multiply the
amount invested by the interest rate to find the
interest earned in the first year.

Interest rate	12%	13%	Total
Amount invested	x	y	\$4800
Interest earned in first year	12%x	13%y	\$604

$\rightarrow$ x + y = 4800

12%x + 13%y = 604

Translate. Adding x and y on the second row, we
get \$4800. This gives us one equation:

x + y = 4800

Adding the amounts of interest in the third row,
we get \$604. This gives us another equation:

12%x + 13%y = 604, or

0.12x + 0.13y = 604   Expressing percents
as decimals

Now we have a system of equations:

x + y = 4800

0.12x + 0.13y = 604

Carry out. We solve the system. First, multiply
the second equation by 100 to clear of decimals:

x + y =   4800

12x + 13y = 60,400

Multiply the first equation by -12 and add it to
the second:

x + y = 4800

y = 2800

Multiply the second equation by -1 and add it to
the first:

x   = 2000

y = 2800

27. (continued)

Check. The solution of the system is (2000, 2800). See if these numbers check in the original problem.

Amount invested: $2000 + $2800 = $4800

Amount of interest earned at 12%: 12%($2000) = 0.12(2000) = $240

Amount of interest earned at 13%: 13%($2800) = 0.13(2800) = $364

Since $240 + $364 = $604, the total interest earned checks.

The numbers check in the original problem.

State. $2000 was invested at 12%, and $2800 was invested at 13%.

29. Familiarize. Organize the information in a table.

Let x = the speed of the boat and y = the speed of the stream.

When the boat goes downstream (or with the current) it travels at a speed of x + y. When it goes upstream (or against the current) it travels at a speed of x - y.

	Rate, or speed	Time	Distance
Downstream	x + y	2 hr	46 km
Upstream	x - y	3 hr	51 km

$\longrightarrow$ 2(x + y) = 46
$\longrightarrow$ 3(x - y) = 51

Translate. Using the distance formula d = rt in the second row we get an equation:

2(x + y) = 46, or

x + y = 23   Multiplying by $\frac{1}{2}$

Using the distance formula d = rt in the third row we get another equation:

3(x - y) = 51, or

x - y = 17   Multiplying by $\frac{1}{3}$

Now we have a system of equations:

x + y = 23
x - y = 17

Carry out. We solve the system.

Multiply the first equation by -1 and add it to the second:

x + y = 23
-2y = -6

Multiply the first equation by 2:

2x + 2y = 46
-2y = -6

Add the second equation to the first:

2x   = 40
-2y = -6

29. (continued)

Solve each equation for its variable:

x   = 20
y = 3

Check. The solution of the system is (20, 3). See if these numbers check in the original problem. The rate of travel downstream is 20 km/hr + 3 km/hr = 23 km/hr. In two hours the boat would travel 23 km/hr · 2 hr = 46 km. The rate of travel upstream is 20 km/hr - 3 km/hr = 17 km/hr. In three hours the boat would travel 17 km/hr · 3 hr = 51 km. The numbers check in the original problem.

State. The speed of the boat is 20 km/hr, and the speed of the stream is 3 km/hr.

31. Familiarize. Organize the information in a table.

Let d = the number of dimes and q = the number of quarters.

Type of coin	Dimes	Quarters	Total
Number	d	q	20
Value	0.10d	0.25q	$3.05

$\longrightarrow$ d + q = 20

0.10d + 0.25q = 3.05

Translate. Adding d and q on the second row we get 20. This gives us one equation:

d + q = 20

The total value of the dimes is 0.10d, and the total value of the quarters is 0.25q. The total value of the coins is $3.05, so we have

0.10d + 0.25q = 3.05

Now we have a system of equations:

d + q = 20
0.10d + 0.25q = 3.05

Carry out. We solve the system. First, multiply the second equation by 100 to clear of decimals.

d + q = 20
10d + 25q = 305

Multiply the first equation by -10 and add it to the second:

d + q = 20
15q = 105

Multiply the first equation by -15:

-15d - 15q = -300
15q = 105

Add the second equation to the first:

-15d   = -195
15q = 105

Solve each equation for its variable:

d   = 13
q = 7

31. (continued)

Check. The solution of the system is (13, 7). See if these numbers check in the original problem.

Number of coins: 13 + 7 = 20

Value of dimes: $0.10(13) = $1.30

Value of quarters: $0.25(7) = $1.75

Since $1.30 + $1.75 = $3.05, the total value of the coins checks.

These numbers check in the original problem.

State. There are 13 dimes and 7 quarters.

33. $\dfrac{x + y}{4} - \dfrac{x - y}{3} = 1$

$\dfrac{x - y}{2} + \dfrac{x + y}{4} = -9$

First, write the equations in standard form $Ax + By = C$. Multiply the first equation by 12 and the second by 4 to clear of fractions.

$3(x + y) - 4(x - y) = 12$

$2(x - y) + x + y = -36$

Multiply to remove parentheses:

$3x + 3y - 4x + 4y = 12$

$2x - 2y + x + y = -36$

Collect like terms on the left side of each equation:

$-x + 7y = 12$

$3x - y = -36$

Now solve the system. Multiply the first equation by 3 and add it to the second:

$-x + 7y = 12$

$20y = 0$

Multiply the first equation by 20:

$-20x + 140y = 240$

$20y = 0$

Multiply the second equation by -7 and add it to the first:

$-20x \qquad = 240$

$20y = 0$

Solve each equation for its variable.

$x \qquad = -12$

$\qquad y = 0$

Check:

$$
\begin{array}{c|c}
\dfrac{x + y}{4} - \dfrac{x - y}{3} = 1 & \dfrac{x - y}{2} + \dfrac{x + y}{4} = -9 \\
\hline
\dfrac{-12 + 0}{4} - \dfrac{-12 - 0}{3} \;\Big|\; 1 & \dfrac{-12 - 0}{2} + \dfrac{-12 + 0}{4} \;\Big|\; -9 \\
\dfrac{-12}{4} - \dfrac{-12}{3} & \dfrac{-12}{2} + \dfrac{-12}{4} \\
-3 - (-4) & -6 + (-3) \\
-3 + 4 & \qquad\qquad -9 \\
1 &
\end{array}
$$

The solution is (-12, 0).

35. $\sqrt{2}x + \pi y = 3$

$\pi x - \sqrt{2}y = 1$

Multiply the second equation by $\sqrt{2}$:

$\sqrt{2}x + \pi y = 3$

$\pi\sqrt{2}x - 2y = \sqrt{2}$

Multiply the first equation by $-\pi$ and add it to the second:

$$
\begin{array}{l|l}
\sqrt{2}x + \pi y = 3 & -\pi\sqrt{2}x - \pi^2 y = -3\pi \\
(-\pi^2 - 2)y = -3\pi + \sqrt{2} \leftarrow & \underline{\pi\sqrt{2}x - 2y = \sqrt{2}} \\
& (-\pi^2 - 2)y = -3\pi + \sqrt{2}
\end{array}
$$

Multiply the first equation by $(-\pi^2 - 2)$:

$(-\pi^2\sqrt{2} - 2\sqrt{2})x + (-\pi^3 - 2\pi)y = -3\pi^2 - 6$

$(-\pi^2 - 2)y = -3\pi + \sqrt{2}$

Multiply the second equation by $-\pi$ and add it to the first:

$$
\begin{array}{l}
(-\pi^2\sqrt{2} - 2\sqrt{2})x + (-\pi^3 - 2\pi)y = -3\pi^2 - 6 \\
\underline{\qquad\qquad\qquad (\pi^3 + 2\pi)y = 3\pi^2 - \pi\sqrt{2}} \\
(-\pi^2\sqrt{2} - 2\sqrt{2})x \qquad\qquad = -6 - \pi\sqrt{2} \\
(-\pi^2\sqrt{2} - 2\sqrt{2})x \qquad\qquad = -6 - \pi\sqrt{2} \\
\qquad\qquad (-\pi^2 - 2)y = -3\pi + \sqrt{2}
\end{array}
$$

Solve each equation for its variable:

$x \qquad = \dfrac{-6 - \pi\sqrt{2}}{-\pi^2\sqrt{2} - 2\sqrt{2}}$

$y = \dfrac{-3\pi + \sqrt{2}}{-\pi^2 - 2}$

Simplify x by rationalizing the denominator:

$x = \dfrac{-6 - \pi\sqrt{2}}{-\pi^2\sqrt{2} - 2\sqrt{2}} \cdot \dfrac{\sqrt{2}}{\sqrt{2}} = \dfrac{-6\sqrt{2} - 2\pi}{-2\pi^2 - 4}$

$= \dfrac{-2(3\sqrt{2} + \pi)}{-2(\pi^2 + 2)} = \dfrac{3\sqrt{2} + \pi}{\pi^2 + 2}$

Simplify y by removing a factor of -1 from the numerator and denominator:

$y = \dfrac{-3\pi + \sqrt{2}}{-\pi^2 - 2} = \dfrac{-1 \cdot (3\pi - \sqrt{2})}{-1 \cdot (\pi^2 + 2)} = \dfrac{3\pi - \sqrt{2}}{\pi^2 + 2}$

The solution is $\left(\dfrac{3\sqrt{2} + \pi}{\pi^2 + 2}, \dfrac{3\pi - \sqrt{2}}{\pi^2 + 2} \right)$.

37. Use the Computer Software Supplement or a calculator to do this exercise.

$$4.026x - 1.448y = 18.32$$

$$0.724y = -9.16 + 2.013x$$

First, write the second equation in standard form $Ax + By = C$.

$$4.026x - 1.448y = 18.32$$

$$-2.013x + 0.724y = -9.16$$

Now interchange the two equations so the x-coefficient of the second will be a multiple of the first:

$$-2.013x + 0.724y = -9.16$$

$$4.026x - 1.448y = 18.32$$

Multiply the first equation by 2 and add it to the second:

$$-2.013x + 0.724y = -9.16$$

$$0 = 0$$

Since the second equation, $0 = 0$, is true for all values of x and y, the system has infinitely many solutions. (See Exercise 13 for an explanation.)

39. **Familiarize.** Organize the information in a table.

Let x = the number of members who ordered one book and y = the number of members who ordered two books. Note that the x members who ordered one book each ordered a total of x books and paid a total of $12x; the y members who ordered two books each ordered a total of 2y books and paid a total of $20y. (The price is one book for $12 or two books for $20.)

Type of order	One book	Two books	Total
Price	$12	$20	
Number of books ordered	x	2y	880
Amount taken in	$12x	$20y	$9840

$\longrightarrow x + 2y = 880$

$\longrightarrow 12x + 20y = 9840$

Translate. Adding x and 2y on the third row, we get 880. This gives us one equation:

$$x + 2y = 880$$

Adding $12x and $20y on the fourth row, we get $9840. This gives us another equation:

$$12x + 20y = 9840$$

Now we have a system of equations:

$$x + 2y = 880$$
$$12x + 20y = 9840$$

Carry out. We solve the system.

Multiply the first equation by −12 and add it to the second:

$$x + 2y = 880$$
$$-4y = -720 \longleftarrow$$

$$\begin{array}{r} -12x - 24y = -10{,}560 \\ 12x + 20y = 9840 \\ \hline -4y = -720 \end{array}$$

39. (continued)

At this point we could solve the second equation for y, the information the exercise asks us to find. However, in order to be able to check our answer we must also find x so we continue with the solution.

Multiply the first equation by 2:

$$2x + 4y = 1760$$

$$-4y = -720$$

Add the second equation to the first:

$$2x = 1040$$

$$-4y = -720$$

Solve each equation for its variable:

$$x = 520$$

$$y = 180$$

Check. The solution of the system is (520, 180). See if these numbers check in the original problem.

Number of books ordered: $520 + 2(180) = 520 + 360 = 880$ (Remember that the 180 members who each ordered two books ordered a total of 2×180 books.)

Money from orders for one book: $12(520) = 6240

Money from orders for two books: $20(180) = 3600

Since $6240 + $3600 = $9840, the total amount taken in checks.

The numbers check in the original problem.

State. 180 members ordered two books.

41. **Familiarize and translate.** Let a = the number of people ahead of you and b = the number of people behind you. The problem tells us that there are two more people ahead of you than there are behind you. This gives us one equation:

$$a = b + 2, \text{ or } a - b = 2$$

The entire line consists of the number of people ahead of you, you, and the number of people behind you or $a + 1 + b$. The problem tells us that this number is three times the number of people behind you. This gives us another equation:

$$a + 1 + b = 3b, \text{ or } a - 2b = -1$$

Now we have a system of equations:

$$a - b = 2$$
$$a - 2b = -1$$

Carry out. We solve the system.

Multiply the first equation by −1 and add it to the second:

$$a - b = 2$$
$$-b = -3$$

Multiply the second equation by −1 and add it to the first:

$$a = 5$$
$$-b = -3$$

41. (continued)

Solve each equation for its variable:

a = 5

b = 3

Check. The solution of the system is (5, 3). See if these numbers check in the original problem. The number of people ahead of you, 5, is two more than the number of people behind you, 3. In the entire line there are 5 + 1 + 3, or 9, people. This is three times the number of people behind you, or 3 × 3. The numbers check in the original problem.

State. There are 5 people ahead of you.

43. Substituting -2 for x and 3 for y gives us an equation in m and b:

3 = -2m + b, or -2m + b = 3

Substituting 4 for x and -5 for y gives us another equation in m and b:

-5 = 4m + b, or 4m + b = -5

Now we have a system of equations:

-2m + b = 3

4m + b = -5

Solve the system to find m and b.

Multiply the first equation by 2 and add it to the second:

$$-2m + b = 3 \quad \left[\begin{array}{l} -4m + 2b = 6 \\ \underline{4m + b = -5} \\ 3b = 1 \end{array} \right.$$
$$3b = 1 \longleftarrow$$

Multiply the first equation by -3:

6m - 3b = -9

3b = 1

Add the second equation to the first:

$$6m = -8 \longleftarrow \left[\begin{array}{l} 6m - 3b = -9 \\ \underline{3b = 1} \\ 6m = -8 \end{array} \right.$$
$$3b = 1$$

Solve each equation for its variable:

$$m = -\frac{4}{3}$$

$$b = \frac{1}{3}$$

Check: For $m = -\frac{4}{3}$ and $b = \frac{1}{3}$, the equation y = mx + b becomes $y = -\frac{4}{3}x + \frac{1}{3}$. Check to see if (-2, 3) and (4, -5) are solutions of this equation.

$y = -\frac{4}{3}x + \frac{1}{3}$		$y = -\frac{4}{3}x + \frac{1}{3}$	
3	$-\frac{4}{3}(-2) + \frac{1}{3}$	-5	$-\frac{4}{3}(4) + \frac{1}{3}$
	$\frac{8}{3} + \frac{1}{3}$		$-\frac{16}{3} + \frac{1}{3}$
	$\frac{9}{3}$		$-\frac{15}{3}$
	3		-5

43. (continued)

Therefore, $m = -\frac{4}{3}$ and $b = \frac{1}{3}$.

45. 3|x| + 5|y| = 30

5|x| + 3|y| = 34

Recall the definition of absolute value:

|a| = a if a ⩾ 0 and

|a| = -a if a < 0

Hence we must consider four possibilities.

Case I: x ⩾ 0 and y ⩾ 0

In this case |x| = x and |y| = y so we can write the system as follows:

3x + 5y = 30

5x + 3y = 34

Solve the system of equations.

Multiply the second equation by 3:

3x + 5y = 30

15x + 9y = 102

Multiply the first equation by -5 and add it to the second:

3x + 5y = 30

-16y = -48

Multiply the first equation by 16:

48x + 80y = 480

-16y = -48

Multiply the second equation by 5 and add it to the first:

48x = 240

-16y = -48

Solve each equation for its variable:

x = 5

y = 3

The solution in this case is (5, 3).

Case II: x ⩾ 0 and y < 0

In this case |x| = x and |y| = -y so we can write the system as follows:

3x - 5y = 30

5x - 3y = 34

Solve the system of equations.

Multiply the second equation by 3:

3x - 5y = 30

15x - 9y = 102

Multiply the first equation by -5 and add it to the second:

3x - 5y = 30

16y = -48

Multiply the first equation by 16:

48x - 80y = 480

16y = -48

45. (continued)

Multiply the second equation by 5 and add it to the first:

$$48x \qquad = 240$$
$$16y = -48$$

Solve each equation for its variable:

$$x \qquad = \quad 5$$
$$y = -3$$

The solution in this case is (5, -3).

Case III: $x < 0$ and $y \geqslant 0$

In this case $|x| = -x$ and $|y| = y$ so we can write the system as follows:

$$-3x + 5y = 30$$
$$-5x + 3y = 34$$

Solve the system of equations.
Multiply the second equation by 3:

$$-3x + 5y = \quad 30$$
$$-15x + 9y = 102$$

Multiply the first equation by -5 and add it to the second:

$$-3x + 5y = \quad 30$$
$$-16y = -48$$

Multiply the first equation by 16:

$$-48x + 80y = 480$$
$$-16y = -48$$

Multiply the second equation by 5 and add it to the first:

$$-48x \qquad = 240$$
$$-16y = -48$$

Solve each equation for its variable:

$$x \qquad = -5$$
$$y = \quad 3$$

The solution in this case is (-5, 3).

Case IV: $x < 0$ and $y < 0$

In this case $|x| = -x$ and $|y| = -y$ so we can write the system as follows:

$$-3x - 5y = 30$$
$$-5x - 3y = 34$$

Solve the system of equations.
Multiply the second equation by 3:

$$-3x - 5y = \quad 30$$
$$-15x - 9y = 102$$

Multiply the first equation by -5 and add it to the second:

$$-3x - 5y = \quad 30$$
$$16y = -48$$

Multiply the first equation by 16:

$$-48x - 80y = 480$$
$$16y = -48$$

45. (continued)

Multiply the second equation by 5 and add it to the first:

$$-48x \qquad = 240$$
$$16y = -48$$

Solve each equation for its variable:

$$x \qquad = -5$$
$$y = -3$$

The solution in this case is (-5, -3).

Thus the system of equations has four solutions:
(5, 3), (5, -3), (-5, 3), (-5, -3)
The check is left to the reader.

47. The equilibrium point is the point of intersection of the graphs of the demand and supply functions, or the solution of the system of equations formed by the demand and supply functions. We solve the system of equations:

$$x + 43p = 800$$
$$x - 16p = 210$$

Multiply the first equation by -1 and add it to the second:

$$x + 43p = \quad 800$$
$$-59p = -590$$

Multiply the first equation by 59:

$$59x + 2537p = 47,200$$
$$-59p = \quad -590$$

Multiply the second equation by 43 and add it to the first:

$$59x \qquad = 21,830$$
$$-59p = \quad -590$$

Solve each equation for its variable:

$$x \qquad = 370$$
$$p = \quad 10$$

The solution is (370, $10). The check is left to the reader.

49. We solve the system of equations. (See Exercise 47 for an explanation.)

$$x = 760 - 13p$$
$$x = 430 + 2p$$

First, we write each equation in standard form $Ax + By = C$.

$$x + 13p = 760$$
$$x - 2p = 430$$

Multiply the first equation by -1 and add it to the second:

$$x + 13p = \quad 760$$
$$-15p = -330$$

47. (continued)

Multiply the first equation by 15:

$$15x + 195p = 11,400$$
$$-15p = -330$$

Multiply the second equation by 13 and add it to the first.

$$15x = 7110$$
$$-15p = -330$$

Solve each equation for its variable:

$$x = 474$$
$$p = 22$$

The solution is (474, $22). The check is left to the reader.

Exercise Set 2.3

1. We substitute -1 for x, 1 for y, and 0 for z in each of the equations:

2x + 3y - 5z = 1		6x - 6y + 10z = 3	
2(-1) + 3(1) - 5(0)	1	6(-1) - 6(1) + 10(0)	3
-2 + 3 - 0		-6 - 6 + 0	
1		-12	

It is not necessary to continue since (-1, 1, 0) is not a solution of the second equation and hence is not a solution of the system of equations.

3.
$$x + 4y = 5$$
$$-3x + 2y = 13$$

Write an echelon tableau.

x	y		1
1*	4		5
-3	2		13

The first pivot is the 1 in the first row, first column. We carry out the pivoting by multiplying the first row by 3 and adding it to the second.

x	y		1	
1	4		5	
0	14*		28	New Row 2 = 3(Row 1) + (Row 2)

The next pivot is the 14 in the second row, second column. To carry out the pivoting we first multiply the first row by 7 to make the element above the pivot a multiple of 14.

x	y		1	
7	28		35	New Row 1 = 7(Row 1)
0	14*		28	

Then we multiply the second row by -2 and add it to the first.

x	y		1	
7	0		-21	New Row 1 = -2(Row 2) + (Row 1)
0	14		28	

3. (continued)

We complete the solution by multiplying to obtain 1's down the main diagonal.

x	y		1	
1	0		-3	New Row 1 = $\frac{1}{7}$(Row 1)
0	1		2	New Row 2 = $\frac{1}{14}$(Row 2)

We consider the corresponding system of equations and read off the solution.

$$x = -3, y = 2$$

The solution is (-3, 2).

5.
$$-x + 3y = 2$$
$$2x - y = 11$$

Write an echelon tableau.

x	y		1
-1*	3		2
2	-1		11

The first pivot is the -1 in the first row, first column. We carry out the pivoting by multiplying the first row by 2 and adding it to the second row.

x	y		1	
-1	3		2	
0	5*		15	New Row 2 = 2(Row 1) + (Row 2)

The next pivot is the 5 in the second row, second column. To carry out the pivoting we first multiply the first row by 5 to make the element above the pivot a multiple of 5.

x	y		1	
-5	15		10	New Row 1 = 5(Row 1)
0	5*		15	

Then we multiply the second row by -3 and add it to the first row.

x	y		1	
-5	0		-35	New Row 1 = -3(Row 2) + (Row 1)
0	5		15	

We complete the solution by multiplying to obtain 1's down the main diagonal.

x	y		1	
1	0		7	New Row 1 = $-\frac{1}{5}$(Row 1)
0	1		3	New Row 2 = $\frac{1}{5}$(Row 2)

We consider the corresponding system of equations and read off the solution.

$$x = 7, y = 3$$

The solution is (7, 3).

7. $2x - 5y = 10$

$4x + 3y = 7$

Write an echelon tableau.

x	y		1
2*	-5		10
4	3		7

The first pivot is the 2 in the first row, first column. We carry out the pivoting by multiplying the first row by -2 and adding it to the second.

x	y		1	
2	-5		10	
0	13*		-13	New Row 2 = -2(Row 1) + (Row 2)

The next pivot is the 13 in the second row, second column. To carry out the pivoting we first multiply the first row by 13 to make the element above the pivot a multiple of 13.

x	y		1	
26	-65		130	New Row 1 = 13(Row 1)
0	13*		-13	

Then we multiply the second row by 5 and add it to the first.

x	y		1	
26	0		65	New Row 1 = 5(Row 2) + (Row 1)
0	13		-13	

We complete the solution by multiplying to obtain 1's down the main diagonal.

x	y		1	
1	0		$\frac{5}{2}$	New Row 1 = $\frac{1}{26}$(Row 1)
0	1		-1	New Row 2 = $\frac{1}{13}$(Row 2)

We consider the corresponding system of equations and read off the solution.

$x = \frac{5}{2}, \ y = -1$

The solution is $\left(\frac{5}{2}, -1\right)$.

9. $x + 6y + 3z = 4$

$2x + y + 2z = 3$

$3x - 2y + z = 0$

Write an echelon tableau.

x	y	z		1
1*	6	3		4
2	1	2		3
3	-2	1		0

The first pivot is the 1 in the first row, first column. We carry out the pivoting by multiplying the first row by -2 and adding it to the second row. Then we multiply the first row by -3 and add it to the third row.

9. (continued)

x	y	z		1	
1	6	3		4	
0	-11*	-4		-5	New Row 2 = -2(Row 1) + (Row 2)
0	-20	-8		-12	New Row 3 = -3(Row 1) + (Row 3)

The next pivot is the -11 in the second row, second column. To carry out the pivoting we first multiply the first and third rows by 11 to make the elements above and below the pivot multiples of -11.

x	y	z		1	
11	66	33		44	New Row 1 = 11(Row 1)
0	-11*	-4		-5	
0	-220	-88		-132	New Row 3 = 11(Row 3)

Now we multiply the second row by 6 and add it to the first row. Then we multiply the second row by -20 and add it to the third row.

x	y	z		1	
11	0	9		14	New Row 1 = 6(Row 2) + (Row 1)
0	-11	-4		-5	
0	0	-8*		-32	New Row 3 = -20(Row 2) + (Row 3)

The next pivot is the -8 in the third row, third column. To carry out the pivoting we first multiply the first row by 8 and the second row by 2 to make the elements above the pivot multiples of -8.

x	y	z		1	
88	0	72		112	New Row 1 = 8(Row 1)
0	-22	-8		-10	New Row 2 = 2(Row 2)
0	0	-8*		-32	

Now we multiply the third row by 9 and add it to the first row. Then we multiply the third row by -1 and add it to the second row.

x	y	z		1	
88	0	0		-176	New Row 1 = 9(Row 3) + (Row 1)
0	-22	0		22	New Row 2 = -1(Row 3) + (Row 2)
0	0	-8		-32	

We complete the solution by multiplying to obtain 1's down the main diagonal.

x	y	z		1	
1	0	0		-2	New Row 1 = $\frac{1}{88}$(Row 1)
0	1	0		-1	New Row 2 = $-\frac{1}{22}$(Row 2)
0	0	1		4	New Row 3 = $-\frac{1}{8}$(Row 3)

We consider the corresponding system of equations and read off the solution.

$x = -2, \ y = -1, \ z = 4$

The solution is $(-2, -1, 4)$.

11. $x + y + z = 6$
   $2x - y - z = -3$
   $x - 2y + 3z = 6$
Write an echelon tableau.

x	y	z	1
1*	1	1	6
2	-1	-1	-3
1	-2	3	6

The first pivot is the 1 in the first row, first column. We carry out the pivoting by multiplying the first row by -2 and adding it to the second row. Then we multiply the first row by -1 and add it to the third row.

x	y	z	1	
1	1	1	6	
0	-3*	-3	-15	New Row 2 = -2(Row 1) + (Row 2)
0	-3	2	0	New Row 3 = -1(Row 1) + (Row 3)

The next pivot is the -3 in the second row, second column. To carry out the pivoting we first multiply the first row by 3 to make the element above the pivot a multiple of -3.

x	y	z	1	
3	3	3	18	New Row 1 = 3(Row 1)
0	-3*	-3	-15	
0	-3	2	0	

Then we add the second row to the first row. We also multiply the second row by -1 and add it to the third row.

x	y	z	1	
3	0	0	3	New Row 1 = (Row 2) + (Row 1)
0	-3	-3	-15	
0	0	5*	15	New Row 3 = -1(Row 2) + (Row 3)

The next pivot is the 5 in the third row, third column. To carry out the pivoting we first multiply the second row by 5 to make the element directly above the pivot a multiple of 5.

x	y	z	1	
3	0	0	3	
0	-15	-15	-75	New Row 2 = 5(Row 2)
0	0	5*	15	

Then we multiply the third row by 3 and add it to the second row.

x	y	z	1	
3	0	0	3	
0	-15	0	-30	New Row 2 = 3(Row 3) + (Row 2)
0	0	5	15	

11. (continued)

We complete the solution by multiplying to obtain 1's down the main diagonal.

x	y	z	1	
1	0	0	1	New Row 1 = $\frac{1}{3}$(Row 1)
0	1	0	2	New Row 2 = $-\frac{1}{15}$(Row 2)
0	0	1	3	New Row 3 = $\frac{1}{5}$(Row 3)

We consider the corresponding system of equations and read off the solution.

   $x = 1, y = 2, z = 3$

The solution is (1, 2, 3).

13. $x + y + z = 6$
   $2x - y + 3z = 9$
   $-x + 2y + 2z = 9$
Write an echelon tableau.

x	y	z	1
1*	1	1	6
2	-1	3	9
-1	2	2	9

The first pivot is the 1 in the first row, first column. To carry out the pivoting we multiply the first row by -2 and add it to the second row. Then we add the first row to the third row.

x	y	z	1	
1	1	1	6	
0	-3*	1	-3	New Row 2 = -2(Row 1) + (Row 2)
0	3	3	15	New Row 3 = (Row 1) + (Row 3)

The next pivot is the 3 in the second row, second column. To carry out the pivoting we first multiply the first row by 3 to make the element above the pivot a multiple of -3.

x	y	z	1
3	3	3	18
0	-3*	1	-3
0	3	3	15

Then we add the second row to the first row and also to the third row.

x	y	z	1	
3	0	4	15	New Row 1 = (Row 2) + (Row 1)
0	-3	1	-3	
0	0	4*	12	New Row 3 = (Row 2) + (Row 3)

The next pivot is the 4 in the third row, third column. To carry out the pivoting we first multiply the second row by 4 to make the element directly above the pivot a multiple of 4.

13. (continued)

x	y	z	1
3	0	4	15
0	-12	4	-12
0	0	4	12

New Row 2 = 4(Row 2)

Then we multiply the third row by -1 and add it to both the first row and the third row.

x	y	z	1
3	0	0	3
0	-12	0	-24
0	0	4	12

New Row 1 = -1(Row 3) + (Row 1)
New Row 2 = -1(Row 3) + (Row 2)

We complete the solution by multiplying to obtain 1's down the main diagonal.

x	y	z	1
1	0	0	1
0	1	0	2
0	0	1	3

New Row 1 = $\frac{1}{3}$(Row 1)
New Row 2 = $-\frac{1}{12}$(Row 2)
New Row 3 = $\frac{1}{4}$(Row 3)

We consider the corresponding system of equations and read off the solution.

$x = 1, y = 2, z = 3$

The solution is (1, 2, 3).

15. $3a - 2b + 7c = 13$
  $a + 8b - 6c = -47$
  $7a - 9b - 9c = -3$

Write an echelon tableau.

a	b	c	1
3	-2	7	13
1	8	-6	-47
7	-9	-9	-3

We first interchange the first and second rows so that each number in the first column below the first number is a multiple of that number.

a	b	c	1
1*	8	-6	-47
3	-2	7	13
7	-9	-9	-3

Interchange Row 1 and Row 2

The first pivot is the 1 in the first row, first column. We carry out the pivoting by multiplying the first row by -3 and adding it to the second row. Then we multiply the first row by -7 and add it to the third row.

a	b	c	1
1	8	-6	-47
0	-26*	25	154
0	-65	33	326

New Row 2 = -3(Row 1) + (Row 2)
New Row 3 = -7(Row 1) + (Row 3)

15. (continued)

The next pivot is the -26 in the second row, second column. To carry out the pivoting we first multiply the first row by 13 (104 is the least common multiple of 8 and 26) and the second row by 2 (130 is the least common multiple of 26 and 65) to make the elements above and below the pivot multiples of -26.

a	b	c	1
13	104	-78	-611
0	-26*	25	154
0	-130	66	652

Now we multiply the second row by 4 and add it to the first row. Then we multiply the second row by -5 and add it to the third row.

a	b	c	1
13	0	22	5
0	-26	25	154
0	0	-59*	-118

New Row 1 = 4(Row 2) + (Row 1)
New Row 3 = -5(Row 2) + (Row 3)

The next point is the -59 in the third row, third column. To carry out the pivoting we first multiply the first and second rows by 59 to make the elements above the pivot multiples of -59.

a	b	c	1
767	0	1298	295
0	-1534	1475	9086
0	0	-59*	-118

New Row 1 = 59(Row 1)
New Row 2 = 59(Row 2)

Now we multiply the third row by 22 and add it to the first row. Then we multiply the third row by 25 and add it to the second row.

a	b	c	1
767	0	0	-2301
0	-1534	0	6136
0	0	-59	-118

New Row 1 = 22(Row 3) + (Row 1)
New Row 2 = 25(Row 3) + (Row 2)

We complete the solution by multiplying to obtain 1's down the main diagonal.

a	b	c	1
1	0	0	-3
0	1	0	-4
0	0	1	2

New Row 1 = $\frac{1}{767}$(Row 1)
New Row 2 = $-\frac{1}{1534}$(Row 2)
New Row 3 = $-\frac{1}{59}$(Row 3)

We consider the corresponding system of equations and read off the solution.

$a = -3, b = -4, c = 2$

The solution is (-3, -4, 2).

17. $3p \quad\;\; + 2r = 11$

$\quad\;\; q - 7r = 4$

$\;\, p - 6q \quad\;\; = 1$

Write an echelon tableau.

p	q	r	1
3	0	2	11
0	1	-7	4
1	-6	0	1

We first interchange the first and third rows so that each number in the first column below the first number is a multiple of that number.

p	q	r	1	
1*	-6	0	1	Interchange Row 1 and Row 3
0	1	-7	4	
3	0	2	11	

The first pivot is the 1 in the first row, first column. We carry out the pivoting by multiplying the first row by -3 and adding it to the third row.

p	q	r	1	
1	-6	0	1	
0	1*	-7	4	
0	18	2	8	New Row 3 = -3(Row 1) + (Row 3)

The next pivot is the 1 in the second row, second column. To carry out the pivoting we multiply the second row by 6 and add it to the first row. Then we multiply the second row by -18 and add it to the third row.

p	q	r	1	
1	0	-42	25	New Row 1 = 6(Row 2) + (Row 1)
0	1	-7	4	
0	0	128*	-64	New Row 3 = -18(Row 2) + (Row 3)

The next pivot is the 128 in the third row, third column. To carry out the pivoting we first multiply the first row by 64 (the least common multiple of 42 and 128 is 2688) and the second row by 128 to make the elements above the pivot multiples of 128.

p	q	r	1	
64	0	-2688	1600	New Row 1 = 64(Row 1)
0	128	-896	512	New Row 2 = 128(Row 2)
0	0	128*	-64	

Now we multiply the third row by 21 and add it to the first. Then we multiply the third row by 7 and add it to the second.

p	q	r	1	
64	0	0	256	New Row 1 = 21(Row 3) + (Row 1)
0	128	0	64	New Row 2 = 7(Row 3) + (Row 2)
0	0	128	-64	

17. (continued)

We complete the solution by multiplying to obtain 1's down the main diagonal.

p	q	r	1	
1	0	0	4	New Row 1 = $\frac{1}{64}$(Row 1)
0	1	0	$\frac{1}{2}$	New Row 2 = $\frac{1}{128}$(Row 2)
0	0	1	$-\frac{1}{2}$	New Row 3 = $\frac{1}{128}$(Row 3)

We consider the corresponding system of equations and read off the solution.

$p = 4, q = \frac{1}{2}, r = -\frac{1}{2}$

The solution is $\left(4, \frac{1}{2}, -\frac{1}{2}\right)$.

19. $w - x + y + z = 0$

$2w + 2x + y - z = 5$

$3w + x - y - z = -4$

$w + x - 3y - 2z = -7$

Write an echelon tableau.

w	x	y	z	1
1*	-1	1	1	0
2	2	1	-1	5
3	1	-1	-1	-4
1	1	-3	-2	-7

The first pivot is the 1 in the first row, first column. To carry out the pivoting we perform the operations indicated below.

w	x	y	z	1	
1	-1	1	1	0	
0	4*	-1	-3	5	New Row 2 = -2(Row 1) + (Row 2)
0	4	-4	-4	-4	New Row 3 = -3(Row 1) + (Row 3)
0	2	-4	-3	-7	New Row 4 = -1(Row 1) + (Row 4)

The next pivot is the 4 in the second row, second column. To carry out the pivoting we first multiply the first row by 4 and the fourth row by 2 to make all the elements above and below the pivot multiples of 4.

w	x	y	z	1	
4	-4	4	4	0	New Row 1 = 4(Row 1)
0	4*	-1	-3	5	
0	4	-4	-4	-4	
0	4	-8	-6	-14	New Row 4 = 2(Row 4)

19. (continued)

Then we perform the operations indicated below.

w	x	y	z	1
4	0	3	1	5
0	4	-1	-3	5
0	0	-3*	-1	-9
0	0	-7	-3	-19

New Row 1 = (Row 2) + (Row 1)

New Row 3 = -1(Row 2) + (Row 3)
New Row 4 = -1(Row 2) + (Row 4)

The next pivot is the -3 in the third row, third column. To carry out the pivoting we first multiply the second and fourth rows by 3 to make all the elements above and below the pivot multiples of -3.

w	x	y	z	1
4	0	3	1	5
0	12	-3	-9	15
0	0	-3*	-1	-9
0	0	-21	-9	-57

New Row 2 = 3(Row 2)

New Row 4 = 3(Row 4)

Then we perform the operations indicated below.

w	x	y	z	1
4	0	0	0	-4
0	12	0	-8	24
0	0	-3	-1	-9
0	0	0	-2*	6

New Row 1 = (Row 3) + (Row 1)
New Row 2 = -1(Row 3) + (Row 2)

New Row 4 = -7(Row 3) + (Row 4)

The next pivot is the -2 in the fourth row, fourth column. To carry out the pivoting we first multiply the third row by -2 to make all the elements above the pivot a multiple of -2.

w	x	y	z	1
4	0	0	0	-4
0	12	0	-8	24
0	0	6	2	18
0	0	0	-2*	6

New Row 3 = -2(Row 3)

Then we perform the operations indicated below.

w	x	y	z	1
4	0	0	0	-4
0	12	0	0	0
0	0	6	0	24
0	0	0	-2	6

New Row 2 = -4(Row 4) + (Row 2)
New Row 3 = (Row 4) + (Row 3)

We complete the solution by multiplying to obtain 1's down the main diagonal.

19. (continued)

w	x	y	z	1
1	0	0	0	-1
0	1	0	0	0
0	0	1	0	4
0	0	0	1	-3

We consider the corresponding system of equations and read off the solution.

$$w = -1, \ x = 0, \ y = 4, \ z = -3$$

The solution is (-1, 0, 4, -3).

21.
$$-w + 2x - 3y + z = -8$$
$$-w + x + y - z = -4$$
$$-w + x - y - z = -14$$
$$w + x + y + z = 22$$

Write an echelon tableau.

w	x	y	z	1
-1*	2	-3	1	-8
-1	1	1	-1	-4
-1	1	-1	-1	-14
1	1	1	1	22

The first pivot is the -1 in the first row, first column. We carry out the pivoting by performing the operations indicated below.

w	x	y	z	1
-1	2	-3	1	-8
0	-1*	4	-2	4
0	-1	2	-2	-6
0	3	-2	2	14

New Row 2 = -1(Row 1) + (Row 2)
New Row 3 = -1(Row 1) + (Row 3)
New Row 4 = (Row 1) + (Row 4)

The next pivot is the -1 in the second row, second column. We carry out the pivoting by performing the operations indicated below.

w	x	y	z	1
-1	0	5	-3	0
0	-1	4	-2	4
0	0	-2*	0	-10
0	0	10	-4	26

New Row 1 = 2(Row 2) + (Row 1)

New Row 3 = -1(Row 2) + (Row 3)
New Row 4 = 3(Row 2) + (Row 4)

The next pivot is the -2 in the third row, third column. To carry out the pivoting we first multiply the first row by 2 to make all the elements above and below the pivot multiples of -2.

w	x	y	z	1
-2	0	10	-6	0
0	-1	4	-2	4
0	0	-2*	0	-10
0	0	10	-4	26

New Row 1 = 2(Row 1)

21. (continued)

Then we perform the operations indicated below.

w	x	y	z		1
-2	0	0	-6		-50
0	-1	0	-2		-16
0	0	-2	0		-10
0	0	0	-4*		-24

New Row 1 = 5(Row 3) + (Row 1)
New Row 2 = 2(Row 3) + (Row 2)
New Row 4 = 5(Row 3) + (Row 4)

The next pivot is the -4 in the fourth row, fourth column. To carry out the pivoting we first multiply the first and second rows by 2 to make all the elements above the pivot multiples of -4.

w	x	y	z		1
-4	0	0	-12		-100
0	-2	0	-4		-32
0	0	-2	0		-10
0	0	0	-4*		-24

New Row 1 = 2(Row 1)
New Row 2 = 2(Row 2)

Then we perform the operations indicated below.

w	x	y	z		1
-4	0	0	0		-28
0	-2	0	0		-8
0	0	-2	0		-10
0	0	0	-4		-24

New Row 1 = -3(Row 4) + (Row 1)
New Row 2 = -1(Row 4) + (Row 2)

We compute the solution by multiplying to obtain 1's down the main diagonal.

w	x	y	z		1
1	0	0	0		7
0	1	0	0		4
0	0	1	0		5
0	0	0	1		6

New Row 1 = $-\frac{1}{4}$(Row 1)
New Row 2 = $-\frac{1}{2}$(Row 2)
New Row 3 = $-\frac{1}{2}$(Row 3)
New Row 4 = $-\frac{1}{4}$(Row 4)

We consider the corresponding system of equations and read off the solution.

$w = 7, x = 4, y = 5, z = 6$

The solution is $(7, 4, 5, 6)$.

23. Familiarize and translate. Let x = the first number, y = the second number, and z = the third number. Use the information given in the problem to write three equations.

The sum of the three numbers is five, so

$x + y + z = 5$.

The first number minus the second plus the third is 1 so

$x - y + z = 1$.

23. (continued)

The first minus the third is 3 more than the second so

$x - z = y + 3$, or $x - y - z = 3$.

Now we have a system of equations:

$x + y + z = 5$
$x - y + z = 1$
$x - y - z = 3$

Carry out. We solve the system. Write an echelon tableau.

x	y	z		1
1*	1	1		5
1	-1	1		1
1	-1	-1		3

The first pivot is the 1 in the first row, first column. Proceed as indicated below.

x	y	z		1
1	1	1		5
0	-2*	0		-4
0	-2	-2		-2

New Row 2 = -1(Row 1) + (Row 2)
New Row 3 = -1(Row 1) + (Row 3)

The next pivot is the -2 in the second row, second column. To carry out the pivoting we proceed as indicated below.

x	y	z		1
2	2	2		10
0	-2*	0		-4
0	-2	-2		-2

New Row 1 = 2(Row 1)

x	y	z		1
2	0	2		6
0	-2	0		-4
0	0	-2*		2

New Row 1 = (Row 2) + (Row 1)
New Row 3 = -1(Row 2) + (Row 3)

The next pivot is the -2 in the third row, third column. We proceed as indicated below.

x	y	z		1
2	0	0		8
0	-2	0		-4
0	0	-2		2

New Row 1 = (Row 3) + (Row 1)

x	y	z		1
1	0	0		4
0	1	0		2
0	0	1		-1

New Row 1 = $\frac{1}{2}$(Row 1)
New Row 2 = $-\frac{1}{2}$(Row 2)
New Row 3 = $-\frac{1}{2}$(Row 3)

The solution of the system is $(4, 2, -1)$.

23. (continued)

Check. First check to see if the sum of the three numbers is 5: $4 + 2 + (-1) = 5$

Next check to see if the first minus the second plus the third is 1: $4 - 2 + (-1) = 1$

Then check to see if the first minus the third is 3 more than the second: $4 - (-1) = 4 + 1 = 5 = 2 + 3$

The numbers check in the original problem.

State. The three numbers are 4, 2, and -1.

25. Familiarize and translate. Let a = the number of lenses polished by machine A in one week, b = the number of lenses polished by machine B in one week, and c = the number of lenses polished by machine C in one week. Use the information given to write three equations.

When all three machines are working, 5700 lenses can be polished in one week:

$a + b + c = 5700$

When only A and B are working, 3400 lenses can be polished in one week:

$a + b = 3400$

When only B and C are working, 4200 lenses can be polished in one week:

$b + c = 4200$

We have a system of equations:

$$a + b + c = 5700$$
$$a + b \quad\;\; = 3400$$
$$\quad\;\; b + c = 4200$$

Carry out. We solve the system. Write an echelon tableau.

a	b	c	1
1*	1	1	5700
1	1	0	3400
0	1	1	4200

The first pivot is the 1 in the first row, first column. We proceed as indicated.

a	b	c	1	
1	1	1	5700	
0	0	-1	-2300	New Row 2 = -1(Row 1) + (Row 2)
0	1	1	4200	

The next pivot location is the second row, second column. We interchange rows two and three to obtain a nonzero number in that position.

a	b	c	1	
1	1	1	5700	
0	1*	1	4200	Interchange Row 2 and Row 3
0	0	-1	-2300	

Then we proceed as indicated.

25. (continued)

a	b	c	1	
1	0	0	1500	New Row 1 = -1(Row 2) + (Row 1)
0	1	1	4200	
0	0	-1*	-2300	

The next pivot is the -1 in the third row, third column. We proceed as indicated.

a	b	c	1	
1	0	0	1500	
0	1	0	1900	New Row 2 = (Row 3) + (Row 2)
0	0	-1	-2300	

a	b	c	1	
1	0	0	1500	
0	1	0	1900	
0	0	1	2300	New Row 3 = -1(Row 3)

The solution of the system is (1500, 1900, 2300).

Check. When all three machines are working the number of lenses polished is $1500 + 1900 + 2300 = 5700$. When only A and B are working, the number of lenses polished is $1500 + 1900 = 3400$. When only B and C are working, the number of lenses polished is $1900 + 2300 = 4200$. The numbers check in the original problem.

State. Machine A polishes 1500 lenses, machine B polishes 1900 lenses, and machine C polishes 2300 lenses in one week.

27. Familiarize and translate. Let a, b, c = the number of gallons per hour pumped by pumps A, B, and C, respectively. Use the information given to write three equations.

When A, B, and C are running together they pump 3700 gallons per hour:

$a + b + c = 3700$

When only A and B are running, 2200 gallons per hour can be pumped:

$a + b = 2200$

When only A and C are running, 2400 gallons per hour can be pumped:

$a + c = 2400$

We have a system of equations:

$$a + b + c = 3700$$
$$a + b \quad\;\; = 2200$$
$$a \quad\;\; + c = 2400$$

Carry out. Solve the system. Write an echelon tableau.

a	b	c	1
1*	1	1	3700
1	1	0	2200
1	0	1	2400

27. (continued)

The first pivot is the 1 in the first row, first column. We proceed as indicated.

a	b	c		1	
1	1	1		3700	
0	0	-1		-1500	New Row 2 = -1(Row 1) + (Row 2)
0	-1	0		-1300	New Row 3 = -1(Row 1) + (Row 3)

The next pivot location is the second row, second column. We interchange rows two and three to obtain a nonzero number in that position.

a	b	c		1	
1	1	1		3700	
0	-1*	0		-1300	Interchange Row 2 and Row 3
0	0	-1		-1500	

Then we proceed as indicated.

a	b	c		1	
1	0	1		2400	New Row 1 = (Row 2) + (Row 1)
0	-1	0		-1300	
0	0	-1*		-1500	

The next pivot is the -1 in the third row, third column. We proceed as indicated.

a	b	c		1	
1	0	0		900	New Row 1 = (Row 3) + (Row 1)
0	-1	0		-1300	
0	0	-1		-1500	

a	b	c		1	
1	0	0		900	
0	1	0		1300	New Row 2 = -1(Row 2)
0	0	1		1500	New Row 3 = -1(Row 3)

The solution of the system is (900, 1300, 1500).

Check. When all three pumps are running, the number of gallons per hour that can be pumped is 900 + 1300 + 1500 = 3700.

When only A and B are running, the number of gallons per hour that can be pumped is 900 + 1300 = 2200.

When only A and C are running, the number of gallons per hour that can be pumped is 900 + 1500 = 2400.

The numbers check in the original problem.

State. The pumping capacities of pumps A, B, and C are 900, 1300, and 1500 gallons per hour, respectively.

29. Familiarize. Recall the formula for simple interest $I = Prt$, where I is interest, P is principal, r is the interest rate, and t is time. Let x = the amount invested at 7% and y = the amount invested at 8%.

Translate. The total amount invested is $8800 so
$$x + y = 8800.$$

The interest from x is 7%x, and the interest from y is 8%y. The total interest earned in one year is $663 so
$$7\%x + 8\%y = 663$$
or $0.07x + 0.08y = 663$

We have a system of equations.
$$x + y = 8800$$
$$0.07x + 0.08y = 663$$

Carry out. Solve the system. Write an echelon tableau.

x	y		1
1	1		8800
0.07	0.08		663

First we multiply the second row by 100 to clear of decimals.

x	y		1	
1*	1		8800	
7	8		66,300	New Row 2 = 100(Row 2)

The first pivot is the 1 in the first row, first column. We proceed as indicated.

x	y		1	
1	1		8800	
0	1*		4700	New Row 2 = -7(Row 1) + (Row 2)

The next pivot is the 1 in the second row, second column. We proceed as indicated.

x	y		1	
1	0		4100	New Row 1 = -1(Row 2) + Row 1
0	1		4700	

The solution of the system is (4100, 4700).

Check. The total amount invested is $4100 + $4700 = $8800.

In one year the $4100 investment earns 7%($4100) = 0.07($4100) = $287, and the $4700 investment earns 8%($4700) = 0.08($4700) = $376. The total interest is $287 + $376 = $663.

The numbers check in the original problem.

State. $4100 is invested at 7%, and $4700 is invested at 8%.

31. Familiarize. The formula for simple interest is explained in Exercise 29 above. Let x = the amount invested at 5% and y = the amount invested at 6%.

Translate. The interest from x is 5%x, and the interest from y is 6%y. The total interest earned in one year is $3900 so

$$5\%x + 6\%y = 3900$$

or $0.05x + 0.06y = 3900$.

The amount invested at 6% is $10,000 more than the amount invested at 5% so

$$y = x + 10,000$$

or $-x + y = 10,000$.

We have a system of equations:

$$0.05x + 0.06y = 3900$$
$$-x + y = 10,000$$

Carry out. Solve the system. Write an echelon tableau.

x	y	1
0.05	0.06	3900
-1	1	10,000

First we multiply the first row by 100 to clear of decimals.

x	y	1	
5	6	390,000	New Row 1 = 100(Row 1)
-1	1	10,000	

Then we interchange the rows so the element below the first pivot will be a multiple of the pivot.

x	y	1	
-1*	1	10,000	Interchange Row 1 and Row 2
5	6	390,000	

The first pivot is the -1 in the first row, first column. We proceed as indicated.

x	y	1	
-1	1	10,000	
0	11*	440,000	New Row 2 = 5(Row 1) + (Row 2)

The next pivot is the 11 in the second row, second column. We proceed as indicated.

x	y	1	
11	-11	-110,000	New Row 1 = -11(Row 1)
0	11*	440,000	

x	y	1	
11	0	330,000	New Row 1 = (Row 2) + (Row 1)
0	11	440,000	

x	y	1	
1	0	30,000	New Row 1 = $\frac{1}{11}$(Row 1)
0	1	40,000	New Row 2 = $\frac{1}{11}$(Row 2)

31. (continued)

The solution of the system is (30,000, 40,000).

Check. In one year the $30,000 investment earns 5%($30,000) = 0.05($30,000) = $1500, and the $40,000 investment earns 6%($40,000) = 0.06($40,000) = $2400. The total interest is $1500 + $2400 = $3900.

The amount invested at 6%, or $40,000, is $10,000 more than the amount invested at 5%, or $30,000.

The numbers check in the original problem.

State. $30,000 is invested at 5%, and $40,000 is invested at 6%.

33. Familiarize. Organize the information in a table. Let x = the amount of soybean meal to be used and y = the amount of cornmeal.

Type of meal	Soybean	Cornmeal	Total
Amount used	x	y	350 lb
Percent of protein	16%	9%	12%
Amount of protein in mixture	16%x	9%y	12%(350 lb), or 42 lb

$$16\%x + 9\% = 42$$
$$x + y = 350$$

Translate. Adding x and y in the second row we get 350. This gives us one equation:

$$x + y = 350$$

Adding the amounts of protein in the fourth row, we get 42. This gives us another equation:

$$16\%x + 9\%y = 42$$

or $0.16x + 0.09y = 42$.

We have a system of equations:

$$x + y = 350$$
$$0.16x + 0.09y = 42$$

Carry out. Solve the system. Write an echelon tableau.

x	y	1
1	1	350
0.16	0.09	42

First, multiply the second row by 100 to clear of decimals.

x	y	1	
1*	1	350	
16	9	4200	New Row 2 = 100(Row 2)

The first pivot is the 1 in the first row, first column. We proceed as indicated.

33. (continued)

x	y		1
1	1		350
0	-7*		-1400

New Row 2 = -16(Row 1) + (Row 2)

The next pivot is the -7 in the second row, second column. We proceed as indicated.

x	y		1
7	7		2450
0	-7*		-1400

New Row 1 = 7(Row 1)

x	y		1
7	0		1050
0	-7		-1400

New Row 1 = (Row 2) + (Row 1)

x	y		1
1	0		150
0	1		200

New Row 1 = $\frac{1}{7}$(Row 1)

New Row 2 = $-\frac{1}{7}$(Row 2)

The solution of the system is (150, 200).

Check. Amount of mixture: 150 lb + 200 lb = 350 lb

Amount of protein in mixture: 16%(150 lb) + 9%(200 lb) = 0.16(150 lb) + 0.09(200 lb) = 24 lb + 18 lb = 42 lb

The numbers check in the original problem.

State. 150 pounds of soybean meal and 200 pounds of cornmeal should be used.

35. Familiarize and translate. Let x = the number of nickels, y = the number of dimes, and z = the number of quarters. Use the information given to write three equations.

The total number of coins is 22:

$x + y + z = 22$

The total value is $2.90:

$0.05x + 0.10y + 0.25z = 2.90$

There are six more nickels than dimes:

$x = y + 6$

or $x - y = 6$

We have a system of equations:

$$x + y + z = 22$$
$$0.05x + 0.10y + 0.25z = 2.90$$
$$x - y = 6$$

Carry out. Solve the system. Write an echelon tableau.

x	y	z		1
1	1	1		22
0.05	0.10	0.25		2.90
1	-1	0		6

35. (continued)

We first multiply the second row by 100 to clear of decimals.

x	y	z		1
1*	1	1		22
5	10	25		290
1	-1	0		6

New Row 2 = 100(Row 2)

The first pivot is the 1 in the first row, first column. We proceed as indicated.

x	y	z		1
1	1	1		22
0	5*	20		180
0	-2	-1		-16

New Row 2 = -5(Row 1) + (Row 2)

New Row 3 = -1(Row 1) + (Row 3)

The next pivot is the 5 in the second row, second column. We proceed as indicated.

x	y	z		1
-5	-5	-5		-110
0	5*	20		180
0	-10	-5		-80

New Row 1 = -5(Row 1)

New Row 3 = 5(Row 3)

x	y	z		1
-5	0	15		70
0	5	20		180
0	0	35*		280

New Row 1 = (Row 2) + (Row 1)

New Row 3 = 2(Row 2) + (Row 3)

The next pivot is the 35 in the third row, third column. We carry out the pivoting by first multiplying the first and second row by 7 (105 is the least common multiple of 15 and 35 and 140 is the least common multiple of 20 and 35).

x	y	z		1
-35	0	105		490
0	35	140		1260
0	0	35*		280

New Row 1 = 7(Row 1)

New Row 2 = 7(Row 2)

Then we proceed as indicated.

x	y	z		1
-35	0	0		-350
0	35	0		140
0	0	35		280

New Row 1 = -3(Row 3) + (Row 1)

New Row 2 = -4(Row 3) + (Row 2)

x	y	z		1
1	0	0		10
0	1	0		4
0	0	1		8

New Row 1 = $-\frac{1}{35}$(Row 1)

New Row 2 = $\frac{1}{35}$(Row 2)

New Row 3 = $\frac{1}{35}$(Row 3)

The solution of the system is (10, 4, 8).

35. (continued)

Check. The total number of coins is 10 + 4 + 8 = 22. The total value of the coins is $0.05(10) + $0.10(4) + $0.25(8) = $0.50 + $0.40 + $2.00 = $2.90.

The number of nickels, 10, is six more than the number of dimes, 4.

The numbers check in the original problem.

State. There are 10 nickels, 4 dimes, and 8 quarters.

37. Familiarize. The formula for simple interest is explained in Exercise 29 above. Let x = the amount invested at 7%, y = the amount invested at 8%, and z = the amount invested at 9%.

Translate. The interest from x is 7%x, the interest from y is 8%y, and the interest from z is 9%z. The total interest earned is $212 so

$$7\%x + 8\%y + 9\%z = 212$$

or $0.07x + 0.08y + 0.09z = 212$.

The total amount invested is $2500 so

$$x + y + z = 2500.$$

There is $1100 more invested at 9% than at 8% so

$$z = y + 1100$$

or $-y + z = 1100$

We have a system of equations:

$$0.07x + 0.08y + 0.09z = 212$$
$$x + y + z = 2500$$
$$-y + z = 1100$$

Carry out. Solve the system. Write an echelon tableau.

x	y	z	1
0.07	0.08	0.09	212
1	1	1	2500
0	-1	1	1100

First, multiply the first row by 100 to clear of decimals.

x	y	z	1	
7	8	9	21,200	New Row 1 = 100(Row 1)
1	1	1	2500	
0	-1	1	1100	

Next, interchange the first and second rows so each element below the first pivot is a multiple of the pivot.

x	y	z	1	
1*	1	1	2500	Interchange Row 1
7	8	9	21,200	and Row 2
0	-1	1	1100	

37. (continued)

The first pivot is the 1 in the first row, first column. We proceed as indicated.

x	y	z	1	
1	1	1	2500	
0	1*	2	3700	New Row 2 = −7(Row 1) + (Row 2)
0	-1	1	1100	

The next pivot is the 1 in the second row, second column. We proceed as indicated.

x	y	z	1	
1	0	-1	-1200	New Row 1 = −1(Row 2) + (Row 1)
0	1	2	3700	
0	0	3*	4800	New Row 3 = (Row 2) + (Row 3)

The next pivot is the 3 in the third row, third column. We proceed as indicated.

x	y	z	1	
3	0	-3	-3600	New Row 1 = 3(Row 1)
0	3	6	11,100	New Row 2 = 3(Row 2)
0	0	3*	4800	

x	y	z	1	
3	0	0	1200	New Row 1 = (Row 3) + (Row 1)
0	3	0	1500	New Row 2 = −2(Row 3) + (Row 2)
0	0	3	4800	

x	y	z	1	
1	0	0	400	New Row 1 = $\frac{1}{3}$(Row 1)
0	1	0	500	New Row 2 = $\frac{1}{3}$(Row 2)
0	0	1	1600	New Row 3 = $\frac{1}{3}$(Row 3)

The solution of the system is (400, 500, 1600).

Check. Interest from 7% investment: 7%($400) = 0.07($400) = $28

Interest from 8% investment: 8%($500) = 0.08($500) = $40

Interest from 9% investment: 9%($1600) = 0.09($1600) = $144

Total interest earned: $28 + 40 + $144 = $212

Total amount invested: $400 + $500 + $1600 = $2500

The amount invested at 9%, $1600, is $1100 more than the amount invested at 8%, $500.

The numbers check in the original problem.

State. $400 is invested at 7%, $500 is invested at 8%, and $1600 is invested at 9%.

39. Use the Computer Software Supplement or a calculator to do this exercise.

$$4.12x - 1.35y - 18.2z = 601.3$$
$$-3.41x + 68.9y + 38.7z = 1777$$
$$0.955x - 0.813y - 6.53z = 160.2$$

Write an echelon tableau.

x	y	z	1
4.12*	-1.35	-18.2	601.3
-3.41	68.9	38.7	1777
0.955	-0.813	-6.53	160.2

The first pivot is the 4.12 in the first row, first column. We proceed as indicated.

x	y	z	1
4.12*	-1.35	-18.2	601.3
-14.0492	283.868	159.444	7321.24
3.9346	-3.34956	-26.9036	660.024

New Row 3 = -4.12(Row 3)
New Row 2 = 4.12(Row 2)

x	y	z	1
4.12	-1.35	-18.2	601.3
0	279.2645*	97.382	9371.673
0	-2.06031	-9.5226	85.7825

New Row 3 = -0.955(Row 1) + (Row 3)
New Row 2 = 3.41(Row 1) + (Row 2)

The next pivot is the 279.2645 in the second row, second column. We proceed as indicated.

x	y	z	1
1150.56974	-377.007075	-5082.6139	167,921.7439
0	279.2645*	97.382	9371.673
0	-575.371442	-2659.324128	23,956.00697

New Row 3 = 279.2645(Row 3)
New Row 1 = 279.2645(Row 1)

x	y	z	1
1150.56974	0	-4951.1482	180,573.5025
0	279.2645	97.382	9371.673
0	0	-2458.68702*	43,264.55857

New Row 3 = 2.06031(Row 2) + (Row 3)
New Row 1 = 1.35(Row 2) + (Row 1)

The next pivot is the -2458.68702 in the third row, third column. We proceed as indicated.

x	y	z	1
-2,828,890.885	0	12,173,328.81	-443,973,726.8
0	-686,624.0013	-239,431.8594	-23,042,010.76
0	0	-2458.68702*	43,264.55857

New Row 2 = -2458.68702(Row 2)
New Row 1 = -2458.68702(Row 1)

39. (continued)

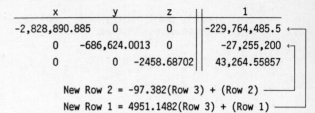

x	y	z	1
-2,828,890.885	0	0	-229,764,485.5
0	-686,624.0013	0	-27,255,200
0	0	-2458.68702	43,264.55857

New Row 2 = -97.382(Row 3) + (Row 2)
New Row 1 = 4951.1482(Row 3) + (Row 1)

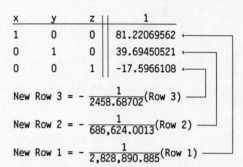

x	y	z	1
1	0	0	81.22069562
0	1	0	39.69450521
0	0	1	-17.5966108

New Row 3 = $-\dfrac{1}{2458.68702}$(Row 3)

New Row 2 = $-\dfrac{1}{686,624.0013}$(Row 2)

New Row 1 = $-\dfrac{1}{2,828,890.885}$(Row 1)

Rounding to the nearest hundredth, we see that the solution of the system is (81.22, 39.69, -17.60).

41. Substituting 1 for x and 4 for y gives us one equation:

$$4 = a(1)^2 + b(1) + c$$

or $4 = a + b + c$

Substituting -1 for x and -2 for y gives us another equation:

$$-2 = a(-1)^2 + b(-1) + c$$

or $-2 = a - b + c$

Substituting 2 for x and 13 for y gives us a third equation:

$$13 = a(2)^2 + b(2) + c$$

or $13 = 4a + 2b + c$

Now we have a system of equations which we will write in standard form:

$$a + b + c = 4$$
$$a - b + c = -2$$
$$4a + 2b + c = 13$$

We solve the system. First, write an echelon tableau.

a	b	c	1
1*	1	1	4
1	-1	1	-2
4	2	1	13

The first pivot is the 1 in the first row, first column. We proceed as indicated.

a	b	c	1	
1	1	1	4	
0	-2*	0	-6	New Row 2 = -1(Row 1) + (Row 2)
0	-2	-3	-3	New Row 3 = -4(Row 1) + (Row 3)

41. (continued)

The next pivot is the -2 in the second row, second column. We proceed as indicated.

a	b	c		1	
2	2	2		8	New Row 1 = 2(Row 1)
0	-2*	0		-6	
0	-2	-3		-3	

a	b	c		1	
2	0	2		2	New Row 1 = (Row 2) + (Row 1)
0	-2	0		-6	
0	0	-3*		3	New Row 3 = -1(Row 2) + (Row 3)

The next pivot is the -3 in the third row, third column. We proceed as indicated.

a	b	c		1	
6	0	6		6	New Row 1 = 3(Row 1)
0	-2	0		-6	
0	0	-3*		3	

a	b	c		1	
6	0	0		12	New Row 1 = 2(Row 3) + (Row 1)
0	-2	0		-6	
0	0	-3		3	

a	b	c		1	
1	0	0		2	New Row 1 = $\frac{1}{6}$(Row 1)
0	1	0		3	New Row 2 = $-\frac{1}{2}$(Row 2)
0	0	1		-1	New Row 3 = $-\frac{1}{3}$(Row 3)

The solution of the system is (2, 3, -1), or a = 2, b = 3, c = -1. Substituting these values in the equation $y = ax^2 + bx + c$, we get $y = 2x^2 + 3x - 1$.

43. a) We must find a quadratic function $y = ax^2 + bx + c$ which contains the points (1, 1000), (2, 2000), and (3, 8000).

Substituting 1 for x and 1000 for y gives us one equation:

$$1000 = a(1)^2 + b(1) + c$$

or $1000 = a + b + c$

Substituting 2 for x and 2000 for y gives us another equation:

$$2000 = a(2)^2 + b(2) + c$$

or $2000 = 4a + 2b + c$

Substituting 3 for x and 8000 for y gives us a third equation:

$$8000 = a(3)^2 + b(3) + c$$

or $8000 = 9a + 3b + c$

Now we have a system of equations:

$$a + b + c = 1000$$
$$4a + 2b + c = 2000$$
$$9a + 3b + c = 8000$$

43. (continued)

We solve the system. First, write an echelon tableau.

a	b	c		1
1*	1	1		1000
4	2	1		2000
9	3	1		8000

The first pivot is the 1 in the first row, first column. We proceed as indicated.

a	b	c		1	
1	1	1		1000	
0	-2*	-3		-2000	New Row 2 = -4(Row 1) + (Row 2)
0	-6	-8		-1000	New Row 3 = -9(Row 1) + (Row 3)

The next pivot is the -2 in the second row, second column. We proceed as indicated.

a	b	c		1	
2	2	2		2000	New Row 1 = 2(Row 1)
0	-2*	-3		-2000	
0	-6	-8		-1000	

a	b	c		1	
2	0	-1		0	New Row 1 = (Row 2) + (Row 1)
0	-2	-3		-2000	
0	0	1*		5000	New Row 3 = -3(Row 2) + (Row 3)

The next pivot is the 1 in the third row, third column. We proceed as indicated.

a	b	c		1	
2	0	0		5000	New Row 1 = (Row 3) + (Row 1)
0	-2	0		13,000	New Row 2 = 3(Row 3) + (Row 2)
0	0	1		5000	

a	b	c		1	
1	0	0		2500	New Row 1 = $\frac{1}{2}$(Row 1)
0	1	0		-6500	New Row 2 = $-\frac{1}{2}$(Row 2)
0	0	1		5000	

We read off the solution, a = 2500, b = -6500, c = 5000. Substituting these values in the quadratic function $y = ax^2 + bx + c$, we get $y = 2500x^2 - 6500x + 5000$.

b) To predict earnings for the fourth month, we substitute 4 for x in the function found in (a):

$$y = 2500(4)^2 - 6500(4) + 5000$$
$$= 2500(16) - 6500(4) + 5000$$
$$= 40,000 - 26,000 + 5000$$
$$= 19,000$$

The predicted earnings for the fourth month are $19,000.

45. Akron's consolidated net income, A, consists of its 80% share of Benson's consolidated net income, B, plus its 70% share of Caskin's consolidated net income, C, plus its own $190,000 net income. Thus,

$$A = 80\%B + 70\%C + 190,000$$

or $A = 0.8B + 0.7C + 190,000$.

Benson's consolidated net income, B, consists of its 15% share of Caskin's consolidated net income plus its own $170,000 net income. Therefore,

$$B = 15\%C + 170,000$$

or $B = 0.15C + 170,000$.

Caskin's consolidated net income, C, consists of its 25% share of Akron's consolidated net income plus its own $230,000 net income. Then,

$$C = 25\%A + 230,000$$

or $C = 0.25A + 230,000$.

47. Use a calculator to do this exercise.

$$1.01x - 0.905y + 2.12z = -2.54$$
$$1.32x + 2.05y + 2.97z = 3.97$$
$$2.21x + 1.35y + 0.001z = -3.15$$

Write an echelon tableau.

x	y	z	1
1.01*	-0.905	2.12	-2.54
1.32	2.05	2.97	3.97
2.21	1.35	0.001	-3.15

The first pivot is 1.01 in the first row, first column. We proceed as indicated.

x	y	z	1
1.01*	-0.905	2.12	-2.54
1.332	2.0705	2.9997	4.0097 ←
2.2321	1.3635	0.00101	-3.1815 ←

New Row 3 = 1.01(Row 3)
New Row 2 = 1.01(Row 2)

x	y	z	1
1.01	-0.905	2.12	-2.54
0	3.2651*	0.2013	7.3625 ←
0	3.36355	-4.68419	2.4319 ←

New Row 3 = -2.21(Row 1) + (Row 3)
New Row 2 = -1.32(Row 1) + (Row 2)

The next pivot is the 3.2651 in the second row, second column. We proceed as indicated.

x	y	z	1
3.297751	-2.9549155	6.922012	-8.293354
0	3.2651*	0.2013	7.3625 ←
0	10.98232711	-15.29434877	7.94039669 ←

New Row 3 = 3.2651(Row 3)
New Row 2 = 3.2651(Row 2)

47. (continued)

x	y	z	1
3.297751	0	7.1041885	-1.6302915 ←
0	3.2651	0.2013	7.3625
0	0	-15.97143139*	-16.82374019 ←

New Row 3 = -3.36355(Row 2) + (Row 3)
New Row 1 = 0.905(Row 2) + (Row 1)

The next pivot is the -15.97143139 in the third row, third column. We proceed as indicated.

x	y	z	1
-52.66980384	0	-113.4640592	26.03808884 ←
0	-52.14832068	-3.215049139	-117.5896636 ←
0	0	-15.97143139*	-16.82374019

New Row 2 = -15.97143139(Row 2)
New Row 1 = -15.97143139(Row 1)

x	y	z	1
-52.66980384	0	0	145.5571104 ←
0	-52.14832068	0	-114.1829147 ←
0	0	-15.97143139	-16.82374019

New Row 2 = -0.2013(Row 3) + (Row 2)
New Row 1 = -7.1041885(Row 3) + (Row 1)

x	y	z	1
1	0	0	-2.763577986 ←
0	1	0	2.189579898 ←
0	0	1	1.053364585 ←

New Row 3 = $-\dfrac{1}{15.97143139}$(Row 3)

New Row 2 = $-\dfrac{1}{52.14832068}$(Row 2)

New Row 1 = $-\dfrac{1}{52.66980384}$(Row 1)

Rounding to the nearest hundredth, we see that the solution is (-2.76, 2.19, 1.05).

49. Familiarize and translate. Let x = the hundred's digit, y = the ten's digit, and z = the unit's digit. Use the given information to write three equations.

The sum of the digits is 14, so

$$x + y + z = 14.$$

The ten's digit is 2 more than the unit's digit, so

$$y = z + 2$$

or $y - z = 2$.

If the digits are reversed the number is unchanged, so

$$100z + 10y + x = 100x + 10y + z$$

or $-99x + 99z = 0$.   (Adding -100x, -10y, and -z)

49. (continued)

Now we have a system of equations:
$$x + y + z = 14$$
$$y - z = 2$$
$$-99x + 99z = 0$$

Carry out. Solve the system. Write an echelon tableau.

x	y	z	1
1*	1	1	14
0	1	-1	2
-99	0	99	0

The first pivot is the 1 in the first row, first column. We proceed as indicated.

x	y	z	1	
1	1	1	14	
0	1*	-1	2	
0	99	198	1386	New Row 3 = 99(Row 1) + (Row 3)

The next pivot is the 1 in the second row, second column. We proceed as indicated.

x	y	z	1	
1	0	2	12	New Row 1 = -1(Row 2) + (Row 1)
0	1	-1	2	
0	0	297*	1188	New Row 3 = -99(Row 2) + (Row 3)

The next pivot is the 297 in the third row, third column. We proceed as indicated.

x	y	z	1	
297	0	594	3564	New Row 1 = 297(Row 1)
0	297	-297	594	New Row 2 = 297(Row 2)
0	0	297*	1188	

x	y	z	1	
297	0	0	1188	New Row 1 = -2(Row 3) + (Row 1)
0	297	0	1782	New Row 2 = (Row 3) + (Row 2)
0	0	297	1188	

x	y	z	1	
1	0	0	4	New Row 1 = $\frac{1}{297}$(Row 1)
0	1	0	6	New Row 2 = $\frac{1}{297}$(Row 2)
0	0	1	4	New Row 3 = $\frac{1}{297}$(Row 3)

The solution of the system is (4, 6, 4).

49. (continued)

Check. See if the number 464 fits the conditions in the original problem.

Sum of the digits: $4 + 6 + 4 = 14$

The ten's digit, 6, is two more than the unit's digit, 4.

If the digits are reversed, 464, the number is unchanged.

State. The number is 464.

51. Familiarize and translate. We know that the thousand's digit must be 1. Since the problem tells us that the thousand's digit is the same as the ten's digit, we also know that the ten's digit is 1. Let x = the hundred's digit and y = the one's digit. Use the remaining information given to write two equations.

The sum of the digits is 13 so
$$1 + x + 1 + y = 13$$
$$\text{or} \quad x + y + 2 = 13$$
$$\text{or} \quad x + y = 11.$$

The hundred's digit is seven more than the one's digit so
$$x = y + 7$$
$$\text{or } x - y = 7.$$

Now we have a system of equations:
$$x + y = 11$$
$$x - y = 7$$

Carry out. Solve the system. Write an echelon tableau.

x	y	1
1*	1	11
1	-1	7

The first pivot is the 1 in the first row, first column. We proceed as indicated.

x	y	1	
1	1	11	
0	-2*	-4	New Row 2 = -1(Row 1) + (Row 2)

The next pivot is the -2 in the second row, second column. We proceed as indicated.

x	y	1	
2	2	22	New Row 1 = 2(Row 1)
0	-2*	-4	

x	y	1	
2	0	18	New Row 1 = (Row 2) + (Row 1)
0	-2	-4	

x	y	z	
1	0	9	New Row 1 = $\frac{1}{2}$(Row 1)
0	1	2	New Row 2 = $-\frac{1}{2}$(Row 2)

51. (continued)

The solution of the system is (9, 2).

Check. See if the year 1912 checks in the original problem.

Sum of the digits: $1 + 9 + 1 + 2 = 13$

The hundred's digit, 9, is seven more than the one's digit, 2.

The thousand's digit is the same as the ten's digit.

State. The year is 1912.

53. Substituting 4 for x and 2 for y, we get an equation:

$$2 = a(4)^2 + b(4) + c$$

or $2 = 16a + 4b + c$

Substituting 2 for x and 0 for y, we get another equation:

$$0 = a(2)^2 + b(2) + c$$

$$0 = 4a + 2b + c$$

Substituting 1 for x and 2 for y, we get a third equation:

$$2 = a(1)^2 + b(1) + c$$

$$2 = a + b + c$$

Now we have a system of equations:

$$16a + 4b + c = 2$$
$$4a + 2b + c = 0$$
$$a + b + c = 2$$

We solve the system. First, write an echelon tableau.

a	b	c	1
16	4	1	2
4	2	1	0
1	1	1	2

Interchange the first and third rows so all the elements below the first pivot are multiples of the pivot.

a	b	c	1
1*	1	1	2
4	2	1	0
16	4	1	2

The first pivot is the 1 in the first row, first column. We proceed as indicated.

a	b	c	1	
1	1	1	2	
0	-2*	-3	-8	New Row 2 = -4(Row 1) + (Row 2)
0	-12	-15	-30	New Row 3 = -16(Row 1) + (Row 3)

The next pivot is the -2 in the second row, second column. We proceed as indicated.

53. (continued)

a	b	c	1	
2	2	2	4	New Row 1 = 2(Row 1)
0	-2*	-3	-8	
0	-12	-15	-30	

a	b	c	1	
2	0	-1	-4	New Row 1 = (Row 2) + (Row 1)
0	-2	-3	-8	
0	0	3*	18	New Row 3 = -6(Row 2) + (Row 3)

The next pivot is the 3 in the third row, third column. We proceed as indicated.

a	b	c	1	
6	0	-3	-12	New Row 1 = 3(Row 1)
0	-2	-3	-8	
0	0	3	18	

a	b	c	1	
6	0	0	6	New Row 1 = (Row 3) + (Row 1)
0	-2	0	10	New Row 2 = (Row 3) + (Row 2)
0	0	3	18	

a	b	c	1	
1	0	0	1	New Row 1 = $\frac{1}{6}$(Row 1)
0	1	0	-5	New Row 2 = $-\frac{1}{2}$(Row 2)
0	0	1	6	New Row 3 = $\frac{1}{3}$(Row 3)

We read off the solution a = 1, b = -5, c = 6. Substituting these values in the equation $y = ax^2 + bx + c$, we get $y = x^2 - 5x + 6$.

55. Substituting 1 for x, $\frac{3}{4}$ for y and 3 for z, we get

$$A + \frac{3}{4}B + 3C = 12.$$

Substituting $\frac{4}{3}$ for x, 1 for y, and 2 for z, we get

$$\frac{4}{3}A + B + 2C = 12.$$

Substituting 2 for x, 1 for y, and 1 for z, we get

$$2A + B + C = 12.$$

Now we have a system of equations:

$$A + \frac{3}{4}B + 3C = 12$$

$$\frac{4}{3}A + B + 2C = 12$$

$$2A + B + C = 12$$

We solve the system. First write an echelon tableau.

55. (continued)

A	B	C	1
1	$\frac{3}{4}$	3	12
$\frac{4}{3}$	1	2	12
2	1	1	12

Multiply the first row by 4 and the second row by 3 to clear of fractions.

A	B	C	1	
4	3	12	48	New Row 1 = 4(Row 1)
4	3	6	36	New Row 2 = 3(Row 2)
2	1	1	12	

Interchange the first and third rows so the elements below the first pivot will be multiples of the pivot.

A	B	C	1	
2*	1	1	12	Interchange Row 1 and Row 3
4	3	6	36	
4	3	12	48	

The first pivot is the 2 in the first row, first column. We proceed as indicated.

A	B	C	1	
2	1	1	12	
0	1*	4	12	New Row 2 = -2(Row 1) + (Row 2)
0	1	10	24	New Row 3 = -2(Row 1) + (Row 3)

The next pivot is the 1 in the second row, second column. We proceed as indicated.

A	B	C	1	
2	0	-3	0	New Row 1 = -1(Row 2) + (Row 1)
0	1	4	12	
0	0	6*	12	New Row 3 = -1(Row 2) + (Row 3)

The next pivot is the 6 in the third row, third column. We proceed as indicated.

A	B	C	1	
4	0	-6	0	New Row 1 = 2(Row 1)
0	3	12	36	New Row 2 = 3(Row 2)
0	0	6*	12	

A	B	C	1	
4	0	0	12	New Row 1 = (Row 3) + (Row 1)
0	3	0	12	New Row 2 = -2(Row 3) + (Row 2)
0	0	6	12	

55. (continued)

A	B	C	1	
1	0	0	3	New Row 1 = $\frac{1}{4}$(Row 1)
0	1	0	4	New Row 2 = $\frac{1}{3}$(Row 2)
0	0	1	2	New Row 3 = $\frac{1}{6}$(Row 3)

We read off the solution, A = 3, B = 4, C = 2. Substituting in the equation Ax + By + Cz = 12, we get 3x + 4y + 2z = 12.

Exercise Set 2.4

1. $9x - 3y = 15$

$6x - 2y = 10$

We translate to the elimination tableau.

x	y	1
9*	-3	15
6	-2	10

Pivoting on the first pivot using the elimination method we obtain

x	y	1	
9*	-3	15	
18	-6	30	New Row 2 = 3(Row 2)

x	y	1	
9	-3	15	
0	0	0	New Row 2 = -2(Row 1) + (Row 2)

We have taken the pivoting as far as possible since both the x and y entries in the bottom row are 0. Translating back to a system of equations we obtain

$9x - 3y = 15$

$0 = 0.$

Since the equation 0 = 0 is obviously true we have a dependent system and every point on the line 9x - 3y = 15 is a solution of the system. We solve for x.

$9x = 3y + 15$

$x = \frac{3y + 15}{9} = \frac{3(y + 5)}{9} = \frac{y + 5}{3}$

The solution is $x = \frac{y + 5}{3}$, y = any real number. We could also solve for y.

$-3y = -9x + 15$

$y = \frac{-9x + 15}{-3} = \frac{-3(3x - 5)}{-3} = 3x - 5$

We could also state the solution as y = 3x - 5, x = any real number.

3.   5c + 2d = 24

30c + 12d = 10

We translate to the elimination tableau.

c	d		1
5*	2		24
30	12		10

Pivoting on the first pivot using the elimination method we obtain

c	d		1	
5	2		24	
0	0		-134	New Row 2 = -6(Row 1) + (Row 2)

We have taken the pivoting as far as possible since both the x and y entries in the bottom row are 0. Translating back to a system of equations we obtain

$$5x + 2y = 24$$
$$0 = -134.$$

Since the second equation is false, the system has no solution.

5.   3x + 2y = 5

4y = 10 - 6x

Write the second equation in standard form, Ax + By = C.

$$3x + 2y = 5$$
$$6x + 4y = 10$$

Write an elimination tableau.

x	y		1
3*	2		5
6	4		10

Pivoting on the first pivot we obtain

x	y		1	
3	2		5	
0	0		0	New Row 2 = -2(Row 1) + (Row 2)

We have taken the pivoting as far as possible since both the x and y entries on the bottom row are 0. Translating back to a system of equations we obtain

$$3x + 2y = 5$$
$$0 = 0.$$

Since the equation 0 = 0 is obviously true, we have a dependent system and every point on the line 3x + 2y = 5 is a solution of the system. We solve for x.

$$3x = 5 - 2y$$
$$x = \frac{5 - 2y}{3}$$

The solution is $x = \frac{5 - 2y}{3}$, y = any real number.

We could also solve for y.

$$2y = 5 - 3x$$
$$y = \frac{5 - 3x}{2}$$

5.   (continued)

We could also state the solution as $y = \frac{5 - 3x}{2}$, x = any real number.

Using the second expression for the solution, we can choose any real value for x and substitute to find the corresponding value for y.

When x = 1, $y = \frac{5 - 3 \cdot 1}{2} = \frac{5 - 3}{2} = \frac{2}{2} = 1$. This gives the solution (1, 1).

When x = 0, $y = \frac{5 - 3 \cdot 0}{2} = \frac{5 - 0}{2} = \frac{5}{2}$. This gives the solution $\left(0, \frac{5}{2}\right)$.

When $x = \frac{5}{3}$, $y = \frac{5 - 3 \cdot \frac{5}{3}}{2} = \frac{5 - 5}{2} = \frac{0}{2} = 0$. This gives the solution $\left(\frac{5}{3}, 0\right)$.

7.   12y - 8x = 6

4x + 3 = 6y

Write the equations in the standard form Ax + By = C.

$$-8x + 12y = 6$$
$$4x - 6y = -3$$

Write an elimination tableau.

x	y		1
-8	12		6
4	-6		-3

Pivoting on the first pivot we obtain

x	y		1	
4*	-6		-3	Interchange Row 1 and Row 2
-8	12		6	

x	y		1	
4	-6		-3	
0	0		0	New Row 2 = 2(Row 1) + (Row 2)

We have taken the pivoting as far as possible since both the x and y entries in the bottom row are 0. Translating back to a system of equations we obtain

$$4x - 6y = -3$$
$$0 = 0.$$

Since the equation 0 = 0 is obviously true we have a dependent system and every point on the line 4x - 6y = -3 is a solution of the system. We solve for x.

$$4x = 6y - 3$$
$$x = \frac{6y - 3}{4}$$

The solution is $x = \frac{6y - 3}{4}$, y = any real number.

We could also solve for y.

$$-6y = -4x - 3$$
$$y = \frac{-4x - 3}{-6} = \frac{-(4x + 3)}{-6} = \frac{4x + 3}{6}$$

7. (continued)

We could also express the solution as $y = \dfrac{4x + 3}{6}$, x = any real number.

Using the first expression for the solution, we can choose any real value for y and substitute to find the corresponding value for x.

When y = 0, $x = \dfrac{6 \cdot 0 - 3}{4} = \dfrac{0 - 3}{4} = -\dfrac{3}{4}$. This gives the solution $\left[-\dfrac{3}{4}, 0\right]$.

When $y = \dfrac{7}{6}$, $x = \dfrac{6 \cdot \frac{7}{6} - 3}{4} = \dfrac{7 - 3}{4} = \dfrac{4}{4} = 1$. This gives the solution $\left[1, \dfrac{7}{6}\right]$.

When y = 1, $x = \dfrac{6 \cdot 1 - 3}{4} = \dfrac{6 - 3}{4} = \dfrac{3}{4}$. This gives the solution $\left[\dfrac{3}{4}, 1\right]$.

9.
$x + 2y - z = -8$
$2x - y + z = 4$
$8x + y + z = 2$

Write an elimination tableau.

x	y	z		1
1*	2	-1		-8
2	-1	1		4
8	1	1		2

Pivoting on the first pivot we obtain

x	y	z		1	
1	2	-1		-8	
0	-5*	3		20	New Row 2 = -2(Row 1) + (Row 2)
0	-15	9		66	New Row 3 = -8(Row 1) + (Row 3)

Pivoting on the next pivot we obtain

x	y	z		1	
5	10	-5		-40	New Row 1 = 5(Row 1)
0	-5*	3		20	
0	-15	9		66	

x	y	z		1	
5	0	1		0	New Row 1 = 2(Row 2) + (Row 1)
0	-5	3		20	
0	0	0		6	New Row 3 = -3(Row 2) + (Row 3)

We have taken the pivoting as far as possible since the x, y, and z entries in the bottom row are all 0. Translating back to a system of equations we obtain

$5x + z = 0$
$-5y + 3z = 20$
$0 = 6.$

Since the third equation is false, the system has no solution.

11.
$2x + y - 3z = 1$
$x - 4y + z = 6$
$4x - 16y + 4z = 24$

Write an elimination tableau.

x	y	z		1
2	1	-3		1
1	-4	1		6
4	-16	4		24

Interchange the first and second rows so the elements below the first pivot are multiples of the pivot.

x	y	z		1
1*	-4	1		6
2	1	-3		1
4	-16	4		24

Pivoting on the first pivot we obtain

x	y	z		1	
1	-4	1		6	
0	9*	-5		-11	New Row 2 = -2(Row 1) + (Row 2)
0	0	0		0	New Row 3 = -4(Row 1) + (Row 3)

Pivoting on the next pivot we obtain

x	y	z		1	
9	-36	9		54	New Row 1 = 9(Row 1)
0	9*	-5		-11	
0	0	0		0	

x	y	z		1	
9	0	-11		10	New Row 1 = 4(Row 2) + (Row 1)
0	9	-5		-11	
0	0	0		0	

We have taken the pivoting as far as possible since the x, y, and z entries in the bottom row are all 0. We translate back to a system of equations.

$9x - 11z = 10$
$9y - 5z = -11$
$0 = 0$

Since the equation 0 = 0 is obviously true we have a dependent system. We solve for x and y. (Since each of the first two equations contains z we can express both x and y in terms of z.)

$9x - 11z = 10$
$9x = 11z + 10$
$x = \dfrac{11z + 10}{9}$

$9y - 5z = -11$
$9y = 5z - 11$
$y = \dfrac{5z - 11}{9}$

The solution is $x = \dfrac{11z + 10}{9}$, $y = \dfrac{5z - 11}{9}$, z = any real number.

11. (continued)

When $z = 0$, $x = \dfrac{11 \cdot 0 + 10}{9} = \dfrac{0 + 10}{9} = \dfrac{10}{9}$, and

$y = \dfrac{5 \cdot 0 - 11}{9} = \dfrac{0 - 11}{9} = -\dfrac{11}{9}$. One solution is

$\left[\dfrac{10}{9}, \ -\dfrac{11}{9}, \ 0 \right]$.

When $z = 1$, $x = \dfrac{11 \cdot 1 + 10}{9} = \dfrac{11 + 10}{9} = \dfrac{21}{9} = \dfrac{7}{3}$, and

$y = \dfrac{5 \cdot 1 - 11}{9} = \dfrac{5 - 11}{9} = -\dfrac{6}{9} = -\dfrac{2}{3}$. Another

solution is $\left[\dfrac{7}{3}, \ -\dfrac{2}{3}, \ 1 \right]$.

When $z = -1$, $x = \dfrac{11(-1) + 10}{9} = \dfrac{-11 + 10}{9} = -\dfrac{1}{9}$, and

$y = \dfrac{5(-1) - 11}{9} = \dfrac{-5 - 11}{9} = -\dfrac{16}{9}$. A third solution

is $\left[-\dfrac{1}{9}, \ -\dfrac{16}{9}, \ -1 \right]$.

13. $2x + y - 3z = 0$

$x - 4y + z = 0$

$4x - 16y + 4z = 0$

Note that this system is the same as the system in Exercise 11 except for the constant on the right side of each equation. Therefore, the results of pivoting to the left of the vertical lines in the elimination tableau will be the same for both systems. Since this system has all 0's to the right of the vertical lines, the pivoting will produce only 0's there also. The final tableau will be

x	y	z		1
9	0	-11		0
0	9	-5		0
0	0	0		0

(See Exercise 11)

Translating back to a system of equations we have

$9x - 11z = 0$

$9y - 5z = 0$

$0 = 0$

Again we have a dependent system since $0 = 0$ is obviously true. We solve for x and y. (Since each of the first two equations contains z, we can express both x and y in terms of z.)

$9x - 11z = 0$

$9x = 11z$

$x = \dfrac{11z}{9}$

$9y - 5z = 0$

$9y = 5z$

$y = \dfrac{5z}{9}$

The solution is $x = \dfrac{11z}{9}$, $y = \dfrac{5z}{9}$, z = any real number.

When $z = -9$, $x = \dfrac{11(-9)}{9} = \dfrac{-99}{9} = -11$, and

$y = \dfrac{5(-9)}{9} = -\dfrac{45}{9} = -5$. One solution is

$(-11, -5, -9)$.

13. (continued)

When $z = 0$, $x = \dfrac{11 \cdot 0}{9} = \dfrac{0}{9} = 0$, and $y = \dfrac{5 \cdot 0}{9} = \dfrac{0}{9} = 0$. Another solution is $(0, 0, 0)$.

When $z = 9$, $x = \dfrac{11 \cdot 9}{9} = 11$, and $y = \dfrac{5 \cdot 9}{9} = 5$. A third solution is $(11, 5, 9)$.

15. $x + y - z = -3$

$x + 2y + 2z = -1$

Write an elimination tableau.

x	y	z		1
1*	1	-1		-3
1	2	2		-1

Pivoting on the first pivot we obtain

x	y	z		1
1	1	-1		-3
0	1*	3		2

Pivoting on the next pivot we obtain

x	y	z		1
1	0	-4		-5
0	1	3		2

We have taken the pivoting as far as possible. We translate back to a system of equations.

$x - 4z = -5$

$y + 3z = 2$

Solve for x and y. (Since each equation contains z we can express both x and y in terms of z.)

$x - 4z = -5$

$x = 4z - 5$

$y + 3z = 2$

$y = -3z + 2$

The solution is $x = 4z - 5$, $y = -3z + 2$, z = any real number.

When $z = 1$, $x = 4 \cdot 1 - 5 = 4 - 5 = -1$, and $y = -3 \cdot 1 + 2 = -3 + 2 = -1$. One solution is $(-1, -1, 1)$.

When $z = 0$, $x = 4 \cdot 0 - 5 = 0 - 5 = -5$, and $y = -3 \cdot 0 + 2 = 0 + 2 = 2$. Another solution is $(-5, 2, 0)$.

When $z = 2$, $x = 4 \cdot 2 - 5 = 8 - 5 = 3$, and $y = -3 \cdot 2 + 2 = -6 + 2 = -4$. A third solution is $(3, -4, 2)$.

17. $2x + y + z = 0$

    $x + y - z = 0$

$x + 2y + 2z = 0$

Write an elimination tableau.

x	y	z	1
2	1	1	0
1	1	-1	0
1	2	2	0

Interchange the first and second rows so the elements below the first pivot will be multiples of the pivot.

x	y	z	1	
1*	1	-1	0	Interchange Row 1 and Row 2
2	1	1	0	
1	2	2	0	

Pivoting on the first pivot we obtain

x	y	z	1	
1	1	-1	0	
0	-1*	3	0	New Row 2 = -2(Row 1) + (Row 2)
0	1	3	0	New Row 3 = -1(Row 1) + (Row 3)

Pivoting on the next pivot we obtain

x	y	z	1	
1	0	2	0	New Row 1 = (Row 2) + (Row 1)
0	-1	3	0	
0	0	6*	0	New Row 3 = (Row 2) + (Row 3)

Pivoting on the next pivot we obtain

x	y	z	1	
-3	0	-6	0	New Row 1 = -3(Row 1)
0	2	-6	0	New Row 2 = -2(Row 2)
0	0	6*	0	

x	y	z	1	
-3	0	0	0	New Row 1 = (Row 3) + (Row 1)
0	2	0	0	New Row 2 = (Row 3) + (Row 2)
0	0	6	0	

x	y	z	1	
1	0	0	0	New Row 1 = $-\frac{1}{3}$(Row 1)
0	1	0	0	New Row 2 = $\frac{1}{2}$(Row 2)
0	0	1	0	New Row 3 = $\frac{1}{6}$(Row 3)

Translating back to a system of equations we get $x = 0$, $y = 0$, $z = 0$. The solution is (0, 0, 0).

19. $2x + 3y - 5z = -8$

$-4x - 6y + 10z = 17$

Write an elimination tableau.

x	y	z	1
2*	3	-5	-8
-4	-6	10	17

Pivoting on the first pivot we obtain

x	y	z	1	
2	3	-5	-8	
0	0	0	1	New Row 2 = 2(Row 1) + (Row 2)

We have taken the pivoting as far as possible since the x, y, and z entries on the bottom row are all 0. Translate back to a system of equations.

    $2x + 3y - 5z = -8$

              $0 = 1$

Since the second equation is false, the system has no solution.

21. $w + x + y + z = 4$

    $w + x + y + z = 3$

    $w + x + y + z = 3$

Write an elimination tableau.

w	x	y	z	1
1*	1	1	1	4
1	1	1	1	3
1	1	1	1	3

Pivoting on the first pivot we obtain

w	x	y	z	1	
1	1	1	1	4	
0	0	0	0	-1	New Row 2 = -1(Row 1) + (Row 2)
0	0	0	0	-1	New Row 3 = -1(Row 1) + (Row 3)

We have taken the pivoting as far as possible since the w, x, y, and z entries on the second and third rows are all 0. We translate back to a system of equations.

    $w + x + y + z = 4$

              $0 = -1$

              $0 = -1$

Since the equation $0 = -1$ is false, the system has no solution.

23. Each system that has a solution is consistent. The consistent systems are those in Exercises 1, 5, 7, 11, 13, 15, and 17.

Each system that has no solution is inconsistent. The inconsistent systems are those in Exercises 3, 9, 19, and 21.

23. (continued)

A system of linear equations is dependent if there is a system with fewer equations which is equivalent to the original system. If we obtain an obviously true statement such as 0 = 0, or if we find that two equations are identical while using the elimination method, then we know that the system is dependent. The dependent systems are those in Exercises 1, 5, 7, 11, 13, 15, and 21.

The systems that are not dependent are independent. Those in Exercises 3, 9, 17, and 19 are independent.

25. $x - 3y = 2$

$-2x + 6y = -4$

Write an elimination tableau.

x	y		1
1*	-3		2
-2	6		-4

Pivoting on the first pivot we obtain

x	y		1
1	-3		2
0	0		0

We have taken the pivoting as far as possible. We translate back to a system of equations.

$x - 3y = 2$

$0 = 0$

The true statement 0 = 0 tells us that the system is dependent. We solve for x. (We could also solve for y.)

$x - 3y = 2$

$x = 3y + 2$

The solution is $x = 3y + 2$, y = any real number.

27. $x - 3y = 2$

$-2x + 6y = -3$

Write an elimination tableau.

x	y		1
1*	-3		2
-2	6		-3

Pivoting on the first pivot we obtain

x	y		1
1	-3		2
0	0		1

We have taken the pivoting as far as possible. Translate back to a system of equations.

$x - 3y = 2$

$0 = 1$

Since the second equation is false, the system has no solution.

29. $4x + 12y + 16z = 4$

$3x + 4y + 7z = 3$

$x + 8y + 9z = 1$

Write an elimination tableau.

x	y	z		1
4	12	16		4
3	4	7		3
1	8	9		1

Interchange the first and third rows so the elements below the first pivot will be multiples of the pivot.

x	y	z		1	
1*	8	9		1	Interchange Row 1 and Row 3
3	4	7		3	
4	12	16		4	

Pivoting on the first pivot we obtain

x	y	z		1	
1	8	9		1	
0	-20*	-20		0	New Row 2 = -3(Row 1) + (Row 2)
0	-20	-20		0	New Row 3 = -4(Row 1) + (Row 3)

Pivoting on the next pivot we obtain

x	y	z		1	
5	40	45		5	New Row 1 = 5(Row 1)
0	-20*	-20		0	
0	-20	-20		0	

x	y	z		1	
5	0	5		5	New Row 1 = 2(Row 2) + (Row 1)
0	-20	-20		0	
0	0	0		0	New Row 3 = -1(Row 2) + (Row 3)

We have taken the pivoting as far as possible. Translate back to a system of equations.

$5x + 5z = 5$

$-20y - 20z = 0$

$0 = 0$

Since the third equation is obviously true we have a dependent system. Solve for x and y.

$5x + 5z = 5$

$5x = 5 - 5z$

$x = \dfrac{5 - 5z}{5} = \dfrac{5(1 - z)}{5}$

$x = 1 - z$

$-20y - 20z = 0$

$-20y = 20z$

$y = -z$

The solution is $x = 1 - z$, $y = -z$, z = any real number.

31. $3x + 4y + 5z = 0$
$4x + 12y + 16z = 0$
$x + 8y + 11z = 0$

Write an elimination tableau.

x	y	z		1
3	4	5		0
4	12	16		0
1	8	11		0

Interchange the first and third rows so all the elements below the first pivot are multiples of the pivot.

x	y	z		1	
1*	8	11		0	Interchange Row 1 and Row 3
4	12	16		0	
3	4	5		0	

Pivoting on the first pivot we obtain

x	y	z		1	
1	8	11		0	
0	-20*	-28		0	New Row 2 = -4(Row 1) + (Row 2)
0	-20	-28		0	New Row 3 = -3(Row 1) + (Row 3)

Pivoting on the next pivot we obtain

x	y	z		1	
5	40	55		0	New Row 1 = 5(Row 1)
0	-20*	-28		0	
0	-20	-28		0	

x	y	z		1	
5	0	1		0	New Row 1 = 2(Row 2) + (Row 1)
0	-20	-28		0	
0	0	0		0	New Row 3 = -1(Row 2) + (Row 3)

We have taken the pivoting as far as possible. Translate back to a system of equations.

$$5x + z = 0$$
$$-20y - 28z = 0$$
$$0 = 0$$

Since the last equation is obviously true, we have a dependent system. We solve for x and y.

$$5x + z = 0$$
$$5x = -z$$
$$x = -\frac{z}{5}$$
$$-20y - 28z = 0$$
$$-20y = 28z$$
$$y = -\frac{28z}{20}$$
$$y = -\frac{7z}{5}$$

The solution is $x = -\frac{z}{5}$, $y = -\frac{7z}{5}$, z = any real number.

33. $x - y - 6z = 0$
$x + y + 13z = 0$

Write an elimination tableau.

x	y	z		1
1*	-1	-6		0
1	1	13		0

Pivoting on the first pivot we obtain

x	y	z		1	
1	-1	-6		0	
0	2*	19		0	New Row 2 = -1(Row 1) + (Row 2)

Pivoting on the next pivot we obtain

x	y	z		1	
2	-2	-12		0	New Row 1 = 2(Row 1)
0	2*	19		0	

x	y	z		1	
2	0	7		0	New Row 1 = (Row 2) + (Row 1)
0	2	19		0	

We have taken the pivoting as far as possible. Translate back to a system of equations.

$$2x + 7z = 0$$
$$2y + 19z = 0$$

Solve for x and y.

$$2x + 7z = 0$$
$$2x = -7z$$
$$x = -\frac{7z}{2}$$
$$2y + 19z = 0$$
$$2y = -19z$$
$$y = -\frac{19z}{2}$$

The solution is $x = -\frac{7z}{2}$, $y = -\frac{19z}{2}$, z = any real number.

35. $3x \qquad -9z = 3$
$2x + y - z = 6$
$w + x + 2y + 7z = 7$

Write an elimination tableau.

w	x	y	z		1
0	3	0	-9		3
0	2	1	-1		6
1	1	2	7		7

Interchange the first and third rows to obtain a nonzero number in the first pivot location.

w	x	y	z		1	
1*	1	2	7		7	Interchange Row 1 and Row 3
0	2	1	-1		6	
0	3	0	-9		3	

35. (continued)

The elements below the first pivot are zero so we move to the next pivot.

w	x	y	z	1
1	1	2	7	7
0	2*	1	-1	6
0	3	0	-9	3

Pivoting on the next pivot we obtain

w	x	y	z	1	
-2	-2	-4	-14	-14	New Row 1 = -2(Row 1)
0	2*	1	-1	6	
0	6	0	-18	6	New Row 3 = 2(Row 3)

w	x	y	z	1	
-2	0	-3	-15	-8	New Row 1 = (Row 2) + (Row 1)
0	2	1	-1	6	
0	0	-3*	-15	-12	New Row 3 = -3(Row 2) + (Row 3)

Pivoting on the next pivot we obtain

w	x	y	z	1	
-2	0	-3	-15	-8	
0	6	3	-3	18	New Row 2 = 3(Row 2)
0	0	-3*	-15	-12	

w	x	y	z	1	
-2	0	0	0	4	New Row 1 = -1(Row 3) + (Row 1)
0	6	0	-18	6	New Row 2 = (Row 3) + (Row 2)
0	0	-3	-15	-12	

We have taken the pivoting as far as possible. Translate back to a system of equations.

$$-2w = 4$$
$$6x - 18z = 6$$
$$-3y - 15z = -12$$

Solving these equations we obtain

$$-2w = 4$$
$$w = -2$$
$$6x - 18z = 6$$
$$6x = 18z + 6$$
$$x = \frac{18z + 6}{6}$$
$$x = \frac{6(3z + 1)}{6}$$
$$x = 3z + 1$$

35. (continued)

$$-3y - 15z = -12$$
$$-3y = -12 + 15z$$
$$y = \frac{-12 + 15z}{-3}$$
$$y = \frac{-3(4 - 5z)}{-3}$$
$$y = 4 - 5z$$

The solution is $w = -2$, $x = 3z + 1$, $y = 4 - 5z$, z = any real number.

37. $$2x - 2y + 18z = -14$$
$$x - 2y + 13z = -4$$
$$-2y + 8z = 4$$
$$2x + y + 36z = 7$$

Write an elimination tableau.

x	y	z	1
2	-2	18	-14
1	-2	13	-4
0	-2	8	4
2	1	36	7

Interchange the first and second rows so all the elements below the first pivot are multiples of the pivot.

x	y	z	1	
1*	-2	13	-4	Interchange Row 1 and Row 2
2	-2	18	-14	
0	-2	8	4	
2	1	36	7	

Pivoting on the first pivot we obtain

x	y	z	1	
1	-2	13	-4	
0	2*	-8	-6	New Row 2 = -2(Row 1) + (Row 2)
0	-2	8	4	
0	5	10	15	New Row 4 = -2(Row 1) + (Row 4)

Pivoting on the next pivot we obtain

x	y	z	1	
1	-2	13	-4	
0	2*	-8	-6	
0	-2	8	4	
0	10	20	30	New Row 4 = 2(Row 4)

x	y	z	1	
1	0	5	-10	New Row 1 = (Row 2) + (Row 1)
0	2	-8	-6	
0	0	0	-2	New Row 3 = (Row 2) + (Row 3)
0	0	60	60	New Row 4 = -5(Row 2) + (Row 4)

37. (continued)

The third row has all 0's to the left of the vertical lines and a nonzero number to the right. It corresponds to a false equation, 0 = -2, so the system has no solution.

39.

w	x	y	z		1
1	2	-4	8		7
0	-3*	9	12		18
0	3	-9	-12		-18

Note that the pivoting in the first column is complete. The next pivot is the -3 in the second row, second column. Pivoting, we obtain

w	x	y	z		1	
3	6	-12	24		21	New Row 1 = 3(Row 1)
0	-3*	9	12		18	
0	3	-9	-12		-18	

w	x	y	z		1	
3	0	6	48		57	New Row 1 = 2(Row 2) + (Row 1)
0	-3	9	12		18	
0	0	0	0		0	New Row 3 = (Row 2) + (Row 3)

We have taken the pivoting as far as possible. Translate back to a system of equations.

$$3w + 6y + 48z = 57$$
$$-3x + 9y + 12z = 18$$
$$0 = 0$$

Since the equation 0 = 0 is obviously true we have a dependent system. The first two equations each contain y and z, so we can solve for w and x in terms of y and z.

$$3w + 6y + 48z = 57$$
$$3w = 57 - 6y - 48z$$
$$w = \frac{57 - 6y - 48z}{3}$$
$$w = \frac{3(19 - 2y - 16z)}{3}$$
$$w = 19 - 2y - 16z$$
$$-3x + 9y + 12z = 18$$
$$-3x = 18 - 9y - 12z$$
$$x = \frac{18 - 9y - 12z}{-3}$$
$$x = \frac{-3(-6 + 3y + 4z)}{-3}$$
$$x = -6 + 3y + 4z$$

The solution is w = 19 - 2y - 16z, x = -6 + 3y + 4z, y = any real number, z = any real number.

41.

x	y	z		1
1	-1	-2		-5
0	0	0		0
0	0	0		0
0	-2	4		-8

Note that the pivoting in the first column is complete. The next pivot location is the second row, second column. Interchange the second and fourth rows to obtain a nonzero number in that location.

x	y	z		1
1	-1	-2		-5
0	-2*	4		-8
0	0	0		0
0	0	0		0

Pivoting we obtain

x	y	z		1	
-2	2	4		10	New Row 1 = -2(Row 1)
0	-2*	4		-8	
0	0	0		0	
0	0	0		0	

x	y	z		1	
-2	0	8		2	New Row 1 = (Row 2) + (Row 1)
0	-2	4		-8	
0	0	0		0	
0	0	0		0	

We have taken the pivoting as far as possible. Translate back to a system of equations.

$$-2x + 8z = 2$$
$$-2y + 4z = -8$$
$$0 = 0$$
$$0 = 0$$

Since the equation 0 = 0 is obviously true, we have a dependent system. Solve for x and y.

$$-2x + 8z = 2$$
$$-2x = 2 - 8z$$
$$x = \frac{2 - 8z}{-2}$$
$$x = \frac{-2(-1 + 4z)}{-2}$$
$$x = -1 + 4z$$
$$-2y + 4z = -8$$
$$-2y = -8 - 4z$$
$$y = \frac{-8 - 4z}{-2}$$
$$y = \frac{-2(4 + 2z)}{-2}$$
$$y = 4 + 2z$$

The solution is x = -1 + 4z, y = 4 + 2z, z = any real number.

43.

v	w	x	y	z	1
1	0	8	-3	0	6
0	1	4	2	0	4
0	0	0	0	-2	10

Note that the pivoting in the first and second columns is complete. The next pivot location is the third row, third column. The only way to get a nonzero pivot there is to interchange the third and fifth columns (including the variables heading the columns). Instead of doing this, we can move to the element -2 in the third row, fifth column and consider it to be the next pivot. Since the rest of the fifth column already has 0's, the pivoting is complete. Translate back to a system of equations.

$$v + 8x - 3y = 6$$
$$w + 4x + 2y = 4$$
$$-2z = 10$$

Since each of the first two equations contains x and y, we can solve for v and w in terms of x and y.

$$v + 8x - 3y = 6$$
$$v = 6 - 8x + 3y$$
$$w + 4x + 2y = 4$$
$$w = 4 - 4x - 2y$$

We can solve the third equation for z.

$$-2z = 10$$
$$z = -5$$

The solution is $v = 6 - 8x + 3y$, $w = 4 - 4x - 2y$, x = any real number, y = any real number, z = -5.

45. Use the Computer Software Supplement or a calculator to do this exercise. Write each equation in standard form $Ax + By = C$.

$$4.026x - 1.448y = 18.32$$
$$-2.013x + 0.724y = -9.16$$

Write an elimination tableau.

x	y	1
4.026	-1.448	18.32
-2.013	0.724	-9.16

The first pivot position is the first row, first column. Proceed as indicated.

x	y	1	
-2.013*	0.724	-9.16	Interchange Row 1
4.026	-1.448	18.32	and Row 2

x	y	1	
-2.013	0.724	-9.16	
0	0	0	New Row 2 = 2(Row 1) + (Row 2)

We have taken the pivoting as far as possible. Translate back to a system of equations.

$$-2.013x + 0.724y = -9.16$$
$$0 = 0$$

45. (continued)

Since the equation 0 = 0 is obviously true, we have a dependent system. We solve for x. (We could also solve for y.)

$$-2.013x + 0.724y = -9.16$$
$$-2013x + 724y = -9160 \quad \text{Clearing of decimals}$$
$$-2013x = -724y - 9160$$
$$x = \frac{-724y - 9160}{-2013} = \frac{-(724y + 9160)}{-2013}$$
$$x = \frac{724y + 9160}{2013}$$

The solution is $x = \frac{724y + 9160}{2013}$, y = any real number.

47. a)
$$2u + 3v - 5w + 2x - 4y + z = 0$$
$$u - 4v + 6w - 2x + 7y - z = -1$$
$$u + v - w + 3x - 5y + z = 7$$

Write an elimination tableau.

u	v	w	x	y	z	1
2	3	-5	2	-4	1	0
1	-4	6	-2	7	-1	-1
1	1	-1	3	-5	1	7

The first pivot location is the first row, first column. We proceed as indicated.

u	v	w	x	y	z	1	
1*	-4	6	-2	7	-1	-1	Interchange Row 1
2	3	-5	2	-4	1	0	and Row 2
1	1	-1	3	-5	1	7	

u	v	w	x	y	z	1
1	-4	6	-2	7	-1	-1
0	11*	-17	6	-18	3	2
0	5	-7	5	-12	2	8

New Row 3 = -1(Row 1) + (Row 3)
New Row 2 = -2(Row 1) + (Row 2)

The next pivot is the 11 in the second row, second column. We proceed as indicated.

u	v	w	x	y	x	1	
11	-44	66	-22	77	-11	-11	New Row 1 = 11(Row 1)
0	11*	-17	6	-18	3	2	
0	55	-77	55	-132	22	88	New Row 3 = 11(Row 3)

u	v	w	x	y	z	1
11	0	-2	2	5	1	-3
0	11	-17	6	-18	3	2
0	0	8*	25	-42	7	78

New Row 3 = -5(Row 2) + (Row 3)
New Row 1 = 4(Row 2) + (Row 1)

47. (continued)

The next pivot is the 8 in the third row, third column. We proceed as indicated.

u	v	w	x	y	z	1
44	0	-8	8	20	4	-12
0	88	-136	48	-144	24	16
0	0	8*	25	-42	7	78

New Row 2 = 8(Row 2)
New Row 1 = 4(Row 1)

u	v	w	x	y	z	1
44	0	0	33	-22	11	66
0	88	0	473	-858	143	1342
0	0	8	25	-42	7	78

New Row 2 = 17(Row 3) + (Row 2)
New Row 1 = (Row 3) + (Row 1)

We have taken the pivoting as far as possible. Translate back to a system of equations.

$$44u + 33x - 22y + 11z = 66$$
$$88v + 473x - 858y + 143z = 1342$$
$$8w + 25x - 42y + 7z = 78$$

Since each equation contains x, y, and z, we can solve for u, v, and w in terms of x, y, and z.

$$44u + 33x - 22y + 11z = 66$$
$$44u = 66 - 33x + 22y - 11z$$
$$u = \frac{66 - 33x + 22y - 11z}{44}$$
$$u = \frac{11(6 - 3x + 2y - z)}{44}$$
$$u = \frac{6 - 3x + 2y - z}{4}$$

$$88v + 473x - 858y + 143z = 1342$$
$$88v = 1342 - 473x + 858y - 143z$$
$$v = \frac{1342 - 473x + 858y - 143z}{88}$$
$$v = \frac{11(122 - 43x + 78y - 13z)}{88}$$
$$v = \frac{122 - 43x + 78y - 13z}{8}$$

$$8w + 25x - 42y + 7z = 78$$
$$8w = 78 - 25x + 42y - 7z$$
$$w = \frac{78 - 25x + 42y - 7z}{8}$$

The solution is $u = \frac{6 - 3x + 2y - z}{4}$, $v = \frac{122 - 43x + 78y - 13z}{8}$, $w = \frac{78 - 25x + 42y - 7z}{8}$, x = any real number, y = any real number, z = any real number.

Note that these expressions for u, v, and w are equivalent to those in the answer section in the text. We can obtain the text expressions for u, v, and w by multiplying ours by -2, -1, and -1, respectively, in both the numerator and the denominator.

47. (continued)

b) The system has a solution, so it is consistent.

c) We cannot find a system with fewer equations which is equivalent to the original. That is, we did not find a tableau with a row consisting of all 0's nor did we find a tableau with two identical rows when we used the elimination method, so the system is independent.

49. $6x - 9y = -3$
$-4x + 6y = k$

Write an elimination tableau.

x	y	1
6*	-9	-3
-4	6	k

Pivoting on the first pivot we obtain

x	y	1
6	-9	-3
-12	18	3k

New Row 2 = 3(Row 2)

x	y	1
6	-9	-3
0	0	-6 + 3k

New Row 2 = 2(Row 1) + (Row 2)

Translate back to a system of equations.

$$6x - 9y = -3$$
$$0 = -6 + 3k$$

The system is dependent if the second equation is true. We solve for k.

$$0 = -6 + 3k$$
$$-3k = -6$$
$$k = 2$$

The system is dependent for k = 2.

51.   <u>Familiarize and translate.</u> Let x = the number of par-3 holes, y = the number of par-4 holes, and z = the number of par-5 holes. Use the information given to write 3 equations.

The golf course has 18 holes so

$$x + y + z = 18.$$

A golfer who shoots par on every hole has a score of 72 so

$$3x + 4y + 5z = 72.$$

There are the same number of par-3 holes as par-5 holes so

$$x = z$$

or $x - z = 0$.

We have a system of equations.

$$x + y + z = 18$$
$$3x + 4y + 5z = 72$$
$$x \quad\quad - z = 0$$

51. (continued)

<u>Carry out</u>. Solve the system. Write an elimination tableau.

x	y	z	1
1*	1	1	18
3	4	5	72
1	0	-1	0

Pivoting on the first pivot we obtain

x	y	z	1	
1	1	1	18	
0	1*	2	18	New Row 2 = -3(Row 1) + (Row 2)
0	-1	-2	-18	New Row 3 = -1(Row 1) + (Row 3)

Pivoting on the next pivot we obtain

x	y	z	1	
1	0	-1	0	New Row 1 = -1(Row 2) + (Row 1)
0	1	2	18	
0	0	0	0	New Row 3 = (Row 2) + (Row 3)

We have taken the pivoting as far as possible. Translate back to a system of equations.

$$x - z = 0$$
$$y + 2z = 18$$
$$0 = 0$$

Since the equation $0 = 0$ is obviously true, we have a dependent system. We solve for x and y.

$$x - z = 0$$
$$x = z$$

$$y + 2z = 18$$
$$y = 18 - 2z$$

The solution of the system is $x = z$, $y = 18 - 2z$, z = any real number. In terms of this application, however, we must restrict z to integer values such that $1 \le z \le 8$ to satisfy the condition that there is at least one of each type of hole.

<u>Check</u>. The check is left to the reader.

<u>State</u>. We can state the possibilities as follows:

Value of z	Par-3 holes(x)	Par-4 holes(y)	Par-5 holes(z)
1	1	16	1
2	2	14	2
3	3	12	3
4	4	10	4
5	5	8	5
6	6	6	6
7	7	4	7
8	8	2	8

Exercise Set 2.5

1.
$$A = \begin{bmatrix} 1 & 2 \\ 4 & 3 \end{bmatrix}$$

A has 2 rows and 2 columns, so it is a 2×2 matrix.

3.
$$J = \begin{bmatrix} -1 & -2 & -3 \\ 3 & 2 & 1 \end{bmatrix}$$

J has 2 rows and 3 columns, so it is a 2×3 matrix.

5.
$$A + B = \begin{bmatrix} 1 & 2 \\ 4 & 3 \end{bmatrix} + \begin{bmatrix} -3 & 5 \\ 2 & -1 \end{bmatrix} = \begin{bmatrix} 1 + (-3) & 2 + 5 \\ 4 + 2 & 3 + (-1) \end{bmatrix} =$$

$$\begin{bmatrix} -2 & 7 \\ 6 & 2 \end{bmatrix}$$

7. When a zero matrix is added to another matrix of the same dimensions, that same matrix is obtained. Therefore, $E + 0 = E$.

9.
$$3F = \begin{bmatrix} 3 \cdot 3 & 3 \cdot 3 \\ 3(-1) & 3(-1) \end{bmatrix} = \begin{bmatrix} 9 & 9 \\ -3 & -3 \end{bmatrix}$$

11. We found 3F in Exercise 9.

$$2A = \begin{bmatrix} 2 \cdot 1 & 2 \cdot 2 \\ 2 \cdot 4 & 2 \cdot 3 \end{bmatrix} = \begin{bmatrix} 2 & 4 \\ 8 & 6 \end{bmatrix}$$

$$3F + 2A = \begin{bmatrix} 9 & 9 \\ -3 & -3 \end{bmatrix} + \begin{bmatrix} 2 & 4 \\ 8 & 6 \end{bmatrix} = \begin{bmatrix} 9 + 2 & 9 + 4 \\ -3 + 8 & -3 + 6 \end{bmatrix} =$$

$$\begin{bmatrix} 11 & 13 \\ 5 & 3 \end{bmatrix}$$

13. $B - A = B + (-A)$

$$= \begin{bmatrix} -3 & 5 \\ 2 & -1 \end{bmatrix} + \begin{bmatrix} -1 & -2 \\ -4 & -3 \end{bmatrix} \quad \text{(Replacing each element of A by its additive inverse)}$$

$$= \begin{bmatrix} -3 + (-1) & 5 + (-2) \\ 2 + (-4) & -1 + (-3) \end{bmatrix}$$

$$= \begin{bmatrix} -4 & 3 \\ -2 & -4 \end{bmatrix}$$

15. We multiply the rows of B by the columns of A.

$$BA = \begin{bmatrix} -3 & 5 \\ 2 & -1 \end{bmatrix} \begin{bmatrix} 1 & 2 \\ 4 & 3 \end{bmatrix} = \begin{bmatrix} -3 \cdot 1 + 5 \cdot 4 & -3 \cdot 2 + 5 \cdot 3 \\ 2 \cdot 1 - 1 \cdot 4 & 2 \cdot 2 - 1 \cdot 3 \end{bmatrix}$$

$$= \begin{bmatrix} 17 & 9 \\ -2 & 1 \end{bmatrix}$$

17. $J - K = J + (-K)$

$$= \begin{bmatrix} -1 & -2 & -3 \\ 3 & 2 & 1 \end{bmatrix} + \begin{bmatrix} 0 & -8 & 4 \\ -1 & 0 & 1 \end{bmatrix} \quad \text{(Replacing each element of A by its additive inverse)}$$

$$= \begin{bmatrix} -1 + 0 & -2 + (-8) & -3 + 4 \\ 3 + (-1) & 2 + 0 & 1 + 1 \end{bmatrix}$$

$$= \begin{bmatrix} -1 & -10 & 1 \\ 2 & 2 & 2 \end{bmatrix}$$

19. J and K are both 2 × 3 matrices. Since the number of columns of J, 3, is not the same as the number of rows of K, 2, it is not possible to find JK.

21.

$$AI = \begin{bmatrix} 1 & 2 \\ 4 & 3 \end{bmatrix} \begin{bmatrix} 1 & 0 \\ 0 & 1 \end{bmatrix}$$

$$= \begin{bmatrix} 1 \cdot 1 + 2 \cdot 0 & 1 \cdot 0 + 2 \cdot 1 \\ 4 \cdot 1 + 3 \cdot 0 & 4 \cdot 0 + 3 \cdot 1 \end{bmatrix}$$

$$= \begin{bmatrix} 1 & 2 \\ 4 & 3 \end{bmatrix} = A$$

23. We found 2A in Exercise 11.
$D - 2A = D + (-2A)$

$$= \begin{bmatrix} 1 & 1 \\ 1 & 1 \end{bmatrix} + \begin{bmatrix} -2 & -4 \\ -8 & -6 \end{bmatrix} \quad \text{(Replacing each element of A by its additive inverse)}$$

$$= \begin{bmatrix} 1 + (-2) & 1 + (-4) \\ 1 + (-8) & 1 + (-6) \end{bmatrix}$$

$$= \begin{bmatrix} -1 & -3 \\ -7 & -5 \end{bmatrix}$$

25. Interchange the rows and columns of A.

$$A = \begin{bmatrix} 1 & 2 \\ 4 & 3 \end{bmatrix} \qquad A^t = \begin{bmatrix} 1 & 4 \\ 2 & 3 \end{bmatrix}$$

27. Interchange the rows and columns of J.

$$J = \begin{bmatrix} -1 & -2 & -3 \\ 3 & 2 & 1 \end{bmatrix} \qquad J^t = \begin{bmatrix} -1 & 3 \\ -2 & 2 \\ -3 & 1 \end{bmatrix}$$

29. We found J^t in Exercise 27.

$$K = \begin{bmatrix} 0 & 8 & -4 \\ 1 & 0 & -1 \end{bmatrix} \qquad K^t = \begin{bmatrix} 0 & 1 \\ 8 & 0 \\ -4 & -1 \end{bmatrix}$$

$$K^t + J^t = \begin{bmatrix} 0 & 1 \\ 8 & 0 \\ -4 & -1 \end{bmatrix} + \begin{bmatrix} -1 & 3 \\ -2 & 2 \\ -3 & 1 \end{bmatrix}$$

$$= \begin{bmatrix} 0 + (-1) & 1 + 3 \\ 8 + (-2) & 0 + 2 \\ -4 + (-3) & -1 + 1 \end{bmatrix}$$

$$= \begin{bmatrix} -1 & 4 \\ 6 & 2 \\ -7 & 0 \end{bmatrix}$$

31.

$$K = \begin{bmatrix} 0 & 8 & -4 \\ 1 & 0 & -1 \end{bmatrix}$$

a_{11} is the element in the first row, first column.
$a_{11} = 0$

a_{12} is the element in the first row, second column.
$a_{12} = 8$

a_{23} is the element in the second row, third column.
$a_{23} = -1$

a_{21} is the element in the second row, first column.
$a_{21} = 1$

33.

$$X = \begin{bmatrix} w \\ x \\ y \\ z \end{bmatrix}$$

The transpose of a column vector is a row vector.
$$X^t = [w \quad x \quad y \quad z]$$

35. A is a 1 × 2 matrix, and B is a 2 × 1 matrix. Since the number of columns of A, 2, is the same as the number of rows of B, 2, we can find AB.

$$AB = [-2 \quad 1] \begin{bmatrix} 3 \\ -4 \end{bmatrix} = [-2 \cdot 3 + 1 \cdot (-4)] = [-10]$$

Since the number of columns of B, 1, is the same as the number of rows of A, 1, we can find BA.

$$BA = \begin{bmatrix} 3 \\ -4 \end{bmatrix} [-2 \quad 1] = \begin{bmatrix} 3(-2) & 3 \cdot 1 \\ -4(-2) & -4 \cdot 1 \end{bmatrix} = \begin{bmatrix} -6 & 3 \\ 8 & -4 \end{bmatrix}$$

37. A is a 1 × 3 matrix, and B is a 3 × 1 matrix. Since the number of columns of A, 3, is the same as the number of rows of B, 3, we can find AB.

$$AB = [2 \quad 0 \quad -4] \begin{bmatrix} 9 \\ -5 \\ \frac{1}{4} \end{bmatrix} = \left[2 \cdot 9 + 0(-5) - 4\left(\frac{1}{4}\right) \right]$$

$$= [17]$$

Since the number of columns of B, 1, is the same as the number of rows of A, 1, we can find BA.

$$BA = \begin{bmatrix} 9 \\ -5 \\ \frac{1}{4} \end{bmatrix} [2 \quad 0 \quad -4] = \begin{bmatrix} 9 \cdot 2 & 9 \cdot 0 & 9(-4) \\ -5 \cdot 2 & -5 \cdot 0 & -5(-4) \\ \frac{1}{4} \cdot 2 & \frac{1}{4} \cdot 0 & \frac{1}{4}(-4) \end{bmatrix}$$

$$= \begin{bmatrix} 18 & 0 & -36 \\ -10 & 0 & 20 \\ \frac{1}{2} & 0 & -1 \end{bmatrix}$$

39.
$$AB = \begin{bmatrix} 1 & 2 & 0 \\ -1 & 0 & 4 \\ 2 & 5 & 6 \end{bmatrix} \begin{bmatrix} 3 & -4 & 1 \\ 2 & -1 & 0 \\ -3 & 2 & 1 \end{bmatrix}$$

$$= \begin{bmatrix} 1 \cdot 3 + 2 \cdot 2 + 0(-3) & 1(-4) + 2(-1) + 0 \cdot 2 & 1 \cdot 1 + 2 \cdot 0 + 0 \cdot 1 \\ -1 \cdot 3 + 0 \cdot 2 + 4(-3) & -1(-4) + 0(-1) + 4 \cdot 2 & -1 \cdot 1 + 0 \cdot 0 + 4 \cdot 1 \\ 2 \cdot 3 + 5 \cdot 2 + 6(-3) & 2(-4) + 5(-1) + 6 \cdot 2 & 2 \cdot 1 + 5 \cdot 0 + 6 \cdot 1 \end{bmatrix}$$

$$= \begin{bmatrix} 7 & -6 & 1 \\ -15 & 12 & 3 \\ -2 & -1 & 8 \end{bmatrix}$$

$$BA = \begin{bmatrix} 3 & -4 & 1 \\ 2 & -1 & 0 \\ -3 & 2 & 1 \end{bmatrix} \begin{bmatrix} 1 & 2 & 0 \\ -1 & 0 & 4 \\ 2 & 5 & 6 \end{bmatrix}$$

$$= \begin{bmatrix} 3 \cdot 1 - 4(-1) + 1 \cdot 2 & 3 \cdot 2 - 4 \cdot 0 + 1 \cdot 5 & 3 \cdot 0 - 4 \cdot 4 + 1 \cdot 6 \\ 2 \cdot 1 - 1(-1) + 0 \cdot 2 & 2 \cdot 2 - 1 \cdot 0 + 0 \cdot 5 & 2 \cdot 0 - 1 \cdot 4 + 0 \cdot 6 \\ -3 \cdot 1 + 2(-1) + 1 \cdot 2 & -3 \cdot 2 + 2 \cdot 0 + 1 \cdot 5 & -3 \cdot 0 + 2 \cdot 4 + 1 \cdot 6 \end{bmatrix}$$

$$= \begin{bmatrix} 9 & 11 & -10 \\ 3 & 4 & -4 \\ -3 & -1 & 14 \end{bmatrix}$$

41.
$$AB = [-4 \quad 1 \quad 3] \begin{bmatrix} -4 & 2 \\ 1 & 0 \\ 6 & -9 \end{bmatrix}$$

$$= [-4(-4) + 1 \cdot 1 + 3 \cdot 6 \quad -4 \cdot 2 + 1 \cdot 0 + 3(-9)]$$

$$= [35 \quad -35]$$

A is a 1 × 3 matrix, and B is a 3 × 2 matrix. Since the number of columns of B is not the same as the number of rows of A, it is not possible to find BA.

43. $3x - 2y + 4z = 17$

$2x + y - 5z = 13$

We write the coefficients on the left in a matrix. We write the product of that matrix by the column matrix containing the variables and set the result equal to the column matrix containing the constants on the right.

$$\begin{bmatrix} 3 & -2 & 4 \\ 2 & 1 & -5 \end{bmatrix} \begin{bmatrix} x \\ y \\ z \end{bmatrix} = \begin{bmatrix} 17 \\ 13 \end{bmatrix}$$

45. $5a + 7b = -4$

$6a - b = 3$

We write matrices as described in Exercise 43 above.

$$\begin{bmatrix} 5 & 7 \\ 6 & -1 \end{bmatrix} \begin{bmatrix} a \\ b \end{bmatrix} = \begin{bmatrix} -4 \\ 3 \end{bmatrix}$$

47.
a) $AB = \begin{bmatrix} 5000 & 3200 \\ 1600 & 6400 \end{bmatrix} \begin{bmatrix} 45 \\ 55 \end{bmatrix}$

$$= \begin{bmatrix} 5000 \cdot 45 + 3200 \cdot 55 \\ 1600 \cdot 45 + 6400 \cdot 55 \end{bmatrix}$$

$$= \begin{bmatrix} 401,000 \\ 424,000 \end{bmatrix}$$

AB gives the total number of men (Row 1) and women (Row 2) reached by 45 newspaper ads and 55 TV ads.

b)
$CB = [\$150 \quad \$400] \begin{bmatrix} 45 \\ 55 \end{bmatrix}$

$= [\$150 \cdot 45 + \$400 \cdot 55]$

$= [\$28,750]$

CB gives the total cost of 45 newspaper ads and 55 TV ads.

49. The coefficients are in the matrix on the left. The variables are in the column matrix on their right and the constants are in the column matrix on the far right.

$$x + 2y = -1$$
$$4x - 3y = 2$$

51.

Two such matrices are $A = \begin{bmatrix} 1 & 0 \\ 5 & 0 \end{bmatrix}$, $B = \begin{bmatrix} 0 & 0 \\ 1 & 3 \end{bmatrix}$.

$A \neq 0$, $B \neq 0$, but

$$AB = \begin{bmatrix} 1 & 0 \\ 5 & 0 \end{bmatrix} \begin{bmatrix} 0 & 0 \\ 1 & 3 \end{bmatrix} = \begin{bmatrix} 1 \cdot 0 + 0 \cdot 1 & 1 \cdot 0 + 0 \cdot 3 \\ 5 \cdot 0 + 0 \cdot 1 & 5 \cdot 0 + 0 \cdot 3 \end{bmatrix}$$

$$= \begin{bmatrix} 0 & 0 \\ 0 & 0 \end{bmatrix} = 0.$$

Answers may vary.

53.

$$A = \begin{bmatrix} 0 & -1 \\ 1 & 2 \end{bmatrix}, \qquad B = \begin{bmatrix} -1 & 1 \\ 3 & 0 \end{bmatrix}$$

$$A + B = \begin{bmatrix} -1 & 0 \\ 4 & 2 \end{bmatrix}, \quad A - B = \begin{bmatrix} 1 & -2 \\ -2 & 2 \end{bmatrix}$$

$$(A + B)(A - B) = \begin{bmatrix} -1 & 0 \\ 4 & 2 \end{bmatrix} \begin{bmatrix} 1 & -2 \\ -2 & 2 \end{bmatrix}$$

$$= \begin{bmatrix} -1 \cdot 1 + 0(-2) & -1(-2) + 0 \cdot 2 \\ 4 \cdot 1 + 2(-2) & 4(-2) + 2 \cdot 2 \end{bmatrix}$$

$$= \begin{bmatrix} -1 & 2 \\ 0 & -4 \end{bmatrix}$$

$$A^2 = AA = \begin{bmatrix} 0 & -1 \\ 1 & 2 \end{bmatrix} \begin{bmatrix} 0 & -1 \\ 1 & 2 \end{bmatrix}$$

$$= \begin{bmatrix} 0 \cdot 0 - 1 \cdot 1 & 0(-1) - 1 \cdot 2 \\ 1 \cdot 0 + 2 \cdot 1 & 1(-1) + 2 \cdot 2 \end{bmatrix}$$

$$= \begin{bmatrix} -1 & -2 \\ 2 & 3 \end{bmatrix}$$

$$B^2 = BB = \begin{bmatrix} -1 & 1 \\ 3 & 0 \end{bmatrix} \begin{bmatrix} -1 & 1 \\ 3 & 0 \end{bmatrix}$$

$$= \begin{bmatrix} -1(-1) + 1 \cdot 3 & -1 \cdot 1 + 1 \cdot 0 \\ 3(-1) + 0 \cdot 3 & 3 \cdot 1 + 0 \cdot 0 \end{bmatrix}$$

$$= \begin{bmatrix} 4 & -1 \\ -3 & 3 \end{bmatrix}$$

53. (continued)

$$A^2 - B^2 = \begin{bmatrix} -1 & -2 \\ 2 & 3 \end{bmatrix} - \begin{bmatrix} 4 & -1 \\ -3 & 3 \end{bmatrix} = \begin{bmatrix} -5 & -1 \\ 5 & 0 \end{bmatrix}$$

$$\neq (A + B)(A - B)$$

55.

$$A + B = \begin{bmatrix} a_{11} & a_{12} \\ a_{21} & a_{22} \end{bmatrix} + \begin{bmatrix} b_{11} & b_{12} \\ b_{21} & b_{22} \end{bmatrix}$$

$$= \begin{bmatrix} a_{11} + b_{11} & a_{12} + b_{12} \\ a_{21} + b_{21} & a_{22} + b_{22} \end{bmatrix} \quad \text{Definition of matrix addition}$$

$$= \begin{bmatrix} b_{11} + a_{11} & b_{12} + a_{12} \\ b_{21} + a_{21} & b_{22} + a_{22} \end{bmatrix} \quad \text{By commutativity of real numbers}$$

$$= \begin{bmatrix} b_{11} & b_{12} \\ b_{21} & b_{22} \end{bmatrix} + \begin{bmatrix} a_{11} & a_{12} \\ a_{21} & a_{22} \end{bmatrix} \quad \text{Definition of matrix addition}$$

$$= B + A$$

57.

$$k(A + B) = \begin{bmatrix} k(a_{11} + b_{11}) & k(a_{12} + b_{12}) \\ k(a_{21} + b_{21}) & k(a_{22} + b_{22}) \end{bmatrix} \quad \text{Definitions of matrix addition and scalar product}$$

$$= \begin{bmatrix} ka_{11} + kb_{11} & ka_{12} + kb_{12} \\ ka_{21} + kb_{21} & ka_{22} + kb_{22} \end{bmatrix} \quad \text{Distributive law for real numbers}$$

$$= \begin{bmatrix} ka_{11} & ka_{12} \\ ka_{21} & ka_{22} \end{bmatrix} + \begin{bmatrix} kb_{11} & kb_{12} \\ kb_{21} & kb_{22} \end{bmatrix} \quad \text{Definition of matrix addition}$$

$$= k\begin{bmatrix} a_{11} & a_{12} \\ a_{21} & a_{22} \end{bmatrix} + k\begin{bmatrix} b_{11} & b_{12} \\ b_{21} & b_{22} \end{bmatrix} \quad \text{Definition of scalar product}$$

$$= kA + kB$$

59.

$$A(BC) = \begin{bmatrix} a_{11} & a_{12} \\ a_{21} & a_{22} \end{bmatrix} \left(\begin{bmatrix} b_{11} & b_{12} \\ b_{21} & b_{22} \end{bmatrix} \begin{bmatrix} c_{11} & c_{12} \\ c_{21} & c_{22} \end{bmatrix} \right)$$

$$= \begin{bmatrix} a_{11} & a_{12} \\ a_{21} & a_{22} \end{bmatrix} \begin{bmatrix} b_{11}c_{11} + b_{12}c_{21} & b_{11}c_{12} + b_{12}c_{22} \\ b_{21}c_{11} + b_{22}c_{21} & b_{21}c_{12} + b_{22}c_{22} \end{bmatrix}$$

Definition of matrix multiplication

59. (continued)

$$
=
\begin{bmatrix}
a_{11}(b_{11}c_{11}+b_{12}c_{21}) + & a_{11}(b_{11}c_{12}+b_{12}c_{22}) + \\
a_{12}(b_{21}c_{11}+b_{22}c_{21}) & a_{12}(b_{21}c_{12}+b_{22}c_{22}) \\
a_{21}(b_{11}c_{11}+b_{12}c_{21}) + & a_{21}(b_{11}c_{12}+b_{12}c_{22}) + \\
a_{22}(b_{21}c_{11}+b_{22}c_{21}) & a_{22}(b_{21}c_{12}+b_{22}c_{22})
\end{bmatrix}
$$

Definition of matrix multiplication

$$
=
\begin{bmatrix}
a_{11}b_{11}c_{11}+a_{11}b_{12}c_{21} + & a_{11}b_{11}c_{12}+a_{11}b_{12}c_{22} + \\
a_{12}b_{21}c_{11}+a_{12}b_{22}c_{21} & a_{12}b_{21}c_{12}+a_{12}b_{22}c_{22} \\
a_{21}b_{11}c_{11}+a_{21}b_{12}c_{21} + & a_{21}b_{11}c_{12}+a_{21}b_{12}c_{22} + \\
a_{22}b_{21}c_{11}+a_{22}b_{22}c_{21} & a_{22}b_{21}c_{12}+a_{22}b_{22}c_{22}
\end{bmatrix}
$$

Distributive law for real numbers

$$
=
\begin{bmatrix}
(a_{11}b_{11}+a_{12}b_{21})c_{11} + & (a_{11}b_{11}+a_{12}b_{21})c_{12} + \\
(a_{11}b_{12}+a_{12}b_{22})c_{21} & (a_{11}b_{12}+a_{12}b_{22})c_{22} \\
(a_{21}b_{11}+a_{22}b_{21})c_{11} + & (a_{21}b_{11}+a_{22}b_{21})c_{12} + \\
(a_{21}b_{12}+a_{22}b_{22})c_{21} & (a_{21}b_{12}+a_{22}b_{22})c_{22}
\end{bmatrix}
$$

Distributive law for real numbers

$$
=
\begin{bmatrix}
a_{11}b_{11}+a_{12}b_{21} & a_{11}b_{12}+a_{12}b_{22} \\
a_{21}b_{11}+a_{22}b_{21} & a_{21}b_{12}+a_{22}b_{22}
\end{bmatrix}
\begin{bmatrix}
c_{11} & c_{12} \\
c_{21} & c_{22}
\end{bmatrix}
$$

Definition of matrix multiplication

$$
=
\left[
\begin{bmatrix}
a_{11} & a_{12} \\
a_{21} & a_{22}
\end{bmatrix}
\begin{bmatrix}
b_{11} & b_{12} \\
b_{21} & b_{22}
\end{bmatrix}
\right]
\begin{bmatrix}
c_{11} & c_{12} \\
c_{21} & c_{22}
\end{bmatrix}
$$
Definition of matrix multipli- cation

$= (AB)C$

Exercise Set 2.6

1.
$A = \begin{bmatrix} 3 & 2 \\ 5 & 3 \end{bmatrix}$

We find the augmented matrix consisting of A and I:

$$
\begin{bmatrix}
3^* & 2 & 1 & 0 \\
5 & 3 & 0 & 1
\end{bmatrix}
$$

Pivot on the first pivot as indicated:

$$
\begin{bmatrix}
3^* & 2 & 1 & 0 \\
15 & 9 & 0 & 3
\end{bmatrix}
$$
New Row 2 = 3(Row 2)

$$
\begin{bmatrix}
3 & 2 & 1 & 0 \\
0 & -1^* & -5 & 3
\end{bmatrix}
$$
New Row 2 = -5(Row 1) + (Row 2)

1. (continued)

Moving down the main diagonal, pivot on the next pivot:

$$
\begin{bmatrix}
3 & 0 & -9 & 6 \\
0 & -1 & -5 & 3
\end{bmatrix}
$$
New Row 1 = 2(Row 2) + (Row 1)

Multiply to get all 1's on the main diagonal:

$$
\begin{bmatrix}
1 & 0 & -3 & 2 \\
0 & 1 & 5 & -3
\end{bmatrix}
$$
New Row 1 = $\frac{1}{3}$(Row 1)
New Row 2 = -1(Row 2)

The matrix I is on the left, and A^{-1} is on the right. Thus

$$
A^{-1} = \begin{bmatrix} -3 & 2 \\ 5 & -3 \end{bmatrix}.
$$

The check is left to the reader.

3.
$A = \begin{bmatrix} 11 & 3 \\ 7 & 2 \end{bmatrix}$

Find the augmented matrix consisting of A and I:

$$
\begin{bmatrix}
11^* & 3 & 1 & 0 \\
7 & 2 & 0 & 1
\end{bmatrix}
$$

Pivot on the first pivot:

$$
\begin{bmatrix}
11^* & 3 & 1 & 0 \\
77 & 22 & 0 & 11
\end{bmatrix}
$$
New Row 2 = 11(Row 2)

$$
\begin{bmatrix}
11 & 3 & 1 & 0 \\
0 & 1^* & -7 & 11
\end{bmatrix}
$$
New Row 2 = -7(Row 1) + (Row 2)

Moving down the main diagonal, pivot on the next pivot:

$$
\begin{bmatrix}
11 & 0 & 22 & -33 \\
0 & 1 & -7 & 11
\end{bmatrix}
$$
New Row 1 = -3(Row 2) + (Row 1)

Multiply to get all 1's on the main diagonal:

$$
\begin{bmatrix}
1 & 0 & 2 & -3 \\
0 & 1 & -7 & 11
\end{bmatrix}
$$
New Row 1 = $\frac{1}{11}$(Row 1)

We have the matrix I on the left, and A^{-1} is on the right. Thus

$$
A^{-1} = \begin{bmatrix} 2 & -3 \\ -7 & 11 \end{bmatrix}.
$$

The check is left to the reader.

<u>5.</u>

$$A = \begin{bmatrix} 4 & -3 \\ 1 & 2 \end{bmatrix}$$

Find the augmented matrix consisting of A and I:

$$\begin{bmatrix} 4 & -3 & 1 & 0 \\ 1 & 2 & 0 & 1 \end{bmatrix}$$

Interchange the rows so the element below the first pivot is a multiple of the pivot:

$$\begin{bmatrix} 1* & 2 & 0 & 1 \\ 4 & -3 & 1 & 0 \end{bmatrix}$$

Pivot on the first pivot:

$$\begin{bmatrix} 1 & 2 & 0 & 1 \\ 0 & -11* & 1 & -4 \end{bmatrix}$$ New Row 2 = -4(Row 1) + (Row 2)

Moving down the main diagonal, pivot on the next pivot:

$$\begin{bmatrix} 11 & 22 & 0 & 11 \\ 0 & -11* & 1 & -4 \end{bmatrix}$$ New Row 1 = 11(Row 1)

$$\begin{bmatrix} 11 & 0 & 2 & 3 \\ 0 & -11 & 1 & -4 \end{bmatrix}$$ New Row 1 = 2(Row 2) + (Row 1)

Multiply to get all 1's on the main diagonal:

$$\begin{bmatrix} 1 & 0 & \frac{2}{11} & \frac{3}{11} \\ 0 & 1 & -\frac{1}{11} & \frac{4}{11} \end{bmatrix}$$ New Row 1 = $\frac{1}{11}$(Row 1)   New Row 2 = $-\frac{1}{11}$(Row 2)

We have the matrix I on the left, and A^{-1} is on the right. Thus

$$A^{-1} = \begin{bmatrix} \frac{2}{11} & \frac{3}{11} \\ -\frac{1}{11} & \frac{4}{11} \end{bmatrix}, \text{ or } \frac{1}{11}\begin{bmatrix} 2 & 3 \\ -1 & 4 \end{bmatrix}.$$

The check is left to the reader.

<u>7.</u>

$$A = \begin{bmatrix} 3 & 1 & 0 \\ 1 & 1 & 1 \\ 1 & -1 & 2 \end{bmatrix}$$

Find the augmented matrix consisting of A and I:

$$\begin{bmatrix} 3 & 1 & 0 & 1 & 0 & 0 \\ 1 & 1 & 1 & 0 & 1 & 0 \\ 1 & -1 & 2 & 0 & 0 & 1 \end{bmatrix}$$

Interchange the first and second rows so each of the elements below the first pivot is a multiple of the pivot:

$$\begin{bmatrix} 1* & 1 & 1 & 0 & 1 & 0 \\ 3 & 1 & 0 & 1 & 0 & 0 \\ 1 & -1 & 2 & 0 & 0 & 1 \end{bmatrix}$$

<u>7.</u> (continued)

Pivot on the first pivot:

$$\begin{bmatrix} 1 & 1 & 1 & 0 & 1 & 0 \\ 0 & -2* & -3 & 1 & -3 & 0 \\ 0 & -2 & 1 & 0 & -1 & 1 \end{bmatrix}$$ New Row 2 = -3(Row 1) + (Row 2)   New Row 3 = -1(Row 1) + (Row 3)

Moving down the main diagonal, pivot on the next pivot:

$$\begin{bmatrix} 2 & 2 & 2 & 0 & 2 & 0 \\ 0 & -2* & -3 & 1 & -3 & 0 \\ 0 & -2 & 1 & 0 & -1 & 1 \end{bmatrix}$$ New Row 1 = 2(Row 1)

$$\begin{bmatrix} 2 & 0 & -1 & 1 & -1 & 0 \\ 0 & -2 & -3 & 1 & -3 & 0 \\ 0 & 0 & 4* & -1 & 2 & 1 \end{bmatrix}$$ New Row 1 = (Row 2) + (Row 1)   New Row 3 = -1(Row 2) + (Row 3)

Moving down the main diagonal, pivot on the next pivot.

$$\begin{bmatrix} 8 & 0 & -4 & 4 & -4 & 0 \\ 0 & -8 & -12 & 4 & -12 & 0 \\ 0 & 0 & 4* & -1 & 2 & 1 \end{bmatrix}$$ New Row 1 = 4(Row 1)   New Row 2 = 4(Row 2)

$$\begin{bmatrix} 8 & 0 & 0 & 3 & -2 & 1 \\ 0 & -8 & 0 & 1 & -6 & 3 \\ 0 & 0 & 4 & -1 & 2 & 1 \end{bmatrix}$$ New Row 1 = (Row 3) + (Row 1)   New Row 2 = 3(Row 3) + (Row 2)

Multiply to get all 1's on the main diagonal:

$$\begin{bmatrix} 1 & 0 & 0 & \frac{3}{8} & -\frac{1}{4} & \frac{1}{8} \\ 0 & 1 & 0 & -\frac{1}{8} & \frac{3}{4} & -\frac{3}{8} \\ 0 & 0 & 1 & -\frac{1}{4} & \frac{1}{2} & \frac{1}{4} \end{bmatrix}$$ New Row 1 = $\frac{1}{8}$(Row 1)   New Row 2 = $-\frac{1}{8}$(Row 2)   New Row 3 = $\frac{1}{4}$(Row 3)

We have I on the left, and A^{-1} is on the right. Thus

$$A^{-1} = \begin{bmatrix} \frac{3}{8} & -\frac{1}{4} & \frac{1}{8} \\ -\frac{1}{8} & \frac{3}{4} & -\frac{3}{8} \\ -\frac{1}{4} & \frac{1}{2} & \frac{1}{4} \end{bmatrix}.$$

The check is left to the reader.

9.
$$A = \begin{bmatrix} 1 & -1 & 2 \\ 0 & 1 & 3 \\ 2 & 1 & -2 \end{bmatrix}$$

Find the augmented matrix consisting of A and I:

$$\begin{bmatrix} 1* & -1 & 2 & 1 & 0 & 0 \\ 0 & 1 & 3 & 0 & 1 & 0 \\ 2 & 1 & -2 & 0 & 0 & 1 \end{bmatrix}$$

Pivot on the first pivot:

$$\begin{bmatrix} 1 & -1 & 2 & 1 & 0 & 0 \\ 0 & 1* & 3 & 0 & 1 & 0 \\ 0 & 3 & -6 & -2 & 0 & 1 \end{bmatrix}$$ New Row 3 = -2(Row 1) + (Row 3)

Moving down the main diagonal, pivot on the next pivot:

$$\begin{bmatrix} 1 & 0 & 5 & 1 & 1 & 0 \\ 0 & 1 & 3 & 0 & 1 & 0 \\ 0 & 0 & -15* & -2 & -3 & 1 \end{bmatrix}$$ New Row 1 = (Row 2) + (Row 1)

New Row 3 = -3(Row 2) + (Row 3)

Moving down the main diagonal, pivot on the next pivot:

$$\begin{bmatrix} 3 & 0 & 15 & 3 & 3 & 0 \\ 0 & 5 & 15 & 0 & 5 & 0 \\ 0 & 0 & -15* & -2 & -3 & 1 \end{bmatrix}$$ New Row 1 = 3(Row 1)
New Row 2 = 5(Row 2)

$$\begin{bmatrix} 3 & 0 & 0 & 1 & 0 & 1 \\ 0 & 5 & 0 & -2 & 2 & 1 \\ 0 & 0 & -15 & -2 & -3 & 1 \end{bmatrix}$$ New Row 1 = (Row 3) + (Row 1)
New Row 2 = (Row 3) + (Row 2)

Multiply to get 1's on the main diagonal:

$$\begin{bmatrix} 1 & 0 & 0 & \frac{1}{3} & 0 & \frac{1}{3} \\ 0 & 1 & 0 & -\frac{2}{5} & \frac{2}{5} & \frac{1}{5} \\ 0 & 0 & 1 & \frac{2}{15} & \frac{1}{5} & -\frac{1}{15} \end{bmatrix}$$ New Row 1 = $\frac{1}{3}$(Row 1)

New Row 2 = $\frac{1}{5}$(Row 2)

New Row 3 = $-\frac{1}{15}$(Row 3)

We have the matrix I on the left, and A^{-1} is on the right. Thus

$$A^{-1} = \begin{bmatrix} \frac{1}{3} & 0 & \frac{1}{3} \\ -\frac{2}{5} & \frac{2}{5} & \frac{1}{5} \\ \frac{2}{15} & \frac{1}{5} & -\frac{1}{15} \end{bmatrix}$$

The check is left to the reader.

11.
$$A = \begin{bmatrix} 1 & -4 & 8 \\ 1 & -3 & 2 \\ 2 & -7 & 10 \end{bmatrix}$$

We find the augmented matrix consisting of A and I:

$$\begin{bmatrix} 1* & -4 & 8 & 1 & 0 & 0 \\ 1 & -3 & 2 & 0 & 1 & 0 \\ 2 & -7 & 10 & 0 & 0 & 1 \end{bmatrix}$$

Pivot on the first pivot:

$$\begin{bmatrix} 1 & -4 & 8 & 1 & 0 & 0 \\ 0 & 1* & -6 & -1 & 1 & 0 \\ 0 & 1 & -6 & -2 & 0 & 1 \end{bmatrix}$$ New Row 2 = -1(Row 1) + (Row 2)
New Row 3 = -2(Row 1) + (Row 3)

Moving down the main diagonal, pivot on the next pivot:

$$\begin{bmatrix} 1 & 0 & -16 & -3 & 4 & 0 \\ 0 & 1 & -6 & -1 & 1 & 0 \\ 0 & 0 & 0 & -1 & -1 & 1 \end{bmatrix}$$ New Row 1 = 4(Row 2) + (Row 1)

New Row 3 = -1(Row 1) + (Row 3)

We have a row of all 0's on the side of the augmented matrix corresponding to A (that is, the left side). Therefore, A^{-1} does not exist.

13.
$$A = \begin{bmatrix} 1 & 2 & 3 & 4 \\ 0 & 1 & 3 & -5 \\ 0 & 0 & 1 & -2 \\ 0 & 0 & 0 & -1 \end{bmatrix}$$

We find the augmented matrix consisting of A and I:

$$\begin{bmatrix} 1 & 2 & 3 & 4 & 1 & 0 & 0 & 0 \\ 0 & 1* & 3 & -5 & 0 & 1 & 0 & 0 \\ 0 & 0 & 1 & -2 & 0 & 0 & 1 & 0 \\ 0 & 0 & 0 & -1 & 0 & 0 & 0 & 1 \end{bmatrix}$$

Notice that all the elements below the 1 in the first row, first column are 0. Therefore, we will pivot on the 1 in the second row, second column. We obtain:

$$\begin{bmatrix} 1 & 0 & -3 & 14 & 1 & -2 & 0 & 0 \\ 0 & 1 & 3 & -5 & 0 & 1 & 0 & 0 \\ 0 & 0 & 1* & -2 & 0 & 0 & 1 & 0 \\ 0 & 0 & 0 & -1 & 0 & 0 & 0 & 1 \end{bmatrix}$$

New Row 1 = -2(Row 2) + (Row 1)

13. (continued)

Pivot on the 1 in the third row, third column:

$$\begin{bmatrix} 1 & 0 & 0 & 8 & 1 & -2 & 3 & 0 \\ 0 & 1 & 0 & 1 & 0 & 1 & -3 & 0 \\ 0 & 0 & 1 & -2 & 0 & 0 & 1 & 0 \\ 0 & 0 & 0 & -1^* & 0 & 0 & 0 & 1 \end{bmatrix}$$

New Row 2 = -3(Row 3) + (Row 2)
New Row 1 = 3(Row 3) + (Row 1)

Pivot on the -1 in the fourth row, fourth column:

$$\begin{bmatrix} 1 & 0 & 0 & 0 & 1 & -2 & 3 & 8 \\ 0 & 1 & 0 & 0 & 0 & 1 & -3 & 1 \\ 0 & 0 & 1 & 0 & 0 & 0 & 1 & -2 \\ 0 & 0 & 0 & -1 & 0 & 0 & 0 & 1 \end{bmatrix}$$

New Row 3 = -2(Row 4) + (Row 3)
New Row 2 = (Row 4) + (Row 2)
New Row 1 = 8(Row 4) + (Row 1)

Multiply to get all 1's on the main diagonal:

$$\begin{bmatrix} 1 & 0 & 0 & 0 & 1 & -2 & 3 & 8 \\ 0 & 1 & 0 & 0 & 0 & 1 & -3 & 1 \\ 0 & 0 & 1 & 0 & 0 & 0 & 1 & -2 \\ 0 & 0 & 0 & 1 & 0 & 0 & 0 & -1 \end{bmatrix}$$
New Row 4 = -1(Row 4)

We have the matrix I on the left, and A^{-1} is on the right. Thus

$$A^{-1} = \begin{bmatrix} 1 & -2 & 3 & 8 \\ 0 & 1 & -3 & 1 \\ 0 & 0 & 1 & -2 \\ 0 & 0 & 0 & -1 \end{bmatrix}$$

The check is left to the reader.

15.
$$A = \begin{bmatrix} -2 & 5 & 3 \\ 4 & -1 & 3 \\ 4 & -10 & -6 \end{bmatrix}$$

We find the augmented matrix consisting of A and I:

$$\begin{bmatrix} -2^* & 5 & 3 & 1 & 0 & 0 \\ 4 & -1 & 3 & 0 & 1 & 0 \\ 4 & -10 & -6 & 0 & 0 & 1 \end{bmatrix}$$

15. (continued)

Pivot on the -2 in the first row, first column:

$$\begin{bmatrix} -2 & 5 & 3 & 1 & 0 & 0 \\ 0 & 9 & 9 & 2 & 1 & 0 \\ 0 & 0 & 0 & 2 & 0 & 1 \end{bmatrix}$$

New Row 3 = 2(Row 1) + (Row 3)
New Row 2 = 2(Row 1) + (Row 2)

We have a row of all 0's on the side of the augmented matrix corresponding to A (that is, the left side). Therefore, A^{-1} does not exist.

17. $11x + 3y = -4$
       $7x + 2y = 5$

Let $\begin{bmatrix} 11 & 3 \\ 7 & 2 \end{bmatrix} = A$, $\begin{bmatrix} x \\ y \end{bmatrix} = X$, and $\begin{bmatrix} -4 \\ 5 \end{bmatrix} = B$.

Then we can write a matrix equation equivalent to the system of equations:

$A \cdot X = B$

To solve the matrix equation we multiply by A^{-1} on the left on each side of the equation:

$X = A^{-1} \cdot B$

Substituting, we have:

$$\begin{bmatrix} x \\ y \end{bmatrix} = \begin{bmatrix} 2 & -3 \\ -7 & 11 \end{bmatrix} \begin{bmatrix} -4 \\ 5 \end{bmatrix}$$

$$= \begin{bmatrix} -23 \\ 83 \end{bmatrix}$$

The solution of the system is (-23, 83).

19. $3x + y \qquad = 2$
       $x - y + 2z = -4$
       $x + y + z = 5$

Let $\begin{bmatrix} 3 & 1 & 0 \\ 1 & -1 & 2 \\ 1 & 1 & 1 \end{bmatrix} = A$, $\begin{bmatrix} x \\ y \\ z \end{bmatrix} = X$, and $\begin{bmatrix} 2 \\ -4 \\ 5 \end{bmatrix} = B$.

Then the system of equations is equivalent to the matrix equation

$A \cdot X = B$,

and the solution is given by

$X = A^{-1} \cdot B$ 　　　Multiplying by A^{-1} on the left on each side

19. (continued)

Substituting, we have

$$\begin{bmatrix} x \\ y \\ z \end{bmatrix} = \frac{1}{8} \begin{bmatrix} 3 & 1 & -2 \\ -1 & -3 & 6 \\ -2 & 2 & 4 \end{bmatrix} \begin{bmatrix} 2 \\ -4 \\ 5 \end{bmatrix}$$

$$= \frac{1}{8} \begin{bmatrix} -8 \\ 40 \\ 8 \end{bmatrix}$$

$$= \begin{bmatrix} -1 \\ 5 \\ 1 \end{bmatrix}$$

The solution of the system is (-1, 5, 1).

21. $4x - 3y = 2$
    $x + 2y = -1$

A matrix equation equivalent to the system is

$$\begin{bmatrix} 4 & -3 \\ 1 & 2 \end{bmatrix} \begin{bmatrix} x \\ y \end{bmatrix} = \begin{bmatrix} 2 \\ -1 \end{bmatrix}.$$
$$\quad\ \downarrow \qquad\ \ \downarrow \qquad\ \ \downarrow$$
$$\quad\ A \quad\ \cdot\ \ X \quad\ =\quad\ B$$

In Exercise 5 we found

$$A^{-1} = \begin{bmatrix} \frac{2}{11} & \frac{3}{11} \\ -\frac{1}{11} & \frac{4}{11} \end{bmatrix} = \frac{1}{11} \begin{bmatrix} 2 & 3 \\ -1 & 4 \end{bmatrix}.$$

To solve the matrix equation multiply by A^{-1} on the left on each side:

$A \cdot X = B$
$\quad X = A^{-1} \cdot B$

$$\begin{bmatrix} x \\ y \end{bmatrix} = \frac{1}{11} \begin{bmatrix} 2 & 3 \\ -1 & 4 \end{bmatrix} \begin{bmatrix} 2 \\ -1 \end{bmatrix}$$

$$= \frac{1}{11} \begin{bmatrix} 1 \\ -6 \end{bmatrix}$$

$$= \begin{bmatrix} \frac{1}{11} \\ -\frac{6}{11} \end{bmatrix}$$

The solution of the system is $\left(\frac{1}{11}, -\frac{6}{11}\right)$.

23. $7x - 2y = -3$
    $9x + 3y = 4$

A matrix equation equivalent to the system is

$$\begin{bmatrix} 7 & -2 \\ 9 & 3 \end{bmatrix} \begin{bmatrix} x \\ y \end{bmatrix} = \begin{bmatrix} -3 \\ 4 \end{bmatrix}$$
$$\quad\ \downarrow \qquad\ \ \downarrow \qquad\ \ \downarrow$$
$$\quad\ A \quad\ \cdot\ \ X \quad\ =\quad\ B$$

To find A^{-1} we first write an augmented matrix:

$$\begin{bmatrix} 7^* & -2 & 1 & 0 \\ 9 & 3 & 0 & 1 \end{bmatrix}$$

Pivot as indicated:

$$\begin{bmatrix} 7^* & -2 & 1 & 0 \\ 63 & 21 & 0 & 7 \end{bmatrix} \quad \text{New Row 2 = 7(Row 2)}$$

$$\begin{bmatrix} 7 & -2 & 1 & 0 \\ 0 & 39^* & -9 & 7 \end{bmatrix} \quad \begin{array}{l}\text{New Row 2 =} \\ \text{-9(Row 1) + (Row 2)}\end{array}$$

$$\begin{bmatrix} 273 & -78 & 39 & 0 \\ 0 & 39^* & -9 & 7 \end{bmatrix} \quad \text{New Row 1 = 39(Row 1)}$$

$$\begin{bmatrix} 273 & 0 & 21 & 14 \\ 0 & 39 & -9 & 7 \end{bmatrix} \quad \begin{array}{l}\text{New Row 1 =} \\ \text{2(Row 2) + (Row 1)}\end{array}$$

$$\begin{bmatrix} 1 & 0 & \frac{1}{13} & \frac{2}{39} \\ 0 & 1 & -\frac{3}{13} & \frac{7}{39} \end{bmatrix} \quad \begin{array}{l}\text{New Row 1 = } \frac{1}{273}\text{(Row 1)} \\ \text{New Row 2 = } \frac{1}{39}\text{(Row 2)}\end{array}$$

We have I on the left and A^{-1} on the right:

$$A^{-1} = \begin{bmatrix} \frac{1}{13} & \frac{2}{39} \\ -\frac{3}{13} & \frac{7}{39} \end{bmatrix}$$

To solve the matrix equation we multiply by A^{-1} on the left on each side:

$$X = \begin{bmatrix} \frac{1}{13} & \frac{2}{39} \\ -\frac{3}{13} & \frac{7}{39} \end{bmatrix} \begin{bmatrix} -3 \\ 4 \end{bmatrix}$$

$$\begin{bmatrix} x \\ y \end{bmatrix} = \begin{bmatrix} -\frac{1}{39} \\ \frac{55}{39} \end{bmatrix}$$

The solution of the system is $\left(-\frac{1}{39}, \frac{55}{39}\right)$.

25. $x \quad\;\; + z = 1$
$\quad\; 2x + y \quad\;\; = 3$
$\quad\;\; x - y + z = 4$

A matrix equation equivalent to this system is

$$\begin{bmatrix} 1 & 0 & 1 \\ 2 & 1 & 0 \\ 1 & -1 & 1 \end{bmatrix} \begin{bmatrix} x \\ y \\ z \end{bmatrix} = \begin{bmatrix} 1 \\ 3 \\ 4 \end{bmatrix}$$
$$\quad\;\;\downarrow \qquad\quad \downarrow \qquad \downarrow$$
$$\quad\; A \qquad\cdot\; X \;=\; B$$

To find A^{-1} we first write an augmented matrix:

$$\begin{bmatrix} 1^* & 0 & 1 & 1 & 0 & 0 \\ 2 & 1 & 0 & 0 & 1 & 0 \\ 1 & -1 & 1 & 0 & 0 & 1 \end{bmatrix}$$

Pivot as indicated:

$$\begin{bmatrix} 1 & 0 & 1 & 1 & 0 & 0 \\ 0 & 1^* & -2 & -2 & 1 & 0 \\ 0 & -1 & 0 & -1 & 0 & 1 \end{bmatrix}$$

New Row 2 =
-2(Row 1) + (Row 2)
New Row 3 =
-1(Row 1) + (Row 3)

$$\begin{bmatrix} 1 & 0 & 1 & 1 & 0 & 0 \\ 0 & 1 & -2 & -2 & 1 & 0 \\ 0 & 0 & -2^* & -3 & 1 & 1 \end{bmatrix}$$

New Row 3 =
(Row 2) + (Row 3)

$$\begin{bmatrix} 2 & 0 & 2 & 2 & 0 & 0 \\ 0 & 1 & -2 & -2 & 1 & 0 \\ 0 & 0 & -2^* & -3 & 1 & 1 \end{bmatrix}$$

New Row 1 = 2(Row 1)

$$\begin{bmatrix} 2 & 0 & 0 & -1 & 1 & 1 \\ 0 & 1 & 0 & 1 & 0 & -1 \\ 0 & 0 & -2 & -3 & 1 & 1 \end{bmatrix}$$

New Row 1 =
(Row 3) + (Row 1)
New Row 2 =
-1(Row 3) + (Row 2)

$$\begin{bmatrix} 1 & 0 & 0 & -\frac{1}{2} & \frac{1}{2} & \frac{1}{2} \\ 0 & 1 & 0 & 1 & 0 & -1 \\ 0 & 0 & 1 & \frac{3}{2} & -\frac{1}{2} & -\frac{1}{2} \end{bmatrix}$$

New Row 1 = $\frac{1}{2}$(Row 1)

New Row 3 = $-\frac{1}{2}$(Row 3)

We have I on the left and A^{-1} on the right:

$$A^{-1} = \begin{bmatrix} -\frac{1}{2} & \frac{1}{2} & \frac{1}{2} \\ 1 & 0 & -1 \\ \frac{3}{2} & -\frac{1}{2} & -\frac{1}{2} \end{bmatrix}, \text{ or } \frac{1}{2}\begin{bmatrix} -1 & 1 & 1 \\ 2 & 0 & -2 \\ 3 & -1 & -1 \end{bmatrix}$$

25. (continued)

To solve the matrix equation we multiply by A^{-1} on the left on each side:

$$A \cdot X = B$$
$$\quad X = A^{-1} \cdot B$$

$$\begin{bmatrix} x \\ y \\ z \end{bmatrix} = \frac{1}{2}\begin{bmatrix} -1 & 1 & 1 \\ 2 & 0 & -2 \\ 3 & -1 & -1 \end{bmatrix}\begin{bmatrix} 1 \\ 3 \\ 4 \end{bmatrix}$$

$$= \frac{1}{2}\begin{bmatrix} 6 \\ -6 \\ -4 \end{bmatrix}$$

$$= \begin{bmatrix} 3 \\ -3 \\ -2 \end{bmatrix}$$

The solution of the system is (3, -3, -2).

27. $2w - 3x + 4y - 5z = 1$
$\quad 3w - 2x + 7y - 3z = -2$
$\quad\; w + x - y + z = 3$
$\quad -w - 3x - 6y + 4z = -5$

A matrix equation equivalent to the system is

$$\begin{bmatrix} 2 & -3 & 4 & -5 \\ 3 & -2 & 7 & -3 \\ 1 & 1 & -1 & 1 \\ -1 & -3 & -6 & 4 \end{bmatrix}\begin{bmatrix} w \\ x \\ y \\ z \end{bmatrix} = \begin{bmatrix} 1 \\ -2 \\ 3 \\ -5 \end{bmatrix}$$
$$\quad\;\downarrow \qquad\qquad\quad \downarrow \qquad \downarrow$$
$$\quad\; A \qquad\qquad\;\cdot\; X \;=\; B$$

To find A^{-1} we first write an augmented matrix:

$$\begin{bmatrix} 2 & -3 & 4 & -5 & 1 & 0 & 0 & 0 \\ 3 & -2 & 7 & -3 & 0 & 1 & 0 & 0 \\ 1 & 1 & -1 & 1 & 0 & 0 & 1 & 0 \\ -1 & -3 & -6 & 4 & 0 & 0 & 0 & 1 \end{bmatrix}$$

Then we proceed as indicated:

$$\begin{bmatrix} 1^* & 1 & -1 & 1 & 0 & 0 & 1 & 0 \\ 3 & -2 & 7 & -3 & 0 & 1 & 0 & 0 \\ 2 & -3 & 4 & -5 & 1 & 0 & 0 & 0 \\ -1 & -3 & -6 & 4 & 0 & 0 & 0 & 1 \end{bmatrix}$$

Interchange
Row 1 and Row 3

$$\begin{bmatrix} 1 & 1 & -1 & 1 & 0 & 0 & 1 & 0 \\ 0 & -5^* & 10 & -6 & 0 & 1 & -3 & 0 \\ 0 & -5 & 6 & -7 & 1 & 0 & -2 & 0 \\ 0 & -2 & -7 & 5 & 0 & 0 & 1 & 1 \end{bmatrix}$$

New Row 4 = (Row 1) + (Row 4)

New Row 3 = -2(Row 1) + (Row 3)

New Row 2 = -3(Row 1) + (Row 2)

27. (continued)

$$\begin{bmatrix} 5 & 5 & -5 & 5 & 0 & 0 & 5 & 0 \\ 0 & -5* & 10 & -6 & 0 & 1 & -3 & 0 \\ 0 & -5 & 6 & -7 & 1 & 0 & -2 & 0 \\ 0 & -10 & -35 & 25 & 0 & 0 & 5 & 5 \end{bmatrix}$$

New Row 1 = 5(Row 1)

New Row 4 = 5(Row 4)

$$\begin{bmatrix} 5 & 0 & 5 & -1 & 0 & 1 & 2 & 0 \\ 0 & -5 & 10 & -6 & 0 & 1 & -3 & 0 \\ 0 & 0 & -4* & -1 & 1 & -1 & 1 & 0 \\ 0 & 0 & -55 & 37 & 0 & -2 & 11 & 5 \end{bmatrix}$$

New Row 4 = -2(Row 2) + (Row 4)
New Row 3 = -1(Row 2) + (Row 3)
New Row 1 = (Row 2) + (Row 1)

$$\begin{bmatrix} 20 & 0 & 20 & -4 & 0 & 4 & 8 & 0 \\ 0 & -10 & 20 & -12 & 0 & 2 & -6 & 0 \\ 0 & 0 & -4* & -1 & 1 & -1 & 1 & 0 \\ 0 & 0 & -220 & 148 & 0 & -8 & 44 & 20 \end{bmatrix}$$

New Row 4 = 4(Row 4)
New Row 2 = 2(Row 2)
New Row 1 = 4(Row 1)

$$\begin{bmatrix} 20 & 0 & 0 & -9 & 5 & -1 & 13 & 0 \\ 0 & -10 & 0 & -17 & 5 & -3 & -1 & 0 \\ 0 & 0 & -4 & -1 & 1 & -1 & 1 & 0 \\ 0 & 0 & 0 & 203* & -55 & 47 & -11 & 20 \end{bmatrix}$$

New Row 4 = -55(Row 3) + (Row 4)
New Row 2 = 5(Row 3) + (Row 2)
New Row 1 = 5(Row 3) + (Row 1)

$$\begin{bmatrix} 4060 & 0 & 0 & -1827 & 1015 & -203 & 2639 & 0 \\ 0 & -2030 & 0 & -3451 & 1015 & -609 & -203 & 0 \\ 0 & 0 & -812 & -203 & 203 & -203 & 203 & 0 \\ 0 & 0 & 0 & 203* & -55 & 47 & -11 & 20 \end{bmatrix}$$

New Row 3 = 203(Row 3)
New Row 2 = 203(Row 2)
New Row 1 = 203(Row 1)

$$\begin{bmatrix} 4060 & 0 & 0 & 0 & 520 & 220 & 2540 & 180 \\ 0 & -2030 & 0 & 0 & 80 & 190 & -390 & 340 \\ 0 & 0 & -812 & 0 & 148 & -156 & 192 & 20 \\ 0 & 0 & 0 & 203 & -55 & 47 & -11 & 20 \end{bmatrix}$$

New Row 3 = (Row 4) + (Row 3)
New Row 2 = 17(Row 4) + (Row 2)
New Row 1 = 9(Row 4) + (Row 1)

27. (continued)

$$\begin{bmatrix} 1 & 0 & 0 & 0 & \frac{26}{203} & \frac{11}{203} & \frac{127}{203} & \frac{9}{203} \\ 0 & 1 & 0 & 0 & -\frac{8}{203} & -\frac{19}{203} & \frac{39}{203} & -\frac{34}{203} \\ 0 & 0 & 1 & 0 & -\frac{37}{203} & \frac{39}{203} & -\frac{48}{203} & -\frac{5}{203} \\ 0 & 0 & 0 & 1 & -\frac{55}{203} & \frac{47}{203} & -\frac{11}{203} & \frac{20}{203} \end{bmatrix}$$

New Row 4 = $\frac{1}{203}$(Row 4)

New Row 3 = $-\frac{1}{812}$(Row 3)

New Row 2 = $-\frac{1}{2030}$(Row 2)

New Row 1 = $\frac{1}{4060}$(Row 1)

We have I on the left and A^{-1} on the right:

$$A^{-1} = \begin{bmatrix} \frac{26}{203} & \frac{11}{203} & \frac{127}{203} & \frac{9}{203} \\ -\frac{8}{203} & -\frac{19}{203} & \frac{39}{203} & -\frac{34}{203} \\ -\frac{37}{203} & \frac{39}{203} & -\frac{48}{203} & -\frac{5}{203} \\ -\frac{55}{203} & \frac{47}{203} & -\frac{11}{203} & \frac{20}{203} \end{bmatrix}$$

$$= \frac{1}{203}\begin{bmatrix} 26 & 11 & 127 & 9 \\ -8 & -19 & 39 & -34 \\ -37 & 39 & -48 & -5 \\ -55 & 47 & -11 & 20 \end{bmatrix}$$

To solve the matrix equation we multiply by A^{-1} on the left on both sides:

A·X = B

X = A^{-1}·B

$$\begin{bmatrix} w \\ x \\ y \\ z \end{bmatrix} = \frac{1}{203}\begin{bmatrix} 26 & 11 & 127 & 9 \\ -8 & -19 & 39 & -34 \\ -37 & 39 & -48 & -5 \\ -55 & 47 & -11 & 20 \end{bmatrix}\begin{bmatrix} 1 \\ -2 \\ 3 \\ -5 \end{bmatrix}$$

$$= \frac{1}{203}\begin{bmatrix} 340 \\ 317 \\ -234 \\ -282 \end{bmatrix}$$

$$= \begin{bmatrix} \frac{340}{203} \\ \frac{317}{203} \\ -\frac{234}{203} \\ -\frac{282}{203} \end{bmatrix}$$

The solution of the system is $\left(\frac{340}{203}, \frac{317}{203}, -\frac{234}{203}, -\frac{282}{203}\right)$.

29.

$$AI = \begin{bmatrix} a & b & c \\ d & e & f \\ g & h & i \end{bmatrix} \begin{bmatrix} 1 & 0 & 0 \\ 0 & 1 & 0 \\ 0 & 0 & 1 \end{bmatrix}$$

$$= \begin{bmatrix} a\cdot1+b\cdot0+c\cdot0 & a\cdot0+b\cdot1+c\cdot0 & a\cdot0+b\cdot0+c\cdot1 \\ d\cdot1+e\cdot0+f\cdot0 & d\cdot0+e\cdot1+f\cdot0 & d\cdot0+e\cdot0+f\cdot1 \\ g\cdot1+h\cdot0+i\cdot0 & g\cdot0+h\cdot1+i\cdot0 & g\cdot0+h\cdot0+i\cdot1 \end{bmatrix}$$

$$= \begin{bmatrix} a & b & c \\ d & e & f \\ g & h & i \end{bmatrix} = A$$

$$IA = \begin{bmatrix} 1 & 0 & 0 \\ 0 & 1 & 0 \\ 0 & 0 & 1 \end{bmatrix} \begin{bmatrix} a & b & c \\ d & e & f \\ g & h & i \end{bmatrix}$$

$$= \begin{bmatrix} 1\cdot a+0\cdot d+0\cdot g & 1\cdot b+0\cdot e+0\cdot h & 1\cdot c+0\cdot f+0\cdot i \\ 0\cdot a+1\cdot d+0\cdot g & 0\cdot b+1\cdot e+0\cdot h & 0\cdot c+1\cdot f+0\cdot i \\ 0\cdot a+0\cdot d+1\cdot g & 0\cdot b+0\cdot e+1\cdot h & 0\cdot c+0\cdot f+1\cdot f \end{bmatrix}$$

$$= \begin{bmatrix} a & b & c \\ d & e & f \\ g & h & i \end{bmatrix} = A$$

Therefore, AI = IA = A.

31. A^{-1} exists if we do not have a row of all 0's on either side of the augmented matrix at any time while we are using the elimination method. Since the elements above and below the main diagonal are 0's, we avoid a row of all 0's when $x \neq 0$ and $y \neq 0$. Under these conditions A^{-1} exists.

To find A^{-1}, first write an augmented matrix.

$$\begin{bmatrix} x & 0 & 1 & 0 \\ 0 & y & 0 & 1 \end{bmatrix}$$

No pivoting needs to be done since the elements above and below the main diagonal are 0. We multiply to get 1's on the main diagonal:

$$\begin{bmatrix} 1 & 0 & \frac{1}{x} & 0 \\ 0 & 1 & 0 & \frac{1}{y} \end{bmatrix}$$

$$A^{-1} = \begin{bmatrix} \frac{1}{x} & 0 \\ 0 & \frac{1}{y} \end{bmatrix}$$

33. A^{-1} exists if we do not have a row of all 0's on either side of the augmented matrix at any time while we are using the elimination method. Since all the elements below the main diagonal in A are 0's, we can avoid having a row of all 0's when $x \neq 0$, $y \neq 0$, $z \neq 0$, and $w \neq 0$.

33. (continued)

To find A^{-1}, first write an augmented matrix.

$$\begin{bmatrix} x & 1 & 1 & 1 & 1 & 0 & 0 & 0 \\ 0 & y & 0 & 0 & 0 & 1 & 0 & 0 \\ 0 & 0 & z & 0 & 0 & 0 & 1 & 0 \\ 0 & 0 & 0 & w & 0 & 0 & 0 & 1 \end{bmatrix}$$

The elements below the x in the first row, first column are all 0, so we do not need to pivot on the x. We pivot first on the y in the second row, second column and proceed as indicated.

$$\begin{bmatrix} xy & y & y & y & y & 0 & 0 & 0 \\ 0 & y^* & 0 & 0 & 0 & 1 & 0 & 0 \\ 0 & 0 & z & 0 & 0 & 0 & 1 & 0 \\ 0 & 0 & 0 & w & 0 & 0 & 0 & 1 \end{bmatrix}$$
New Row 1 = y(Row 1)

$$\begin{bmatrix} xy & 0 & y & y & y & -1 & 0 & 0 \\ 0 & y & 0 & 0 & 0 & 1 & 0 & 0 \\ 0 & 0 & z^* & 0 & 0 & 0 & 1 & 0 \\ 0 & 0 & 0 & w & 0 & 0 & 0 & 1 \end{bmatrix}$$
New Row 1 = -1(Row 2) + (Row 1)

$$\begin{bmatrix} xyz & 0 & yz & yz & yz & -z & 0 & 0 \\ 0 & y & 0 & 0 & 0 & 1 & 0 & 0 \\ 0 & 0 & z^* & 0 & 0 & 0 & 1 & 0 \\ 0 & 0 & 0 & w & 0 & 0 & 0 & 1 \end{bmatrix}$$
New Row 1 = z(Row 1)

$$\begin{bmatrix} xyz & 0 & 0 & yz & yz & -z & -y & 0 \\ 0 & y & 0 & 0 & 0 & 1 & 0 & 0 \\ 0 & 0 & z & 0 & 0 & 0 & 1 & 0 \\ 0 & 0 & 0 & w^* & 0 & 0 & 0 & 1 \end{bmatrix}$$

New Row 1 = -y(Row 3) + (Row 1)

$$\begin{bmatrix} xyzw & 0 & 0 & yzw & yzw & -zw & -yw & 0 \\ 0 & y & 0 & 0 & 0 & 1 & 0 & 0 \\ 0 & 0 & z & 0 & 0 & 0 & 1 & 0 \\ 0 & 0 & 0 & w^* & 0 & 0 & 0 & 1 \end{bmatrix}$$
New Row 1 = w(Row 1)

$$\begin{bmatrix} xyzw & 0 & 0 & 0 & yzw & -zw & -yw & -yz \\ 0 & y & 0 & 0 & 0 & 1 & 0 & 0 \\ 0 & 0 & z & 0 & 0 & 0 & 1 & 0 \\ 0 & 0 & 0 & w & 0 & 0 & 0 & 1 \end{bmatrix}$$

New Row 1 = -yz(Row 4) + (Row 1)

33. (continued)

$$\begin{bmatrix} 1 & 0 & 0 & 0 & \frac{1}{x} & -\frac{1}{xy} & -\frac{1}{xz} & -\frac{1}{xw} \\ 0 & 1 & 0 & 0 & 0 & \frac{1}{y} & 0 & 0 \\ 0 & 0 & 1 & 0 & 0 & 0 & \frac{1}{z} & 0 \\ 0 & 0 & 0 & 1 & 0 & 0 & 0 & \frac{1}{w} \end{bmatrix} \longleftarrow$$

New Row 4 $= \frac{1}{w}$(Row 4)

New Row 3 $= \frac{1}{z}$(Row 3)

New Row 2 $= \frac{1}{y}$(Row 2)

New Row 1 $= \frac{1}{xyzw}$(Row 1)

$$A^{-1} = \begin{bmatrix} \frac{1}{x} & -\frac{1}{xy} & -\frac{1}{xz} & -\frac{1}{xw} \\ 0 & \frac{1}{y} & 0 & 0 \\ 0 & 0 & \frac{1}{z} & 0 \\ 0 & 0 & 0 & \frac{1}{w} \end{bmatrix}$$

Exercise Set 2.7

1. We can write the data in a table.

Industry j / Input Product i	Input to agricultural industry (j = 1)	Input to manufacturing industry (j = 2)
Produce (i = 1)	0.2 bushels of produce per bushel of produce	0.5 bushels of produce per lb of of hardware
Hardware (i = 2)	0.25 lb of hardware per bushel of produce	0.2 lb of hardware per lb of hardware
Labor (i = 0)	5 man-days of labor per bushel of produce	3 man-days of labor per lb of hardware

a) We can use the information in the first two rows of the table to set up the I/O matrix A.

$$A = \begin{bmatrix} 0.2 & 0.5 \\ 0.25 & 0.2 \end{bmatrix}$$

b) We use the information in the third row of the table to set up the matrix A_0.

$$A_0 = [5 \quad 3]$$

1. (continued)

c) First find I - A:

$$I - A = \begin{bmatrix} 1 & 0 \\ 0 & 1 \end{bmatrix} - \begin{bmatrix} 0.2 & 0.5 \\ 0.25 & 0.2 \end{bmatrix}$$

$$= \begin{bmatrix} 0.8 & -0.5 \\ -0.25 & 0.8 \end{bmatrix}$$

Use the elimination method to find $(I - A^{-1})$:

$$\begin{bmatrix} 0.8 & -0.5 & 1 & 0 \\ -0.25 & 0.8 & 0 & 1 \end{bmatrix} \text{ Augmented matrix}$$

Multiply to clear of decimals:

$$\begin{bmatrix} 8* & -5 & 10 & 0 \\ -25 & 80 & 0 & 100 \end{bmatrix} \begin{array}{l} \text{New Row 1 = 10(Row 1)} \\ \text{New Row 2 = 100(Row 2)} \end{array}$$

$$\begin{bmatrix} 8* & -5 & 10 & 0 \\ -200 & 640 & 0 & 800 \end{bmatrix} \text{ New Row 2 = 8(Row 2)}$$

$$\begin{bmatrix} 8 & -5 & 10 & 0 \\ 0 & 515* & 250 & 800 \end{bmatrix} \begin{array}{l} \text{New Row 2 =} \\ \text{25(Row 1) + (Row 2)} \end{array}$$

$$\begin{bmatrix} 824 & -515 & 1030 & 0 \\ 0 & 515* & 250 & 800 \end{bmatrix} \text{ New Row 1 = 103(Row 1)}$$

$$\begin{bmatrix} 824 & 0 & 1280 & 800 \\ 0 & 515 & 250 & 800 \end{bmatrix} \begin{array}{l} \text{New Row 1 =} \\ \text{(Row 2) + (Row 1)} \end{array}$$

$$\begin{bmatrix} 1 & 0 & \frac{160}{103} & \frac{100}{103} \\ 0 & 1 & \frac{50}{103} & \frac{160}{103} \end{bmatrix} \begin{array}{l} \text{New Row 1 = } \frac{1}{824}\text{(Row 1)} \\ \text{New Row 2 = } \frac{1}{515}\text{(Row 2)} \end{array}$$

$$(I - A)^{-1} = \begin{bmatrix} \frac{160}{103} & \frac{100}{103} \\ \frac{50}{103} & \frac{160}{103} \end{bmatrix}, \text{ or } \frac{10}{103}\begin{bmatrix} 16 & 10 \\ 5 & 16 \end{bmatrix}$$

d) The demand matrix $D = [d_1, d_2 \ldots d_n]^t$ where d_i represents the net production of industry i. Therefore,

$$D = [530 \quad 106]^t.$$

1. (continued)

e) $X = (I - A)^{-1}D$

$$= \frac{10}{103} \begin{bmatrix} 16 & 10 \\ 5 & 16 \end{bmatrix} \begin{bmatrix} 530 \\ 106 \end{bmatrix}$$

$$= \frac{10}{103} \begin{bmatrix} 9540 \\ 4346 \end{bmatrix}$$

$$= \begin{bmatrix} 926.2 \\ 421.9 \end{bmatrix}$$

f) $x_0 = A_0 X$

$$= [5 \quad 3] \begin{bmatrix} 926.2 \\ 421.9 \end{bmatrix}$$

$$= [5896.7]$$

$x_0 = 5896.7$ man-days

(Answers may vary in the last digit due to rounding differences.)

3. We can write the data in a table.

Industry j / Input Product i	Input to mining industry (j = 1)	Input to agricultural industry (j = 2)	Input to manufacturing industry (j = 3)
Metal (i = 1)	0.2K$ per 1K$ of metal	0.1K$ per 1K$ of produce	0.5K$ per 1K$ of hardware
Produce (i = 2)	0K$ per 1K$ of metal	0.1K$ per 1K$ of produce	0K$ per 1K$ of hardware
Hardware (i = 3)	0.2K$ per 1K$ of metal	0.2K$ per 1K$ of produce	0.3K$ per 1K$ of hardware
Labor (i = 0)	0.4K$ per 1K$ of metal	0.2K$ per 1K$ of produce	0.3K$ per 1K$ of hardware

a) We can use the information in the first three rows of the table to set up the I/O matrix A.

$$A = \begin{bmatrix} 0.2 & 0.1 & 0.5 \\ 0 & 0.1 & 0 \\ 0.2 & 0.2 & 0.3 \end{bmatrix}$$

b) We use the information in the fourth row of the table to set up the matrix A_0.

$$A_0 = [0.4 \quad 0.2 \quad 0.3]$$

3. (continued)

c) First find I - A:

$$I - A = \begin{bmatrix} 1 & 0 & 0 \\ 0 & 1 & 0 \\ 0 & 0 & 1 \end{bmatrix} - \begin{bmatrix} 0.2 & 0.1 & 0.5 \\ 0 & 0.1 & 0 \\ 0.2 & 0.2 & 0.3 \end{bmatrix}$$

$$= \begin{bmatrix} 0.8 & -0.1 & -0.5 \\ 0 & 0.9 & 0 \\ -0.2 & -0.2 & 0.7 \end{bmatrix}$$

Use the elimination method to find $(I - A)^{-1}$:

$$\begin{bmatrix} 0.8 & -0.1 & -0.5 & 1 & 0 & 0 \\ 0 & 0.9 & 0 & 0 & 1 & 0 \\ -0.2 & -0.2 & 0.7 & 0 & 0 & 1 \end{bmatrix}$$ Augmented matrix

Multiply each row by 10 to clear of decimals:

$$\begin{bmatrix} 8 & -1 & -5 & 10 & 0 & 0 \\ 0 & 9 & 0 & 0 & 10 & 0 \\ -2 & -2 & 7 & 0 & 0 & 10 \end{bmatrix}$$

Proceed with the pivoting:

$$\begin{bmatrix} -2* & -2 & 7 & 0 & 0 & 10 \\ 0 & 9 & 0 & 0 & 10 & 0 \\ 8 & -1 & -5 & 10 & 0 & 0 \end{bmatrix}$$ Interchange Row 1 and Row 3

$$\begin{bmatrix} -2 & -2 & 7 & 0 & 0 & 10 \\ 0 & 9* & 0 & 0 & 10 & 0 \\ 0 & -9 & 23 & 10 & 0 & 40 \end{bmatrix}$$ New Row 3 = 4(Row 1) + (Row 3)

$$\begin{bmatrix} -18 & -18 & 63 & 0 & 0 & 90 \\ 0 & 9* & 0 & 0 & 10 & 0 \\ 0 & -9 & 23 & 10 & 0 & 40 \end{bmatrix}$$ New Row 1 = 9(Row 1)

$$\begin{bmatrix} -18 & 0 & 63 & 0 & 20 & 90 \\ 0 & 9 & 0 & 0 & 10 & 0 \\ 0 & 0 & 23* & 10 & 10 & 40 \end{bmatrix}$$ New Row 1 = 2(Row 2) + (Row 1); New Row 3 = (Row 2) + (Row 3)

$$\begin{bmatrix} -414 & 0 & 1449 & 0 & 460 & 2070 \\ 0 & 9 & 0 & 0 & 10 & 0 \\ 0 & 0 & 23* & 10 & 10 & 40 \end{bmatrix}$$ New Row 1 = 23(Row 1)

$$\begin{bmatrix} -414 & 0 & 0 & -630 & -170 & -450 \\ 0 & 9 & 0 & 0 & 10 & 0 \\ 0 & 0 & 23 & 10 & 10 & 40 \end{bmatrix}$$

New Row 1 = -63(Row 3) + (Row 1)

3. (continued)

$$\begin{bmatrix} 1 & 0 & 0 & \frac{35}{23} & \frac{85}{207} & \frac{25}{23} \\ 0 & 1 & 0 & 0 & \frac{10}{9} & 0 \\ 0 & 0 & 1 & \frac{10}{23} & \frac{10}{23} & \frac{40}{23} \end{bmatrix}$$

New Row 3 = $\frac{1}{23}$(Row 3)

New Row 2 = $\frac{1}{9}$(Row 2)

New Row 1 = $-\frac{1}{414}$(Row 1)

$$(I - A)^{-1} = \begin{bmatrix} \frac{35}{23} & \frac{85}{207} & \frac{25}{23} \\ 0 & \frac{10}{9} & 0 \\ \frac{10}{23} & \frac{10}{23} & \frac{40}{23} \end{bmatrix}$$

or $\frac{5}{207}\begin{bmatrix} 63 & 17 & 45 \\ 0 & 46 & 0 \\ 18 & 18 & 72 \end{bmatrix}$ (Factoring $\frac{5}{207}$ out of each element)

d) The net production of each industry is 1K$. Therefore, the demand matrix is D = [1 1 1]^t.

e) X = (I - A^{-1})D

$$= \frac{5}{207}\begin{bmatrix} 63 & 17 & 45 \\ 0 & 46 & 0 \\ 18 & 18 & 72 \end{bmatrix}\begin{bmatrix} 1 \\ 1 \\ 1 \end{bmatrix}$$

$$= \frac{5}{207}\begin{bmatrix} 125 \\ 46 \\ 108 \end{bmatrix}$$

$$= \begin{bmatrix} 3.019 \\ 1.111 \\ 2.609 \end{bmatrix}$$

f) $x_0 = A_0 X$

$$= [0.4 \quad 0.2 \quad 0.3]\begin{bmatrix} 3.019 \\ 1.111 \\ 2.609 \end{bmatrix}$$

$$= [2.213]$$
$$= 2.213K\$$$

5. To find $(I - A)^{-1}$, we first find I - A:

$$I - A = \begin{bmatrix} 1 & 0 \\ 0 & 1 \end{bmatrix} - \begin{bmatrix} 0.2 & 0.4 \\ 0.4 & 0 \end{bmatrix} = \begin{bmatrix} 0.8 & -0.4 \\ -0.4 & 1 \end{bmatrix}$$

5. (continued)

Use the elimination method to find $(I - A)^{-1}$:

$$\begin{bmatrix} 0.8 & -0.4 & 1 & 0 \\ -0.4 & 1 & 0 & 1 \end{bmatrix} \quad \text{Augmented matrix}$$

$$\begin{bmatrix} 8 & -4 & 10 & 0 \\ -4 & 10 & 0 & 10 \end{bmatrix} \quad \begin{array}{l}\text{Multiply each row by 10 to} \\ \text{clear of decimals}\end{array}$$

$$\begin{bmatrix} -4^* & 10 & 0 & 10 \\ 8 & -4 & 10 & 0 \end{bmatrix} \quad \text{Interchange Row 1 and Row 2}$$

$$\begin{bmatrix} -4 & 10 & 0 & 10 \\ 0 & 16^* & 10 & 20 \end{bmatrix} \quad \text{New Row 2 = 2(Row 1) + (Row 2)}$$

$$\begin{bmatrix} -32 & 80 & 0 & 80 \\ 0 & 16^* & 10 & 20 \end{bmatrix} \quad \text{New Row 1 = 8(Row 1)}$$

$$\begin{bmatrix} -32 & 0 & -50 & -20 \\ 0 & 16 & 10 & 20 \end{bmatrix} \quad \begin{array}{l}\text{New Row 1 =} \\ -5\text{(Row 2) + (Row 1)}\end{array}$$

$$\begin{bmatrix} 1 & 0 & \frac{25}{16} & -\frac{5}{8} \\ 0 & 1 & \frac{5}{8} & \frac{5}{4} \end{bmatrix} \quad \begin{array}{l}\text{New Row 1 = } -\frac{1}{32}\text{(Row 1)} \\[4pt] \text{New Row 2 = } \frac{1}{16}\text{(Row 2)}\end{array}$$

$$(I - A)^{-1} = \begin{bmatrix} \frac{25}{16} & \frac{5}{8} \\ \frac{5}{8} & \frac{5}{4} \end{bmatrix}$$

or $\frac{5}{16}\begin{bmatrix} 5 & 2 \\ 2 & 4 \end{bmatrix}$ (Factoring $\frac{5}{16}$ out of each element)

X = (I - A)$^{-1}$D

$$= \frac{5}{16}\begin{bmatrix} 5 & 2 \\ 2 & 4 \end{bmatrix}\begin{bmatrix} 70 \\ 56 \end{bmatrix}$$

$$= \frac{5}{16}\begin{bmatrix} 462 \\ 364 \end{bmatrix}$$

$$= \begin{bmatrix} 144.4 \\ 113.8 \end{bmatrix}$$

$x_0 = A_0 X$

$$= [0.3 \quad 0.7]\begin{bmatrix} 144.4 \\ 113.8 \end{bmatrix}$$

$$= [122.98], \text{ or } 122.98$$

7. To find $(I - A)^{-1}$, we first find $I - A$:

$$I - A = \begin{bmatrix} 1 & 0 & 0 \\ 0 & 1 & 0 \\ 0 & 0 & 1 \end{bmatrix} - \begin{bmatrix} 0.2 & 0.1 & 0 \\ 0 & 0.3 & 0.2 \\ 0.1 & 0 & 0.9 \end{bmatrix}$$

$$= \begin{bmatrix} 0.8 & -0.1 & 0 \\ 0 & 0.7 & -0.2 \\ -0.1 & 0 & 0.1 \end{bmatrix}$$

Use the elimination method to find $(I - A)^{-1}$:

$$\begin{bmatrix} 0.8 & -0.1 & 0 & 1 & 0 & 0 \\ 0 & 0.7 & -0.2 & 0 & 1 & 0 \\ -0.1 & 0 & 0.1 & 0 & 0 & 1 \end{bmatrix}$$ Augmented matrix

$$\begin{bmatrix} 8 & -1 & 0 & 10 & 0 & 0 \\ 0 & 7 & -2 & 0 & 10 & 0 \\ -1 & 0 & 1 & 0 & 0 & 10 \end{bmatrix}$$ Multiply each row by 10 to clear of decimals

$$\begin{bmatrix} -1* & 0 & 1 & 0 & 0 & 10 \\ 0 & 7 & -2 & 0 & 10 & 0 \\ 8 & -1 & 0 & 10 & 0 & 0 \end{bmatrix}$$ Interchange Row 1 and Row 3

$$\begin{bmatrix} -1 & 0 & 1 & 0 & 0 & 10 \\ 0 & 7 & -2 & 0 & 10 & 0 \\ 0 & -1 & 8 & 10 & 0 & 80 \end{bmatrix}$$ New Row 3 = 8(Row 1) + (Row 3)

$$\begin{bmatrix} -1 & 0 & 1 & 0 & 0 & 10 \\ 0 & -1* & 8 & 10 & 0 & 80 \\ 0 & 7 & -2 & 0 & 10 & 0 \end{bmatrix}$$ Interchange Row 2 and Row 3

$$\begin{bmatrix} -1 & 0 & 1 & 0 & 0 & 10 \\ 0 & -1 & 8 & 10 & 0 & 80 \\ 0 & 0 & 54* & 70 & 10 & 560 \end{bmatrix}$$ New Row 3 = 7(Row 2) + (Row 3)

$$\begin{bmatrix} -54 & 0 & 54 & 0 & 0 & 540 \\ 0 & -54 & 432 & 540 & 0 & 4320 \\ 0 & 0 & 54* & 70 & 10 & 560 \end{bmatrix}$$ New Row 1 = 54(Row 1) New Row 2 = 54(Row 2)

$$\begin{bmatrix} -54 & 0 & 0 & -70 & -10 & -20 \\ 0 & -54 & 0 & -20 & -80 & -160 \\ 0 & 0 & 54 & 70 & 10 & 560 \end{bmatrix}$$

New Row 2 = -8(Row 3) + (Row 2)
New Row 1 = -1(Row 3) + (Row 1)

7. (continued)

$$\begin{bmatrix} 1 & 0 & 0 & \frac{35}{27} & \frac{5}{27} & \frac{10}{27} \\ 0 & 1 & 0 & \frac{10}{27} & \frac{40}{27} & \frac{80}{27} \\ 0 & 0 & 1 & \frac{35}{27} & \frac{5}{27} & \frac{280}{27} \end{bmatrix}$$ New Row 1 = $-\frac{1}{54}$(Row 1) New Row 2 = $-\frac{1}{54}$(Row 2) New Row 3 = $\frac{1}{54}$(Row 3)

$$(I - A)^{-1} = \begin{bmatrix} \frac{35}{27} & \frac{5}{27} & \frac{10}{27} \\ \frac{10}{27} & \frac{40}{27} & \frac{80}{27} \\ \frac{35}{27} & \frac{5}{27} & \frac{280}{27} \end{bmatrix}$$

or $\frac{5}{27} \begin{bmatrix} 7 & 1 & 2 \\ 2 & 8 & 16 \\ 7 & 1 & 56 \end{bmatrix}$ (Factoring $\frac{5}{27}$ out of each element)

$$X = (I - A)^{-1}D$$

$$= \frac{5}{27} \begin{bmatrix} 7 & 1 & 2 \\ 2 & 8 & 16 \\ 7 & 1 & 56 \end{bmatrix} \begin{bmatrix} 48 \\ 96 \\ 168 \end{bmatrix}$$

$$= \frac{5}{27} \begin{bmatrix} 768 \\ 3552 \\ 9840 \end{bmatrix}$$

$$= \begin{bmatrix} 142.2 \\ 657.8 \\ 1822.2 \end{bmatrix}$$

Exercise Set 3.1

1. Graph y < x.

    First, graph the related equation y = x. We draw a dashed line since the inequality symbol is <.

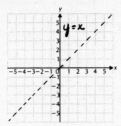

    The line y = x divides the plane into two half-planes. To decide which half-plane contains the solutions to the inequality and should be shaded, we check to see if some point off the line is a solution of the inequality. The point (0, 0) is the easiest to use, but since this line passes through the origin we choose another point, say (1, 0). Substitute 1 for x and 0 for y in the inequality.

    $$\begin{array}{c|c} y & < & x \\ \hline 0 & & 1 \end{array}$$

    Since 0 < 1 is true, (1, 0) is a solution of the inequality and so are all the points in the half-plane containing (1, 0). We shade that half-plane. The graph consists of the shaded half-plane but not the line.

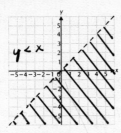

3. Graph y + x ≥ 0.

    First graph the related equation y + x = 0, or y = -x. We draw a solid line since the inequality symbol is ≥.

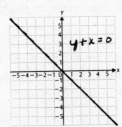

    Check to see if some point off the line is a solution of the inequality. We choose (0, 1). Substitute 0 for x and 1 for y in the inequality.

    $$\begin{array}{c|c} y + x & \geq & 0 \\ \hline 1 + 0 & & 0 \\ 1 & & \end{array}$$

3. (continued)

    Since 1 ≥ 0 is true, (0, 1) is on the graph. We shade the half-plane containing (0, 1). The graph consists of the shaded half-plane and the line.

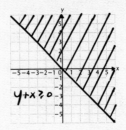

5. Graph 5x - 2y < 10.

    First graph the related equation 5x - 2y = 10. (We can use the intercepts (0, -5) and (2, 0) to draw the graph.) We draw a dashed line since the inequality symbol is <.

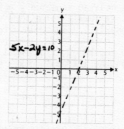

    Next we check to see if some point off the line is a solution of the inequality. We choose (0, 0). Substitute 0 for x and 0 for y.

    $$\begin{array}{c|c} 5x - 2y & < & 10 \\ \hline 5(0) - 2(0) & & 10 \\ 0 - 0 & & \\ 0 & & \end{array}$$

    Since 0 < 10 is true (0, 0) is on the graph. We shade the half-plane containing (0, 0). The graph consists of the shaded half-plane but not the line.

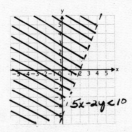

7. Graph 3x + 2y ≥ 6.

    First graph the related equation 3x + 2y = 6. (We can use the intercepts (0, 3) and (2, 0) to draw the graph.) We draw a solid line since the inequality symbol is ≥. Use (0, 0) as a check point. That is, check to see if (0, 0) is a solution of the inequality. Substitute 0 for x and 0 for y.

    $$\begin{array}{c|c} 3x + 2y & \geq & 6 \\ \hline 3(0) + 2(0) & & 6 \\ 0 + 0 & & \\ 0 & & \end{array}$$

7. (continued)

Since $0 \geqslant 6$ is false, (0, 0) is not a solution. Therefore, we shade the half-plane that does <u>not</u> contain (0, 0). The graph consists of the shaded half-plane and the line.

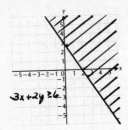

9. Graph $y \geqslant 0$.

This is considered to be a linear inequality in two variables. We can express it as $0x + y \geqslant 0$.

First graph the related equation $y = 0$. This is the x-axis. We draw a solid line since the inequality symbol is $\geqslant$. Then check to see if some point off the line is a solution of the inequality. We choose (0, 1). Substitute 0 for x and 1 for y in the inequality $0x + y \geqslant 0$.

$$
\begin{array}{c|c}
0x + y & \geqslant 0 \\
\hline
0(0) + 1 & 0 \\
0 + 1 & \\
1 & \\
\end{array}
$$

Since $1 \geqslant 0$ is true, (0, 1) is a solution. We shade the half-plane containing (0, 1). The graph consists of the shaded half-plane and the line.

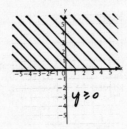

11. Graph $x < -4$.

This is considered to be a linear inequality in two variables. We can express it as $x + 0y < -4$.

First graph the related equation $x = -4$. This is a vertical line through the point (-4, 0). We draw a dashed line since the inequality symbol is <. Then check to see if some point off the line is a solution of the inequality. We choose (0, 0). Substitute 0 for x and 0 for y in the inequality $x + 0y < -4$.

$$
\begin{array}{c|c}
x + 0y & < -4 \\
\hline
0 + 0(0) & -4 \\
0 + 0 & \\
0 & \\
\end{array}
$$

Since $0 < -4$ is false, (0, 0) is not a solution. We shade the half-plane that does <u>not</u> contain (0, 0). The graph consists of the half-plane but not the line.

11. (continued)

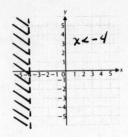

13. Graph $x + y \leqslant 1$

$x - y \leqslant 2$.

The separate graphs are shown below. A pair of arrows points in the direction of the half-plane that is part of the solution set of each inequality.

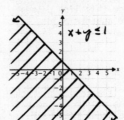 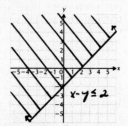

The region satisfying <u>both</u> inequalities, or constraints, is shown below.

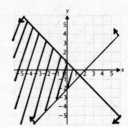

The solution set has one corner point which is the point of intersection of the two lines. It is found by solving the system of equations

$x + y = 1$

$x - y = 2$.

Using the elimination method or any other algebraic technique, we find that the solution is $\left(\frac{3}{2}, -\frac{1}{2}\right)$. This is the corner point.

15. Graph y - 2x > 1
          y - 2x < 3.

The separate graphs are shown below.

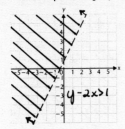 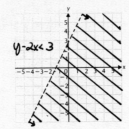

The region satisfying both inequalities, or constraints, is shown below.

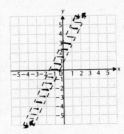

This region has no corner points, since the lines do not intersect.

17. Graph x + y ≤ 6   (1)
               y ≤ 5   (2)
           x, y ≥ 0   (3), (4)

The four separate graphs are shown below. Note that x, y ≥ 0 means x ≥ 0 and y ≥ 0.

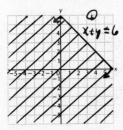 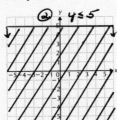

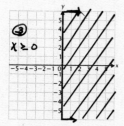 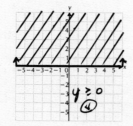

17. (continued)

The region satisfying all four constraints is shown below.

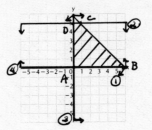

This region has four corner points. Corner point A is the point of intersection of lines (3) and (4). We find it by solving the system composed of equations (3) and (4):

    x = 0
    y = 0

Point A has coordinates (0, 0).

Corner point B is the point of intersection of lines (1) and (4). We find it by solving the system composed of equations (1) and (4):

    x + y = 6
        y = 0

Point B is (6, 0).

Corner point C is the point of intersection of lines (1) and (2). We find it by solving the system composed of equations (1) and (2):

    x + y = 6
        y = 5

Point C is (1, 5).

Corner point D is the point of intersection of lines (2) and (3). We find it by solving the system composed of equations (2) and (3).

    y = 5
    x = 0

Point D is (0, 5).

The four corner points are (0, 0), (6, 0), (1, 5), and (0, 5).

19. Graph $3x + 2y \leqslant 12$ (1)

$x + y \leqslant 5$ (2)

$x, y \geqslant 0$ (3), (4)

The four separate graphs are shown below.

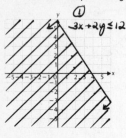 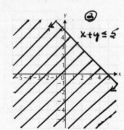

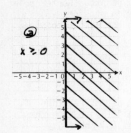 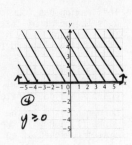

The region satisfying all four constraints is shown below.

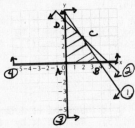

The region has four corner points. We find corner point A by solving the system composed of equations (3) and (4):

$x = 0$

$y = 0$

Point A is (0, 0).

Corner point B is found by solving the system composed of equations (1) and (4):

$3x + 2y = 12$

$y = 0$

Point B is (4, 0).

Corner point C is found by solving the system composed of equations (1) and (2):

$3x + 2y = 12$

$x + y = 5$

Point C is (2, 3).

Corner point D is found by solving the system composed of equations (1) and (3):

$x + y = 5$

$x = 0$

Point D is (0, 5).

The four corner points are (0, 0), (4, 0), (2, 3), and (0, 5).

21. Graph $x + 2y \leqslant 14$ (1)

$4x + 3y \leqslant 26$ (2)

$2x + y \leqslant 12$ (3)

$x, y \geqslant 0$ (4), (5)

The five separate graphs are shown below.

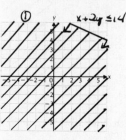 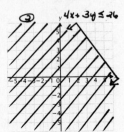

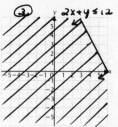 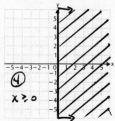

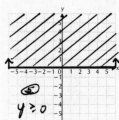

The region satisfying all five constraints is shown below.

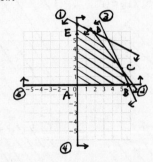

There are five corner points. Corner point A is found by solving the system composed of equations (4) and (5):

$x = 0$

$y = 0$

Point A is (0, 0).

Corner point B is found by solving the system composed of equations (3) and (5):

$2x + y = 12$

$y = 0$

Point B is (6, 0).

21. (continued)

Corner point C is found by solving the system composed of equations (2) and (3):

$$4x + 3y = 26$$
$$2x + y = 12$$

Point C is (5, 2).

Corner point D is found by solving the system composed of equations (1) and (2):

$$x + 2y = 14$$
$$4x + 3y = 26$$

Point D is (2, 6).

Corner point E is found by solving the system composed of equations (1) and (4):

$$x + 2y = 14$$
$$x = 0$$

Point E is (0, 7).

The corner points are (0, 0), (6, 0), (5, 2), (2, 6), and (0, 7).

23. Graph $3u + v \geqslant 9$ (1)
$$u + v \geqslant 7 \qquad (2)$$
$$u + 2v \geqslant 8 \qquad (3)$$
$$u, v \geqslant 0 \qquad (4), (5)$$

The five separate graphs are shown below.

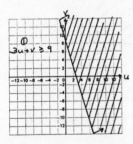
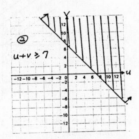
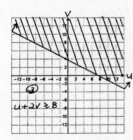
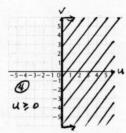
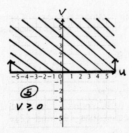

23. (continued)

The region satisfying all five constraints is shown below.

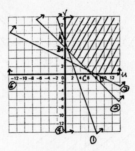

The region has four corner points. Corner point A is found by solving the system composed of equations (1) and (4):

$$3u + v = 9$$
$$u = 0$$

Point A is (0, 9).

Corner point B is found by solving the system composed of equations (1) and (2):

$$3u + v = 9$$
$$u + v = 7$$

Point B is (1, 6).

Corner point C is found by solving the system composed of equations (2) and (3):

$$u + v = 7$$
$$u + 2v = 8$$

Point C is (6, 1).

Corner point D is found by solving the system composed of equations (3) and (5):

$$u + 2v = 8$$
$$v = 0$$

Point D is (8, 0).

The corner points are (0, 9), (1, 6), (6, 1), and (8, 0).

25. Graph $2a + b \geqslant 9$ (1)
$$4a + 3b \geqslant 23 \qquad (2)$$
$$a + 3b \geqslant 8 \qquad (3)$$
$$a, b \geqslant 0 \qquad (4), (5)$$

The five separate graphs are shown below.

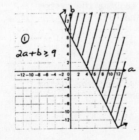
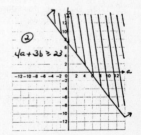

25. (continued)

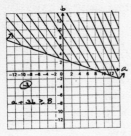

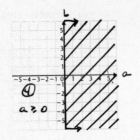

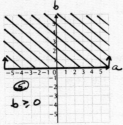

The region satisfying all five constraints is shown below.

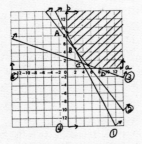

There are four corner points. Corner point A is found by solving the system composed of equations (1) and (4):

$$2a + b = 9$$
$$a = 0$$

Point A is (0, 9).

Corner point B is found by solving the system composed of equations (1) and (2):

$$2a + b = 9$$
$$4a + 3b = 23$$

Point B is (2, 5).

Corner point C is found by solving the system composed of equations (2) and (3):

$$4a + 3b = 23$$
$$a + 3b = 8$$

Point C is (5, 1).

Corner point D is found by solving the system composed of equations (3) and (5):

$$a + 3b = 8$$
$$b = 0$$

Point D is (8, 0).

The corner points are (0, 9), (2, 5), (5, 1), and (8, 0).

27. Graph $4x + y \geqslant 9$ (1)
$$3x + 2y \geqslant 13$$ (2)
$$2x + 5y \geqslant 16$$ (3)
$$x, y \geqslant 0$$ (4), (5)

The separate graphs are shown below.

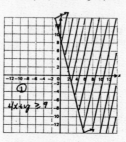

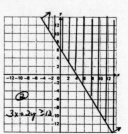

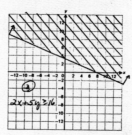

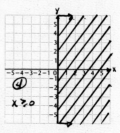

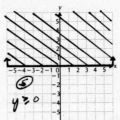

The region satisfying all five constraints is shown below.

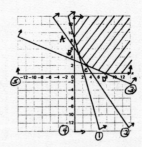

There are four corner points. Corner point A is found by solving the system composed of equations (1) and (4):

$$4x + y = 9$$
$$x = 0$$

Point A is (0, 9).

Corner point B is found by solving the system composed of equations (1) and (2):

$$4x + y = 9$$
$$3x + 2y = 13$$

Point B is (1, 5).

27. (continued)

Corner point C is found by solving the system composed of equations (2) and (3):

$$3x + 2y = 13$$
$$2x + 5y = 16$$

Point C is (3, 2).

Corner point D is found by solving the system composed of equations (3) and (5):

$$2x + 5y = 16$$
$$y = 0$$

Point D is (8, 0).

The corner points are (0, 9), (1, 5), (3, 2), and (8, 0).

29. a) Graph x + 2y ≤ 6   (1)
        x ≥ 0   (2)
        x ≤ 5   (3)
        y ≥ -2   (4)

The separate graphs are shown below.

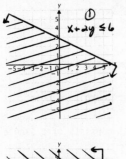 

 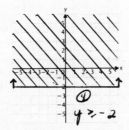

The region satisfying all the constraints is shown below.

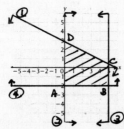

b) The shaded region and its boundary are the solution set. It is nonempty.

c) The solution set is confined to a quadrilateral and its interior. It is bounded.

29. (continued)

d) Each of the constraints affects the solution set. Therefore, there are no redundant constraints.

e) No corner point is the solution of three or more equations. Therefore, there are no degenerate corners.

31. a) Graph   x ≥ -3   (1)
        x - 2y ≤ 4   (2)
        y - 3x ≤ 9   (3)
        3x + y ≤ 10   (4)

The separate graphs are shown below.

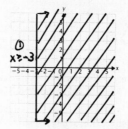 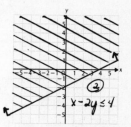

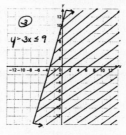 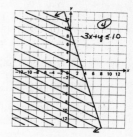

The region satisfying all the constraints is shown below.

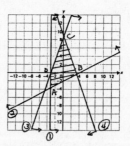

b) The solution set is the shaded region and its boundary. It is nonempty.

c) The solution set is confined to a quadrilateral and its interior. It is bounded.

d) Each of the constraints affects the solution set. There are no redundant constraints.

e) No corner point is the solution of three or more equations. There are no degenerate corners.

33. a) Graph $-3x + 2y \geqslant 6$ (1)

             $2x + y \leqslant -2$ (2)

               $x + y \geqslant 4$ (3)

            $2x + 7y \leqslant 21$ (4)

We graph each inequality using the same set of axes. As usual, a pair of arrows points in the direction of the half-plane that is part of the solution set of each inequality.

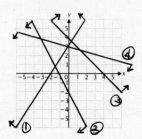

b) There are no points in common to all the constraints. Therefore, the solution set is empty.

c) An empty set is bounded. Therefore, the solution set is bounded.

d) If either constraint (1) or constraint (2) were omitted, the solution set of the system would still be empty. Therefore, constraints (1) and (2) are redundant. (If either (3) or (4) were omitted, however, the solution set would be nonempty. Thus, (3) and (4) are not redundant.)

e) Since the solution set is empty, there are no corner points and, hence, no degeneracies.

35. a) Graph $x \geqslant 0$ (1)

                $y \geqslant 0$ (2)

           $x + y \geqslant 2$ (3)

           $x - y \leqslant 2$ (4)

               $y \leqslant 6$ (5)

We graph each inequality using the same set of axes.

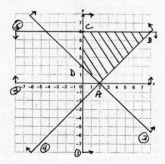

b) The solution set is the shaded region and its boundary. It is nonempty.

c) The solution set is confined to a quadrilateral and its interior. It is bounded.

35. (continued)

d) Constraint (2) does not affect the solution set. It is redundant.

e) Corner point A with coordinates (2, 0) is the solution to three equations, $y = 0$, $x + y = 2$, and $x - y = 2$. It is degenerate.

37. a) Graph $x + y \leqslant 0$ (1)

          $2x - 3y \leqslant 15$ (2)

               $y \leqslant 5$ (3)

               $x \geqslant 0$ (4)

         $2x + y \geqslant 3$ (5)

Graph each inequality using the same set of axes.

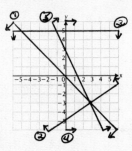

Note that the solution set consists of a single point, $(3, -3)$.

b) Since the solution set contains a point, it is nonempty.

c) Since the solution set consists of a single point, it is bounded.

d) Constraints (3) and (4) do not affect the solution set. They are redundant.

e) The point $(3, -3)$ is the solution to three equations, $x + y = 0$, $2x - 3y = 15$, and $2x + y = 3$. It is degenerate.

39. a) Graph $x \geqslant 0$ (1)

                $y \geqslant 0$ (2)

         $5y - 3x \leqslant 15$ (3)

              $x \leqslant 4y$ (4)

        $2x - 5y \leqslant 10$ (5)

Graph each inequality using the same set of axes.

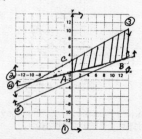

b) The solution set is the shaded region and portions of the lines $5y - 3x = 15$, $x = 4y$, and $2x - 5y = 10$. It is nonempty.

<u>39.</u> (continued)

    c) The solution set is <u>not</u> confined to a polygon and its interior. It is unbounded.

    d) Constraint (2) does not affect the solution set. It is redundant.

    e) The point (0, 0) is the solution of three equations, x = 0, y = 0, and x = 4y. It is degenerate.

<u>41.</u> Graph $y \geqslant x^2 - 4$.

First graph the related equation $y = x^2 - 4$.

The graph divides the planes into two parts, one part "inside" the parabola and the other part "outside" the parabola. To decide which part of the plane to shade, we check to see if some point off the parabola is a solution of the inequality. We choose (0, 0).

$$
\begin{array}{c|c}
y^2 \geqslant & x^2 - 4 \\ \hline
0^2 & 0^2 - 4 \\
0 & 0 - 4 \\
& -4
\end{array}
$$

Since $0 \geqslant -4$ is true, (0, 0) is a solution of the inequality and so are all the other points "inside" the parabola. We shade that region. The graph consists of the parabola and the shaded region.

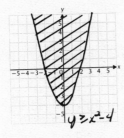

<u>43.</u> Graph $|x - y| > 0$.

Remember that the absolute value of a number is always greater than or equal to zero. Therefore, the solution set will contain all pairs (x, y) except those for which the absolute value is equal to zero; that is, those pairs for which

    x - y = 0, or

    x = y.

The solution set is all points (x, y) except those on the line x = y.

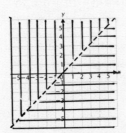

Exercise Set 3.2

<u>1.</u> Maximize and minimize

    F = 4x + 28y

Subject to

    5x + 3y ≤ 34   (1)

    3x + 5y ≤ 30   (2)

    x, y ≥ 0   (3), (4)

We draw the solution set of the system of constraints and find the corner points.

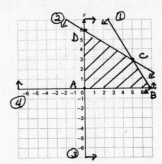

To find corner point A we solve the system

    x = 0

    y = 0.

Point A is (0, 0).

To find corner point B we solve the system

    5x + 3y = 34

    y = 0.

Point B is $\left(\frac{34}{5}, 0\right)$.

To find corner point C we solve the system

    5x + 3y = 34

    3x + 5y = 30.

Point C is (5, 3).

To find corner point D we solve the system

    3x + 5y = 30

    x = 0.

Point D is (0, 6).

Now compute the function values at the corner points.

Corner points	Function F = 4x + 28y	
(0, 0)	$4 \cdot 0 + 28 \cdot 0 = 0 + 0 = 0$	← Minimum
$\left(\frac{34}{5}, 0\right)$	$4 \cdot \frac{34}{5} + 28 \cdot 0 = \frac{136}{5} + 0 = 27\frac{1}{5}$	
(5, 3)	$4 \cdot 5 + 28 \cdot 3 = 20 + 84 = 104$	
(0, 6)	$4 \cdot 0 + 28 \cdot 6 = 0 + 168 = 168$	← Maximum

The maximum value of F is 168 when x = 0 and y = 6. The minimum value of F is 0 when x = 0 and y = 0. We write this as:

Max = 168 at x = 0, y = 6; Min = 0 at x = 0, y = 0

3. Maximize and minimize

   $P = 16x - 2y + 40$

   Subject to

   $6x + 8y \leqslant 48$ (1)

   $y \leqslant 4$ (2)

   $y \geqslant 0$ (3)

   $x \leqslant 7$ (4)

   $x \geqslant 0$ (5)

   Draw the graph of the solution set of the system of constraints and find the corner points.

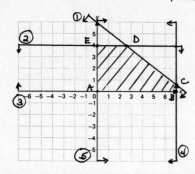

   Corner point A is the solution of the system

   $x = 0$

   $y = 0.$

   Point A is $(0, 0)$.

   Corner point B is the solution of the system

   $y = 0$

   $x = 7.$

   Point B is $(7, 0)$.

   Corner point C is the solution of the system

   $6x + 8y = 48$

   $x = 7.$

   Point C is $\left[7, \frac{3}{4}\right]$.

   Corner point D is the solution of the system

   $6x + 8y = 48$

   $y = 4.$

   Point D is $\left[\frac{8}{3}, 4\right]$.

   Corner point E is the solution of the system

   $y = 4$

   $x = 0.$

   Point E is $(0, 4)$.

   Now we compute the function values at the corner points.

3. (continued)

Corner points	Function $P = 16x - 2y + 40$	
$(0, 0)$	$16 \cdot 0 - 2 \cdot 0 + 40 =$ $0 - 0 + 40 = 40$	
$(7, 0)$	$16 \cdot 7 - 2 \cdot 0 + 40 =$ $112 - 0 + 40 = 152$	← Maximum
$\left[7, \frac{3}{4}\right]$	$16 \cdot 7 - 2 \cdot \frac{3}{4} + 40 =$ $112 - \frac{3}{2} + 40 = 150\frac{1}{2}$	
$\left[\frac{8}{3}, 4\right]$	$16 \cdot \frac{8}{3} - 2 \cdot 4 + 40 =$ $\frac{128}{3} - 8 + 40 = 74\frac{2}{3}$	
$(0, 4)$	$16 \cdot 0 - 2 \cdot 4 + 40 =$ $0 - 8 + 40 = 32$	← Minimum

Max = 152 at $x = 7$, $y = 0$;

Min = 32 at $x = 0$, $y = 4$

5. We graphed the solution set of the system of constraints in Exercise 17, Set 3.1 and found the corner points to be $(0, 0)$, $(6, 0)$, $(1, 5)$, and $(0, 5)$. Now we compute the function values at the corner points.

Corner points	Function $F = x + 2y$	
$(0, 0)$	$0 + 2 \cdot 0 = 0 + 0 = 0$	
$(6, 0)$	$6 + 2 \cdot 0 = 6 + 0 = 6$	
$(1, 5)$	$1 + 2 \cdot 5 = 1 + 10 = 11$	← Maximum
$(0, 5)$	$0 + 2 \cdot 5 = 0 + 10 = 10$	

Max = 11 at $x = 1$, $y = 5$

7. We graphed the solution set of the system of constraints in Exercise 19, Set 3.1 and found the corner points to be $(0, 0)$, $(4, 0)$, $(2, 3)$, and $(0, 5)$. Now we compute the function values at the corner points.

Corner points	Function $f = 5x + 4y$	
$(0, 0)$	$5 \cdot 0 + 4 \cdot 0 = 0 + 0 = 0$	
$(4, 0)$	$5 \cdot 4 + 4 \cdot 0 = 20 + 0 = 20$	
$(2, 3)$	$5 \cdot 2 + 4 \cdot 3 = 10 + 12 = 22$	← Maximum
$(0, 5)$	$5 \cdot 0 + 4 \cdot 5 = 0 + 20 = 20$	

Max = 22 at $x = 2$, $y = 3$

9. We graphed the solution set of the system of constraints in Exercise 21, Set 3.1 and found the corner points to be (0, 0), (6, 0), (5, 2), (2, 6), and (0, 7). Now we compute the function values at the corner points.

Corner points	Function $G = 3x + 4y$	
(0, 0)	$3 \cdot 0 + 4 \cdot 0 = 0 + 0 = 0$	
(6, 0)	$3 \cdot 6 + 4 \cdot 0 = 18 + 0 = 18$	
(5, 2)	$3 \cdot 5 + 4 \cdot 2 = 15 + 8 = 23$	
(2, 6)	$3 \cdot 2 + 4 \cdot 6 = 6 + 24 = 30$	⟵ Maximum
(0, 7)	$3 \cdot 0 + 4 \cdot 7 = 0 + 28 = 28$	

Max = 30 at $x = 2$, $y = 6$

11. We graphed the solution set of the system of constraints in Exercise 23, Set 3.1 and found the corner points to be (0, 9), (1, 6), (6, 1) and (8, 0). Now we compute the function values at the corner points.

Corner points	Function $Z = 3u + 4v$	
(0, 9)	$3 \cdot 0 + 4 \cdot 9 = 0 + 36 = 36$	
(1, 6)	$3 \cdot 1 + 4 \cdot 6 = 3 + 24 = 27$	
(6, 1)	$3 \cdot 6 + 4 \cdot 1 = 18 + 4 = 22$	⟵ Minimum
(8, 0)	$3 \cdot 8 + 4 \cdot 0 = 24 + 0 = 24$	

Min = 22 at $u = 6$, $v = 1$

13. We graphed the solution set of the system of constraints in Exercise 25, Set 3.1 and found the corner points to be (0, 9), (2, 5), (5, 1), and (8, 0). Now we find the function values at the corner points.

Corner points	Function $A = 2a + 5b$	
(0, 9)	$2 \cdot 0 + 5 \cdot 9 = 0 + 45 = 45$	
(2, 5)	$2 \cdot 2 + 5 \cdot 5 = 4 + 25 = 29$	
(5, 1)	$2 \cdot 5 + 5 \cdot 1 = 10 + 5 = 15$	⟵ Minimum
(8, 0)	$2 \cdot 8 + 5 \cdot 0 = 16 + 0 = 16$	

Min = 15 at $a = 5$, $b = 1$

15. We graphed the solution set of the system of constraints in Exercise 27, Set 3.1 and found the corner points to be (0, 9), (1, 5), (3, 2), and (8, 0). Now we find the function values at the corner points.

Corner points	Function $W = 3x + 5y$	
(0, 9)	$3 \cdot 0 + 5 \cdot 9 = 0 + 45 = 45$	
(1, 5)	$3 \cdot 1 + 5 \cdot 5 = 3 + 25 = 28$	
(3, 2)	$3 \cdot 3 + 5 \cdot 2 = 9 + 10 = 19$	⟵ Minimum
(8, 0)	$3 \cdot 8 + 5 \cdot 0 = 24 + 0 = 24$	

Min = 19 at $x = 3$, $y = 2$

17. Maximize and minimize

$$f = x - y$$

Subject to

$$x + 2y \leqslant 6 \qquad (1)$$
$$x \geqslant 0 \qquad (2)$$
$$x \leqslant 5 \qquad (3)$$
$$y \geqslant -2 \qquad (4)$$

We graphed the solution set of the system of constraints in Exercise 29, Set 3.1. The graph is shown below.

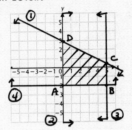

Now we find the corner points.

Corner point A is the solution of the system

$$x = 0$$
$$y = -2.$$

Point A is (0, -2).

Corner point B is the solution of the system

$$x = 5$$
$$y = -2.$$

Point B is (5, -2).

Corner point C is the solution of the system

$$x + 2y = 6$$
$$x = 5.$$

Point C is $\left(5, \frac{1}{2}\right)$.

Corner point D is the solution of the system

$$x + 2y = 6$$
$$x = 0.$$

Point D is (0, 3).

Now we compute the function values at the corner points.

Corner points	Function $f = x - y$	
(0, -2)	$0 - (-2) = 0 + 2 = 2$	
(5, -2)	$5 - (-2) = 5 + 2 = 7$	⟵ Maximum
$\left(5, \frac{1}{2}\right)$	$5 - \frac{1}{2} = 4\frac{1}{2}$	
(0, 3)	$0 - 3 = -3$	⟵ Minimum

Max = 7 at $x = 5$, $y = -2$;
Min = -3 at $x = 0$, $y = 3$

19. Maximize and minimize

    f = x - y

Subject to

    x ⩾ -3

    x - 2y ⩽ 4

    y - 3x ⩽ 9

    3x + y ⩽ 10

We graphed the solution set of the system of constraints in Exercise 31, Set 3.1. The graph is shown below.

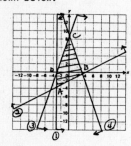

Now we find the corner points.

Corner point A is the solution of the system

    x = -3

    x - 2y = 4.

Point A is $\left[-3, -\frac{7}{2}\right]$.

Corner point B is the solution of the system

    x - 2y = 4

    3x + y = 10.

Point B is $\left[\frac{24}{7}, -\frac{2}{7}\right]$.

Corner point C is the solution of the system

    y - 3x = 9

    3x + y = 10.

Point C is $\left[\frac{1}{6}, \frac{19}{2}\right]$.

Corner point D is the solution of the system

    x = -3

    y - 3x = 9.

Point D is (-3, 0).

Now we compute the function values at the corner points.

Corner points	Function f = x - y	
$\left[-3, -\frac{7}{2}\right]$	$-3 - \left[-\frac{7}{2}\right] = -3 + \frac{7}{2} = \frac{1}{2}$	
$\left[\frac{24}{7}, -\frac{2}{7}\right]$	$\frac{24}{7} - \left[-\frac{2}{7}\right] = \frac{24}{7} + \frac{2}{7} = \frac{26}{7}$	← Maximum
$\left[\frac{1}{6}, \frac{19}{2}\right]$	$\frac{1}{6} - \frac{19}{2} = -\frac{56}{6} = -\frac{28}{3}$	← Minimum
(-3, 0)	$-3 - 0 = -3$	

Max = $\frac{26}{7}$ at x = $\frac{24}{7}$, y = $-\frac{2}{7}$;

Min = $-\frac{28}{3}$ at x = $\frac{1}{6}$, y = $\frac{19}{2}$

21. Maximize and minimize

    Q = 3x - y

Subject to

    3x ⩾ y   (1)

    3y ⩾ x   (2)

    x + y ⩾ 5   (3)

    2x + 3y ⩽ 24 (4)

The system of constraints is the system in Exercise 32, Set 3.1. We graph it and find the corner points.

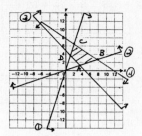

Corner point A is the solution of the system

    3y = x

    x + y = 5.

Point A = $\left[\frac{15}{4}, \frac{5}{4}\right]$.

Corner point B is the solution of the system

    3y = x

    2x + 3y = 24.

Point B is $\left[8, \frac{8}{3}\right]$.

Corner point C is the solution of the system

    3x = y

    2x + 3y = 24.

Point C is $\left[\frac{24}{11}, \frac{72}{11}\right]$.

Corner point D is the solution of the system

    3x = y

    x + y = 5.

Point D is $\left[\frac{5}{4}, \frac{15}{4}\right]$.

Now we compute the function values at the corner points.

Corner points	Function Q = 3x - y	
$\left[\frac{15}{4}, \frac{5}{4}\right]$	$3 \cdot \frac{15}{4} - \frac{5}{4} = \frac{45}{4} - \frac{5}{4} = \frac{40}{4} = 10$	
$\left[8, \frac{8}{3}\right]$	$3 \cdot 8 - \frac{8}{3} = 24 - \frac{8}{3} = \frac{64}{3}$	← Maximum
$\left[\frac{24}{11}, \frac{72}{11}\right]$	$3 \cdot \frac{24}{11} - \frac{72}{11} = \frac{72}{11} - \frac{72}{11} = 0$	← Minimum
$\left[\frac{5}{4}, \frac{15}{4}\right]$	$3 \cdot \frac{5}{4} - \frac{15}{4} = \frac{15}{4} - \frac{15}{4} = 0$	← Minimum

21. (continued)

Since the minimum value, 0, occurs at two corner points, $\left(\frac{24}{11}, \frac{72}{11}\right)$ and $\left(\frac{5}{4}, \frac{15}{4}\right)$, it will also occur at any point on the line segment between these points. Therefore, we say Max = $\frac{64}{3}$ at x = 8, y = $\frac{8}{3}$; Min = 0 at any point on the line between $\left(\frac{24}{11}, \frac{72}{11}\right)$ and $\left(\frac{5}{4}, \frac{15}{4}\right)$.

23. We graphed the solution set of the system of constraints in Exercise 33, Set 3.1 and found the solution set to be empty. Since there is no feasible region there is no maximum or minimum value of the function.

25. Maximize and minimize

F = 2x + 3y + 24

Subject to

x ⩾ 0	(1)
y ⩾ 0	(2)
5y − 3x ⩽ 15	(3)
x ⩽ 4y	(4)
2x − 5y ⩽ 10	(5)

We graphed the solution set of the system of constraints in Exercise 39, Set 3.1. The graph is shown below.

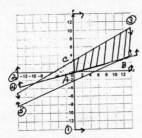

Next we find the corner points.

Corner point A is the solution of the system

x = 0

x = 4y.

Point A is (0, 0).

Corner point B is the solution of the system

x = 4y

2x − 5y = 10.

Point B is $\left(\frac{40}{3}, \frac{10}{3}\right)$.

Corner point C is the solution of the system

x = 0

5y − 3x = 15.

Point C is (0, 3).

Now we compute the function values at the corner points.

25. (continued)

Corner points	Function F = 2x + 3y + 24
(0, 0)	2·0 + 3·0 + 24 = 0 + 0 + 24 = 24 ← Minimum
$\left(\frac{40}{3}, \frac{10}{3}\right)$	2 · $\frac{40}{3}$ + 3 · $\frac{10}{3}$ + 24 = $\frac{80}{3}$ + 10 + 24 = $\frac{182}{3}$
(0, 3)	2·0 + 3·3 + 24 = 0 + 9 + 24 = 33

The feasible region is unbounded, so F has no maximum value; Min = 24 at x = 0, y = 0.

27. We graphed the solution set of the system of constraints in Exercise 37, Set 3.1 and found that it contained a single point (3, −3). Therefore, both the maximum and minimum values of the function must occur at this point. Find the function value at (3, −3):

P = 4x + 3y

P = 4·3 + 3(−3) = 12 − 9 = 3

Max = Min = 3 at x = 3, y = −3

29. Maximize and minimize

F = 2x + y

Subject to

x + 2y ⩽ 30	(1)
2x + 3y ⩾ 35	(2)
x ⩾ 4	(3)
x ⩽ 10	(4)

Graph the solution set of the system of constraints.

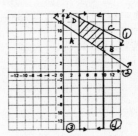

Find the corner points.

Corner point A is the solution of the system

2x + 3y = 35

x = 4.

Point A is (4, 9).

Corner point B is the solution of the system

2x + 3y = 35

x = 10.

Point B is (10, 5).

<u>29.</u> (continued)

Corner point C is the solution of the system

$$x + 2y = 30$$
$$x = 10.$$

Point C is (10, 10).

Corner point D is the solution of the system

$$x + 2y = 30$$
$$x = 4.$$

Point D is (4, 13).

Now compute the function values at the corner points.

Corner points	Function $F = 2x + y$	
(4, 9)	$2 \cdot 4 + 9 = 8 + 9 = 17$	← Minimum
(10, 5)	$2 \cdot 10 + 5 = 20 + 5 = 25$	
(10, 10)	$2 \cdot 10 + 10 = 20 + 10 = 30$	← Maximum
(4, 13)	$2 \cdot 4 + 13 = 8 + 13 = 21$	

Max = 30 at x = 10, y = 10;

Min = 17 at x = 4, y = 9.

<u>31.</u> Maximize and minimize Q = x - y subject to the constraints in Exercise 29 above. In Exercise 29 we graphed the solution set of the system of constraints and found the corner points. Now we compute the function values at the corner points.

Corner points	Function $Q = x - y$	
(4, 9)	$4 - 9 = -5$	
(10, 5)	$10 - 5 = 5$	← Maximum
(10, 10)	$10 - 10 = 0$	
(4, 13)	$4 - 13 = -9$	← Minimum

Max = 5 at x = 10, y = 5;

Min = -9 at x = 4, y = 13

<u>33.</u>
$$2X + Y \geqslant 10 \quad (1)$$
$$-3X + 2Y \leqslant 6 \quad (2)$$
$$X + Y \geqslant 6 \quad (3)$$
$$X \geqslant 0 \quad (4)$$
$$Y \geqslant 0 \quad (5)$$

Graph the solution set of the system of constraints and find the corner points.

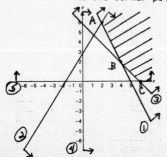

<u>33.</u> (continued)

Corner point A is the solution of the system

$$2X + Y = 10$$
$$-3X + 2Y = 6.$$

Point A is (2, 6).

Corner point B is the solution of the system

$$2X + Y = 10$$
$$X + Y = 6.$$

Point B is (4, 2).

Corner point C is the solution of the system

$$X + Y = 6$$
$$Y = 0.$$

Point C is (6, 0).

The correct answer is choice D.

Exercise Set 3.3

<u>1.</u> a) Let x = number of suits to be made, y = number of dresses to be made and I = income ($) from the sale of the suits and dresses. We head the columns under "Composition" "Suits" and "Dresses" and put x under suits and y under dresses in the "Number of Units" row of the table. Each suit requires 1 yd of polyester, each dress requires 2 yd of polyester, and the clothier's supply of polyester is 60 yd. We enter these figures in the "Polyester" row. Each suit requires 4 yd of wool, each dress requires 3 yd of wool, and the clothier's supply of wool is 120 yd. We enter these figures in the "Wool" row. A suit sells for $120 and a dress for $75. Enter these figures in the "Price per Unit ($)" row. The completed table is shown below.

	Composition		Number of Units of Supply
	Suits	Dresses	
Number of Units	x	y	
Polyester	1	2	60
Wool	4	3	120
Price per Unit ($)	$120	$75	Objective Function Maximize I

b) Noting that the amount of fabric (supply) available cannot be exceeded, we obtain the following constraints:

$$x + 2y \leqslant 60$$
$$4x + 3y \leqslant 120$$

Since the clothier cannot produce a negative number of suits or dresses, we have non-negativity constraints:

$$x, y \geqslant 0$$

We can read the objective function from the bottom row:

$$\text{Maximize } I = 120x + 75y$$

1. (continued)

We can express this formulation as follows:

Maximize I = 120x + 75y

Subject to x + 2y ⩽ 60   (1)

4x + 3y ⩽ 120   (2)

x, y ⩾ 0   (3), (4)

c) Carry out: We solve the problem by graphing the feasible region, determining the corner points, and finding the function values at the corner points.

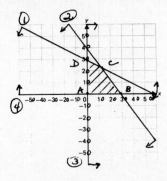

Point A is the solution of the system of equations related to (3) and (4):

x = 0

y = 0

Point A is (0, 0). Point B is the solution of the system of equations related to (2) and (4):

4x + 3y = 120

y = 0

Point B is (30, 0). Point C is the solution of the system of equations related to (1) and (2):

x + 2y = 60

4x + 3y = 120

Point C is (12, 24). Point D is the solution of the system of equations related to (1) and (3):

x + 2y = 60

x = 0

Point D is (0, 30).

Corner Points	I = 120x + 75y
(0, 0)	0
(30, 0)	3600 ← Maximum
(12, 24)	3240
(0, 30)	2250

Check: The best way to check is to go over your work to be sure your algebra and arithmetic are correct.

State: The maximum income will be $3600 when the clothier produces 30 suits and 0 dresses.

3. Familiarize: Let

x = lbs of Mixture I to be made,

y = lbs of Mixture II to be made,

and I = income from the sale of the mixtures.

Set up a table using the information given.

	Composition		Number of lbs Available
	Mixture I	Mixture II	
Number of Units	x	y	
Peanuts	0.6	0.2	1800
Cashews	0.3	0.5	1500
Almonds	0.1	0.3	750
Price per Unit ($)	$0.75	$2.00	Objective Function Maximize I

Note that the percents are converted to decimals, and the price is expressed in dollars.

Translate: Noting that the supply available cannot be exceeded and that a negative amount of a mixture cannot be produced we obtain the following constraints:

0.6x + 0.2y ⩽ 1800

0.3x + 0.5y ⩽ 1500

0.1x + 0.3y ⩽ 750

x, y ⩾ 0

We can read the objective function from the bottom row of the table:

I = 0.75x + 2y

Multiplying by 10 to clear the inequalities of decimals we obtain

Maximize I = 0.75x + 2y

Subject to 6x + 2y ⩽ 18,000   (1)

3x + 5y ⩽ 15,000   (2)

x + 3y ⩽ 7500   (3)

x, y ⩾ 0   (4), (5)

Carry out: We graph the feasible region, determine the corner points, and find the function values at the corner points.

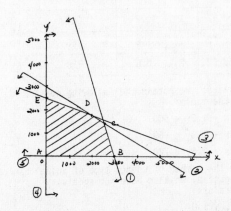

3. (continued)

We find the corner points by solving the systems of equations related to the constraints indicated in the table below.

Corner point	System related to	Coordinates
A	(4), (5)	(0, 0)
B	(1), (5)	(3000, 0)
C	(1), (2)	(2500, 1500)
D	(2), (3)	(1875, 1875)
E	(3), (4)	(0, 2500)

Next we find the function values at the corner points.

Corner point	$I = 0.75x + 2y$
(0, 0)	0
(3000, 0)	2250
(2500, 1500)	4875
(1875, 1875)	5156.25 ← Maximum
(0, 2500)	5000

Check: The best way to check is to go over your work to be sure your algebra and arithmetic are correct.

State: The maximum income will be $5156.25 when 1875 lb of Mixture I and 1875 lb of Mixture II are made.

5. Familiarize: Let

  x = number of animal A1,

  y = number of animal A2,

and I = number of animals that the forest can support.

Set up a table using the information given.

	Composition		Number of Units of Supply
	A1	A2	
Number of Units	x	y	
Food F1	1	1.2	600
Food F2	2	1.8	960
Food F3	2	0.6	720

Translate: We obtain the following constraints from the table:

  $x + 1.2y \leqslant 600$

  $2x + 1.8y \leqslant 960$

  $2x + 0.6y \leqslant 720$

Since the number of either type of animal cannot be negative, we also have nonnegativity constraints:

  $x, y \geqslant 0$

5. (continued)

The objective function gives the total number of animals, or $I = x + y$.

Multiplying by 10 to clear the inequalities of decimals we have

Maximize $I = x + y$

Subject to $10x + 12y \leqslant 6000$ (1)

  $20x + 18y \leqslant 9600$ (2)

  $20x + 6y \leqslant 7200$ (3)

  $x, y \geqslant 0$ (4), (5)

Carry out: We graph the feasible region, determine the corner points, and find the function values at the corner points.

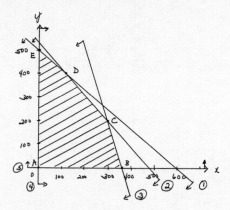

We find the corner points by solving the systems of equations related to the constraints indicated in the table below.

Corner point	System related to	Coordinates
A	(4), (5)	(0, 0)
B	(3), (5)	(360, 0)
C	(2), (3)	(300, 200)
D	(1), (2)	(120, 400)
E	(1), (4)	(0, 500)

Next we find the function values at the corner points.

Corner point	$I = x + y$
(0, 0)	0
(360, 0)	360
(300, 200)	500
(120, 400)	520 ← Maximum
(0, 500)	500

Check: Go over your work to be sure your algebra and arithmetic are correct.

State: The maximum number of animals that the forest can support is 520 for 120 of A1 and 400 of A2.

7. __Familiarize:__ We define the variables as in Exercise 5 above and change the table to show that the number of units of supply of foods F1 and F3 are 720 and 600, respectively.

	Composition		Number of Units of Supply
	A1	A2	
Number of Units	x	y	
Food F1	1	1.2	720
Food F2	2	1.8	960
Food F3	2	0.6	600

__Translate:__ Putting all the information together we obtain a linear program.

Maximize $I = x + y$

Subject to $x + 1.2y \leqslant 720$

$2x + 1.8y \leqslant 960$

$2x + 0.6y \leqslant 600$

$x, y \geqslant 0.$

Multiplying the inequalities by 10 to clear of decimals we have

Maximize $I = x + y$

Subject to $10x + 12y \leqslant 7200$ (1)

$20x + 18y \leqslant 9600$ (2)

$20x + 6y \leqslant 6000$ (3)

$x, y \geqslant 0.$ (4), (5)

__Carry out:__ We graph the feasible region, determine the corner points, and find the function values at the corner points.

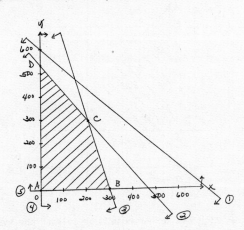

We find the corner points by solving the systems of equations related to the constraints indicated in the table below.

7. (continued)

Corner point	System related to	Coordinates
A	(4), (5)	(0, 0)
B	(3), (5)	(300, 0)
C	(2), (3)	(210, 300)
D	(2), (4)	$\left(0, 533\frac{1}{3}\right)$

Next we find the function values at the corner points.

Corner point	$I = x + y$
(0, 0)	0
(300, 0)	300
(210, 300)	510
$\left(0, 533\frac{1}{3}\right)$	$533\frac{1}{3}$ ⟵ Maximum

__Check:__ Go over your work to be sure your algebra and arithmetic are correct.

__State:__ The maximum number of animals that the forest can support is $533\frac{1}{3}$ for 0 of A1 and $533\frac{1}{3}$ of A2. This means species A1 would not survive. (Note that a fractional answer is given even though it is not totally practical.)

9. __Familiarize:__ Let

L = amount in loans, in millions,

and S = amount in securities, in millions.

Recall the simple interest formula $I = Prt$, where I = simple interest, P = principal, r = interest rate, and t = time. Then in one year the L dollars invested in loans earn L(0.14)(1), or 0.14L, interest, and the S dollars invested in savings earn S(0.1)(1), or 0.1S. Letting

I = total interest, or income

we have

$I = 0.14L + 0.1S.$

The total amount invested is $L + S$. The institution does not have to invest all its funds, so

$L + S \leqslant 600.$

At least $\frac{1}{3}$ of the total amount must be in securities, so

$S \geqslant \frac{1}{3}(L + S)$

or $3S \geqslant L + S$

$2S \geqslant L.$

No more than $480 million can be invested in securities, so

$S \leqslant 480.$

Nonnegativity constraints $L, S \geqslant 0$ are also implied.

9. (continued)

Translate: Putting all the information together we have

Maximize I = 0.14L + 0.1S

Subject to	L + S ≤ 600	(1)
	L - 2S ≤ 0	(2)
	S ≤ 480	(3)
	L, S ≥ 0.	(4), (5)

Carry out: We graph the feasible region, determine the corner points, and find the function values at the corner points.

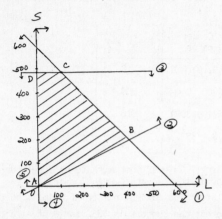

We find the corner points by solving the systems of equations related to the constraints indicated in the table below.

Corner point	System related to	Coordinates
A	(4), (5) or (2), (4) or (2), (5)	(0, 0)
B	(1), (2)	(400, 200)
C	(1), (3)	(120, 480)
D	(3), (4)	(0, 480)

Note that A is degenerate.

Next we find the function values at the corner points.

Corner point	I = 0.14L + 0.1S	
(0, 0)	0	
(400, 200)	76	← Maximum
(120, 480)	64.8	
(0, 480)	48	

Check: Go over your work to be sure your algebra and arithmetic are correct.

State: The revenue will be maximized by investing $400 million in loans and $200 million in securities. The maximum revenue is $76 million.

11. Familiarize: Let

x = the number of Type A questions to be done,

y = the number of Type B questions to be done,

and S = the test score.

Type A questions are worth 10 points and Type B worth 25 points, so

S = 10x + 25y.

You can do no more than 12 questions of Type A and 15 questions of Type B, so

x ≤ 12

and y ≤ 15.

You must do at least 3 questions of Type A and 4 questions of Type B. This provides "nonnegativity" constraints

x ≥ 3

and y ≥ 4.

You can do no more than a total of 20 questions, so

x + y ≤ 20.

Translate: Putting all the information together we get

Maximize S = 10x + 25y

Subject to	x ≤ 12	(1)
	y ≤ 15	(2)
	x + y ≤ 20	(3)
	x ≥ 3	(4)
	y ≥ 4.	(5)

Carry out: We graph the feasible region, determine the corner points, and find the function values at the corner points.

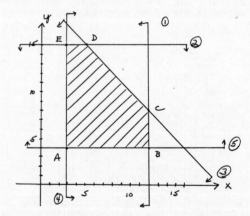

We find the corner points by solving the systems of equations related to the constraints indicated in the table below.

11. (continued)

Corner point	System related to	Coordinates
A	(4), (5)	(3, 4)
B	(1), (5)	(12, 4)
C	(1), (3)	(12, 8)
D	(2), (3)	(5, 15)
E	(2), (4)	(3, 15)

Next we find the function values at the corner points.

Corner point	$S = 10x + 25y$	
(3, 4)	130	
(12, 4)	220	
(12, 8)	320	
(5, 15)	425	← Maximum
(3, 15)	405	

Check: Go over your work to be sure your algebra and arithmetic are correct.

State: You must do 5 questions of Type A and 15 questions of Type B in order to maximize your score. The maximum score is 425.

13. Familiarize: Let

x = the amount invested in corporate bonds,

y = the amount invested in municipal bonds,

and I = total interest, or income.

Using the formula for simple interest $I = Prt$ (see Exercise 9 above), we have

$I = 0.08x + 0.075y$.

The total amount invested is $x + y$. The woman plans to invest up to $40,000 so

$x + y \leqslant 40,000$

She does not want to invest more than $22,000 in corporate bonds or more than $30,000 in municipal bonds, so we have

$x \leqslant 22,000$

and $y \leqslant 30,000$.

The least she is allowed to invest in corporate bonds is $6000, and she cannot invest a negative amount in municipal bonds. This gives us the nonnegativity constraints:

$x \geqslant 6000$

$y \geqslant 0$

Translate: Putting all the information together we have a linear program.

Maximize $I = 0.08x + 0.075y$

Subject to $x + y \leqslant 40,000$ (1)

$x \leqslant 22,000$ (2)

$y \leqslant 30,000$ (3)

$x \geqslant 6000$ (4)

$y \geqslant 0$ (5)

13. (continued)

Carry out: We graph the feasible region, determine the corner points, and find the function values at the corner points.

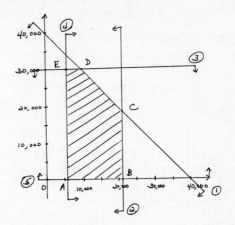

We find the corner points by solving the systems of equations related to the constraints indicated in the table below.

Corner point	System related to	Coordinates
A	(4), (5)	(6000, 0)
B	(2), (5)	(22,000, 0)
C	(1), (2)	(22,000, 18,000)
D	(1), (3)	(10,000, 30,000)
E	(3), (4)	(6000, 30,000)

Next we find the function values at the corner points.

Corner point	$I = 0.08x + 0.075y$	
(6000, 0)	480	
(22,000, 0)	1760	
(22,000, 18,000)	3110	← Maximum
(10,000, 30,000)	3050	
(6000, 30,000)	2730	

Check: Go over your work to be sure your algebra and arithmetic are correct.

State: She should invest $22,000 in corporate bonds and $18,000 in municipal bonds to maximize her income. The maximum income is $3110.

15. Familiarize: Let

    x = the number of batches of SMELLO to be made,

    y = the number of batches of ROPPO to be made,

and I = profit ($) from the sale of the tobacco.

Set up a table using the information given.

	Composition		Number of lbs Available
	SMELLO	ROPPO	
Number of Batches	x	y	
English tobacco	12	8	3000
Virginia tobacco	0	8	2000
Latakia tobacco	4	0	500
Profit per Batch	$10.56	$6.40	Objective Function Maximize I

Translate: We obtain the following constraints from the table:

    $12x + 8y \leqslant 3000$

    $8y \leqslant 2000$

    $4x \leqslant 500$

Since the number of batches made cannot be negative, we also have nonnegativity constraints:

    $x, y \geqslant 0$

We can read the objective function from the bottom row:

    $I = 10.56x + 6.40y$

The translation can be expressed as a linear program.

Maximize $I = 10.56x + 6.40y$

Subject to $12x + 8y \leqslant 3000$ (1)

             $8y \leqslant 2000$ (2)

             $4x \leqslant 500$ (3)

             $x, y \geqslant 0$ (4), (5)

Carry out: We graph the feasible region, determine the corner points, and find the function values at the corner points.

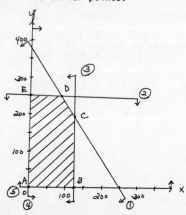

15. (continued)

We find the corner points by solving the systems of equations related to the constraints indicated in the table below.

Corner point	System related to	Coordinates
A	(4), (5)	(0, 0)
B	(3), (5)	(125, 0)
C	(1), (3)	(125, 187.5)
D	(1), (2)	$\left[83\frac{1}{3}, 250\right]$
E	(2), (4)	(0, 250)

Next we find the function values at the corner points.

Corner point	$I = 10.56x + 6.40y$
(0, 0)	0
(125, 0)	1320
(125, 187.5)	2520 ←——— Maximum
$\left[83\frac{1}{3}, 250\right]$	2480
(0, 250)	1600

Check: Go over your work to be sure your algebra and arithmetic are correct.

State: The maximum profit will be yielded when 125 batches of SMELLO and 187.5 batches of ROPPO are made. The maximum profit is $2520.

17. Familiarize: Let

    x = acreage planted in corn,

    y = acreage planted in oats,

and I = profit ($) from the corn and oats.

Set up a table using the information given.

	Composition		Number of Units Available
	Corn	Oats	
Number of Acres	x	y	240
Labor per Acre	2	1	320
Profit per Acre	$40	$30	Objective Function Maximize I

Using the information in the table and adding the nonnegativity constraints we have a linear program.

Maximize $I = 40x + 30y$

Subject to $x + y \leqslant 240$ (1)

           $2x + y \leqslant 320$ (2)

             $x, y \geqslant 0$ (3), (4)

17. (continued)

Carry out: We graph the feasible region, determine the corner points, and find the function values at the corner points.

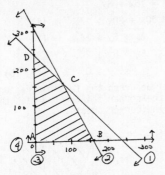

We find the corner points by solving the systems of equations related to the constraints indicated in the table below.

Corner point	System related to	Coordinates
A	(3), (4)	(0, 0)
B	(2), (4)	(160, 0)
C	(1), (2)	(80, 160)
D	(1), (3)	(0, 240)

Next we find the function values at the corner points.

Corner point	I = 40x + 30y
(0, 0)	0
(160, 0)	6400
(80, 160)	8000 ← Maximum
(0, 240)	7200

Check: Go over your work to be sure your algebra and arithmetic are correct.

State: Planting 80 acres of corn and 160 acres of oats will yield the maximum profit of $8000.

Exercise Set 3.4

1. Familiarize: Let

    x = number of sacks of soybean meal,

    y = number of sacks of oats,

and f = cost of the animal feed, in dollars.

Organize the information given in a table.

1. (continued)

	Composition (per 100-lb sack)		Pounds Required
	Soybean Meal	Oats	
Number of 100-lb Sacks	x	y	
Protein	50	15	120
Fat	8	5	24
Mineral Ash	5	1	10
Cost per 100-lb Sack	$15	$5	Objective Function Minimize f

Translate: Use the table for translation.

The amount of protein must be at least 120 pounds, so

    $50x + 15y \geqslant 120$.

Similarly, for fat and mineral ash,

    $8x + 5y \geqslant 24$

    and $5x + y \geqslant 10$.

We have two nonnegativity constraints

    $x, y \geqslant 0$.

We can read the objective function from the bottom row:

    $f = 15x + 5y$

The translation can be expressed as follows:

Minimize $f = 15x + 5y$

Subject to $50x + 15y \geqslant 120$ (1)

        $8x + 5y \geqslant 24$ (2)

        $5x + y \geqslant 10$ (3)

        $x, y \geqslant 0$ (4), (5)

Carry out: We graph the feasible region, determine the corner points, and find the function values at the corner points.

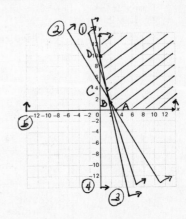

1. (continued)

We find the corner points as follows:

(2) and (5) yield the coordinates of A which are
(3, 0),

(1) and (2) yield the coordinates of B which are
$\left[1\frac{11}{13}, 1\frac{11}{13}\right]$,

(1) and (3) yield the coordinates of C which are
$\left[1\frac{1}{5}, 4\right]$,

(3) and (4) yield the coordinates of C which are
(0, 10).

Compute the function values at the corner points.

Corner point	f = 15x + 5y	
(3, 0)	45	
$\left[1\frac{11}{13}, 1\frac{11}{13}\right]$	$36\frac{12}{13}$	← Minimum
$\left[1\frac{1}{5}, 4\right]$	38	
(0, 10)	50	

Check: Go over your work.

State: $1\frac{11}{13}$ sacks of soybean meal and $1\frac{11}{13}$ sacks of oats should be used. (The minimum cost is $\$36\frac{12}{13}$.)

3. Familiarize: Let

x = number of sacks of soybean meal,

y = number of sacks of alfalfa,

and f = cost of the animal feed, in dollars.

Organize the information given in a table. The table is the same as in Exercise 1 except that the "Oats" column will now be headed "Alfalfa" and will show the nutrient content of the alfalfa, and the number of pounds of mineral ash required will be doubled (that is, increased to 20 pounds).

	Composition (per 100-lb sack)		Pounds Required
	Soybean Meal	Alfalfa	
Number of 100-lb Sacks	x	y	
Protein	50	20	120
Fat	8	6	24
Mineral Ash	5	8	20
Cost per 100-lb Sack	$15	$8	Objective Function Minimize f

3. (continued)

Translate: Using the information from the table and adding the nonnegativity constraints as in Exercise 1 we have a linear program.

Minimize f = 15x + 8y

Subject to 50x + 20y ≥ 120   (1)

8x + 6y ≥ 24   (2)

5x + 8y ≥ 20   (3)

x, y ≥ 0   (4), (5)

Carry out: We graph the feasible region, determine the corner points, and find the function values at the corner points.

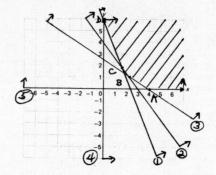

We find the corner points as follows:

(3) and (5) yield the coordinates of A which are
(4, 0),

(2) and (3) yield the coordinates of B which are
$\left[2\frac{2}{17}, 1\frac{3}{17}\right]$,

(1) and (2) yield the coordinates of C which are
$\left[1\frac{5}{7}, 1\frac{5}{7}\right]$,

(1) and (4) yield the coordinates of D which are
(0, 6).

Compute the function values at the corner points.

Corner point	f = 15x + 8y	
(4, 0)	60	
$\left[2\frac{2}{17}, 1\frac{3}{17}\right]$	$41\frac{3}{17}$	
$\left[1\frac{5}{7}, 1\frac{5}{7}\right]$	$39\frac{3}{7}$	← Minimum
(0, 6)	48	

Check: Go over your work.

State: $1\frac{5}{7}$ sacks of soybean meal and $1\frac{5}{7}$ sacks of oats should be used. (The minimum cost is $\$39\frac{3}{7}$.)

5. Familiarize: Let

x = number of type P1 to be used,

y = number of type P2 to be used,

and C = total operating cost, in thousands of dollars.

Organize the information given in a table.

	Composition		Number of Passengers
	Airplane P1	Airplane P2	
Number of Airplanes	x	y	
First-class	40	80	2000
Tourist	40	30	1500
Economy	120	40	2400
Cost per Airplane (in thousands of dollars)	$12	$10	Objective Function Minimize C

Translate: Use the table for translation.

The number of first-class passengers must be at least 2000, so

$40x + 80y \geqslant 2000$.

Similarly, for tourist and economy-class passengers,

$40x + 30y \geqslant 1500$

and $120x + 40y \geqslant 2400$.

We have two nonnegativity constraints:

$x, y \geqslant 0$

We can read the objective function from the bottom row:

$C = 12x + 10y$

The translation can be expressed as follows:

Minimize $C = 12x + 10y$

Subject to $\quad 40x + 80y \geqslant 2000 \qquad (1)$

$\qquad\qquad 40x + 30y \geqslant 1500 \qquad (2)$

$\qquad\qquad 120x + 40y \geqslant 2400 \qquad (3)$

$\qquad\qquad\qquad x, y \geqslant 0 \qquad\qquad (4), (5)$

Carry out: We graph the feasible region, determine the corner points, and find the function values at the corner points.

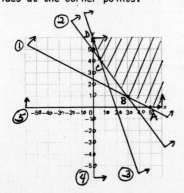

5. (continued)

We find the corner points as follows:

(1) and (5) yield the coordinates of A which are (50, 0),

(1) and (2) yield the coordinates of B which are (30, 10),

(2) and (3) yield the coordinates of C which are (6, 42),

(3) and (4) yield the coordinates of D which are (0, 60).

Compute the function values at the corner points.

Corner point	$C = 12x + 10y$
(50, 0)	600
(30, 10)	460 ← Minimum
(6, 42)	492
(0, 60)	600

Check: Go over your work.

State: The airline should use 30 airplanes of type P1 and 10 airplanes of P2. (The minimum cost is $460 thousand.)

7. Familiarize: Let

x = number of type P1 to be used,

y = number of type P3 to be used,

and C = total operating cost, in thousands of dollars.

Organize the information given in a table. Use the table in Exercise 5, replacing the information in the P2 column with the information pertaining to P3.

	Composition		Number of Passengers
	Airplane P1	Airplane P3	
Number of Airplanes	x	y	
First-class	40	40	2000
Tourist	40	80	1500
Economy	120	80	2400
Cost per Airplane (in thousands of dollars)	$12	$15	Objective Function Minimize C

Translate: Using the information from the table and adding the nonnegativity constraints as in Exercise 5 we have a linear program.

Minimize $C = 12x + 15y$

Subject to $\quad 40x + 40y \geqslant 2000 \qquad (1)$

$\qquad\qquad 40x + 80y \geqslant 1500 \qquad (2)$

$\qquad\qquad 120x + 80y \geqslant 2400 \qquad (3)$

$\qquad\qquad\qquad x, y \geqslant 0 \qquad\qquad (4), (5)$

7. (continued)

Carry out: We graph the feasible region, determine the corner points, and find the function values at the corner points.

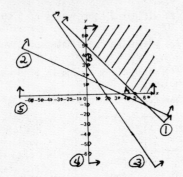

We find the corner points as follows:

(1) and (5) yield the coordinates of A which are (50, 0),

(1) and (4) yield the coordinates of B which are (0, 50).

Compute the function values at the corner points.

Corner point	C = 12x + 15y	
(50, 0)	600	← Minimum
(0, 50)	750	

Check: Go over your work.

State: The airline should use 50 airplanes of type P1 and 0 of type P3. (The minimum cost is $600 thousand.)

9. Familiarize: Define the variables as in Exercise 5. Use the table in Exercise 5, replacing the contract requirements (number of passengers) in the right-hand column with the new requirements given in this Exercise.

	Composition		Number of Passengers
	Airplane P1	Airplane P2	
Number of Airplanes	x	y	
First-class	40	80	1600
Tourist	40	30	2100
Economy	120	40	2400
Cost per Airplane (in thousands of dollars)	$12	$10	Objective Function Minimize C

Translate: Using the information from the table and adding the nonnegativity constraints as in Exercise 5 we have a linear program.

9. (continued)

Minimize C = 12x + 10y

Subject to   40x + 80y ⩾ 1600   (1)

40x + 30y ⩾ 2100   (2)

120x + 40y ⩾ 2400   (3)

x, y ⩾ 0   (4), (5)

Carry out: We graph the feasible region, determine the corner points, and find the function values at the corner points.

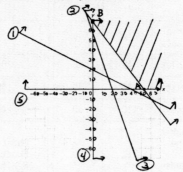

We find the corner points as follows:

(2) and (5) yield the coordinates of A which are (52.5, 0),

(2) and (4) yield the coordinates of B which are (0, 70)

Compute the function values at the corner points.

Corner point	C = 12x + 10y	
(52.5, 0)	630	← Minimum
(0, 70)	700	

Check: Go over your work.

State: The airline should use 52.5 airplanes of type P1 and 0 of type P2. (The minimum cost is $630 thousand.)

11. Familiarize: Define the variables as in Exercise 7. Use the table in Exercise 9, replacing the information in the P2 column with the information pertaining to P3.

	Composition		Number of Passengers
	Airplane P1	Airplane P3	
Number of Airplanes	x	y	
First-class	40	40	1600
Tourist	40	80	2100
Economy	120	80	2400
Cost per Airplane (in thousands of dollars)	$12	$15	Objective Function Minimize C

11. (continued)

Translate: Using the information from the table and adding the nonnegativity constraints as in Exercise 9 we have a linear program.

Minimize $C = 12x + 15y$

Subject to $40x + 40y \geqslant 1600$ (1)

$40x + 80y \geqslant 2100$ (2)

$120x + 80y \geqslant 2400$ (3)

$x, y \geqslant 0$ (4), (5)

Carry out: We graph the feasible region, determine the corner points, and find the function values at the corner points.

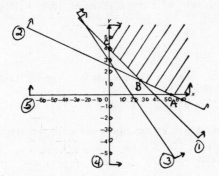

We find the corner points as follows:

(2) and (5) yield the coordinates of A which are (52.5, 0),

(1) and (2) yield the coordinates of B which are (27.5, 12.5),

(1) and (4) yield the coordinates of C which are (0, 40).

Compute the function values at the corner points.

Corner point	$C = 12x + 15y$	
(52.5, 0)	630	
(27.5, 12.5)	517.5	← Minimum
(0, 40)	600	

Check: Go over your work.

State: The airline should use 27.5 airplanes of type P1 and 12.5 of type P3. (The minimum cost is $517.5 thousand.)

13. Familiarize: Let

x = number of days Refinery JR is operated,

y = number of days Refinery Bobby is operated,

and C = production cost, in dollars.

Organize the information given in a table.

13. (continued)

	Composition		Number of Barrels Required
	Refinery JR	Refinery Bobby	
Number of Days	x	y	
Low-grade	6000	3000	30,000
High-grade	2500	5000	20,000
Cost per Day	$20,000	$15,000	Objective Function Minimize C

Translate: Using the information from the table and adding nonnegativity constraints we have a linear program.

Minimize $C = 20{,}000x + 15{,}000y$

Subject to $6000x + 3000y \geqslant 30{,}000$ (1)

$2500x + 5000y \geqslant 20{,}000$ (2)

$x, y \geqslant 0$ (3), (4)

Carry out: We graph the feasible region, determine the corner points and find the function values at the corner points.

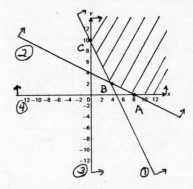

We find the corner points as follows:

(2) and (4) yield the coordinates of A which are (8, 0),

(1) and (2) yield the coordinates of B which are (4, 2),

(1) and (3) yield the coordinates of C which are (0, 10).

Compute the function values at the corner points.

Corner point	$C = 20{,}000x + 15{,}000y$	
(8, 0)	160,000	
(4, 2)	110,000	← Minimum
(0, 10)	150,000	

Check: Go over your work.

State: The company should operate Refinery JR for 4 days and Refinery Bobby for 2 days.

15. Familiarize: Let

    x = number of acres reclaimed for urban purposes,

    y = number of acres reclaimed for agricultural purposes,

and C = reclamation cost, in dollars.

Use the information given to write constraints. At least 4000 acres must be reclaimed for urban purposes, so

    $x \geq 4000$.

At least 5000 acres must be reclaimed for agricultural purposes, so

    $y \geq 5000$.

At least 10,000 acres must be reclaimed. We have

    $x + y \geq 10{,}000$.

The reclamation costs are $400 per acre for urban purposes and $300 per acre for agricultural purposes. This leads to the objective function

    $C = 400x + 300y$.

Translate: We can now write a linear program.

Minimize $C = 400x + 300y$

Subject to $x \geq 4000$ (1)

              $y \geq 5000$ (2)

         $x + y \geq 10{,}000$ (3)

Carry out: We graph the feasible region, determine the corner points, and find the function values at the corner points.

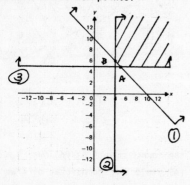

(Units on axes represent thousands.)

We find the corner points as follows:

(1) and (3) yield the coordinates of A which are (5000, 5000),

(1) and (2) yield the coordinates of B which are (4000, 6000).

Compute the function values at the corner points.

Corner point	C = 400x + 300y
(5000, 5000)	3,500,000
(4000, 6000)	3,400,000 ← Minimum

Check: Go over your work.

State: 4000 acres should be reclaimed for urban purposes and 6000 acres for agricultural purposes. (The minimum cost is $3,400,000.)

17. Familiarize: Let

    x = number of teachers hired,

    y = number of aides hired,

and C = salary costs, in dollars.

Use the information given to write constraints. There can be no more than 50 faculty members, so

    $x + y \leq 50$.

There must be at least 20 faculty members, so

    $x + y \geq 20$.

There must be at least 12 teachers. Thus,

    $x \geq 12$.

The number of aides must be at least half the number of teachers. We have

    $y \geq 6$

and $y \geq \frac{1}{2}x$, or $2y \geq x$.

We can use the average annual salaries to find the objective function

    $C = 30{,}000x + 20{,}000y$

Translate: We can now write a linear program.

Minimize $C = 30{,}000x + 20{,}000y$

Subject to $x + y \leq 50$ (1)

              $x + y \geq 20$ (2)

                  $x \geq 12$ (3)

                  $y \geq 6$ (4)

        $-x + 2y \geq 0$ (5)

Carry out: We graph the feasible region, determine the corner points, and find the function values at the corner points.

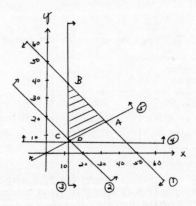

We find the corner points as follows:

(1) and (5) yield the coordinates of A which are $\left(33\frac{1}{3}, 16\frac{2}{3}\right)$,

(1) and (3) yield the coordinates of B which are (12, 38),

(2) and (3) yield the coordinates of C which are (12, 8),

(2) and (5) yield the coordinates of D which are $\left(13\frac{1}{3}, 6\frac{2}{3}\right)$.

17. (continued)

Compute the function values at the corner points.

Corner point	C = 30,000x + 20,000y
$\left(33\frac{1}{3},\ 16\frac{2}{3}\right)$	$1,333,333\frac{1}{3}$
(12, 38)	1,120,000
(12, 8)	520,000 ←——— Minimum
$\left(13\frac{1}{3},\ 6\frac{2}{3}\right)$	$533,333\frac{1}{3}$

Check: Go over your work.

State: 12 teachers and 8 aides should be hired in order to minimize salary costs. (The minimum cost is $520,000.)

19. In Exercise 1 of Set 3.3 we found that the production of 30 suits and 0 dresses would maximize income. The objective function in this Exercise was I = 120x + 75y. If we let b = the price of a dress, then we obtain a modified objective function

$$I_1 = 120x + by,$$

where we must determine b. Consider again the graph of the constraints in Exercise 1. We add the objective function I, which touches the feasible region at B.

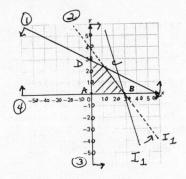

If we solve I_1 for y we obtain

$$Y = \frac{-120}{b}x + \frac{I_1}{b}.$$

For positive values of the price b, the slope, $-120/b$, is negative. As we increase b the slope increases and thus rotates counter-clockwise about the outside of the feasible region until it touches point C. Then point C, point B, and any point on the line between them are all maximum solutions. That is, the maximum value of I_1 is $3600 as it was for the function I in Exercise 1. The objective function I_1 contains the points B(30, 0) and C(12, 24) and has the same value at those points. Then

$$120(30) + b(0) = 120(12) + b(24)$$
$$3600 = 1440 + 24b$$
$$2160 = 24b$$
$$90 = b.$$

If the price of a dress is at least $90, dresses will be profitable.

21. In Exercise 3 of Set 3.3 the nut dealer had 1500 lb of cashews, giving us constraint (2), $0.3x + 0.5y \leqslant 1500$ (before clearing of decimals). If we let b = the number of lb of cashews available, then we obtain a modified constraint

$$0.3x + 0.5y \leqslant b \qquad (2A),$$

where we must determine b. Consider again the graph of the constraints in Exercise 3. We add the modified constraint which is parallel to the equation related to (2). For b = 1500 it touches the feasible region at D, the corner point that yielded the maximum value of the objective function.

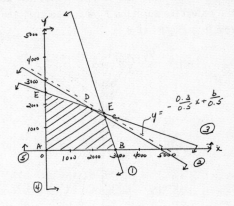

If we solve the equation related to (2A) for y we obtain

$$y = -\frac{0.3}{0.5}x + \frac{b}{0.5}.$$

For positive values of b the y-intercept, b/0.5, is positive. As b increases, the y-intercept increases and the line related to (2A) moves upward, remaining parallel to the equation related to constraint (3) until it touches point E. If constraint (2) were allowed to vary in this manner, then E would be a corner point of the feasible region. The coordinates of E are found by solving the system of equations related to (1) and (3) and are (2437.5, 1687.5). The equation $y = -\frac{0.3}{0.5}x + \frac{b}{0.5}$ must contain point E, so

$$1687.5 = -\frac{0.3}{0.5}(2437.5) + \frac{b}{0.5}$$
$$843.75 = -0.3(2437.5) + b$$
$$843.75 = -731.25 + b$$
$$1575 = b$$

If the dealer has 1575 lb of cashews he can still use the same mixtures profitably without buying more of the other nuts. Since he originally had 1500 lb of cashews, he can buy 1575 - 1500, or 75 lb more of cashews.

23. In Exercise 1 of this set we found that a mixture using $1\frac{11}{13}$ sacks of soybean meal and $1\frac{11}{13}$ sacks of oats will satisfy the minimum nutrition requirements at minimum cost. In this Exercise the objective function was $f = 15x + 5y$. If we let b = the cost of soybean meal per 100-lb sack, then we obtain a modified objective function

$$f_1 = bx + 5y,$$

where we must determine b. Consider again the graph of the constraints in Exercise 1. We add the objective function f_1 which touches the feasible region at B.

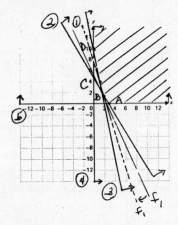

If we solve f_1 for y we obtain

$$y = -\frac{b}{5}x + \frac{f_1}{5}.$$

For positive values of the cost b, the slope, $-b/5$, is negative. As b increases the slope decreases and thus rotates clockwise about the outside of the region until it touches point C. Then point C, point B, and any point on the line segment between them are all minimum solutions. The objective function goes through points $B\left[1\frac{11}{13}, 1\frac{11}{13}\right]$ and $C\left[1\frac{1}{5}, 4\right]$, so

$$b\left[\frac{24}{13}\right] + 5\left[\frac{24}{13}\right] = b\left[\frac{6}{5}\right] + 5(4)$$

$$5b(24) + 25(24) = 13b(6) + 325(4)$$

Multiply by 65 to clear of fractions

$$120b + 600 = 78b + 1300$$

$$42b = 700$$

$$b = \frac{50}{3}, \text{ or } 16\frac{2}{3}.$$

If the price of soybean meal goes beyond $16\frac{2}{3}$, it becomes more economical to use a mixture with less soybean meal than oats.

25. In Exercise 5 of this set we found that operating costs are minimized when 30 P1's and 10 P2's are used. In this Exercise the objective function was $C = 12x + 10y$. If we let b = the cost of operating a plane of type P2, then we obtain a modified objective function

$$C_1 = 12x + by,$$

where we must determine b. Consider again the graph of the constraints in Exercise 5. We add the objective function C_1 which touches the feasible region at B.

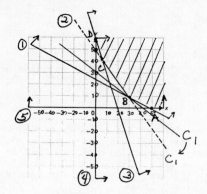

If we solve the equation $C_1 = 12x + by$ for y we obtain

$$y = \frac{-12}{b}x + \frac{C_1}{b}.$$

For positive values of b, the slope, $-12/b$, is negative. As b decreases, the slope decreases and the line rotates clockwise about the outside of the region until it touches point C. Then point C, point B, and any point on the line between them are all minimum solutions. The objective function C_1 contains points B(30, 10) and C(6, 42) and has the same value at those points, so

$$12(30) + b(10) = 12(6) + b(42)$$

$$360 + 10b = 72 + 42b$$

$$288 = 32b$$

$$9 = b.$$

When the operating cost of plane P2 is lowered to $9 thousand, it would be more economical to use more P2's.

Exercise Set 4.1

1. Maximize

$F = x + 2y$ ⟵ Objective function

Subject to

$x + y \leqslant 6$
$y \leqslant 5$ ⟵ Structural constraints

$x, y \geqslant 0$ ⟵ Nonnegativity constraints

a) Add a different nonnegative variable to each structural constraint to take up the slack in the inequality.

$$x + y + u = 6 \quad (u \geqslant 0)$$
$$y + v = 5 \quad (v \geqslant 0)$$

b) Add -x and -2y to both sides of the objective function:

$$-x - 2y + F = 0$$

c) We have the following system of equations:

$$x + y + u = 6$$
$$y + v = 5$$
$$-x - 2y + F = 0$$

The initial simplex tableau is

x	y	u	v	F	1
1	1	1	0	0	6
0	1	0	1	0	5
-1	-2	0	0	1	0

d) The nonbasic variables are those which we need to set equal to 0 in order to compute values for the rest of the variables. In this tableau they are x and y. There are 2 nonbasic variables. (Remember, also, that the number of nonbasic variables = the number of structural variables = 2 in this case.)

e) There are as many basic variables as there are slack variables. Thus, there are 2 basic variables. In this tableau they are u and v.

3. Maximize

$f = 5x + 4y$ ⟵ Objective function

Subject to

$3x + 2y \leqslant 12$
$x + y \leqslant 5$ ⟵ Structural constraints
$4x + 6y \leqslant 13$

$x, y \geqslant 0$ ⟵ Nonnegativity constraints

a) Add a different nonnegative slack variable to each structural constraint.

$$3x + 2y + u = 12 \quad (u \geqslant 0)$$
$$x + y + v = 5 \quad (v \geqslant 0)$$
$$4x + 6y + w = 13 \quad (w \geqslant 0)$$

b) Add -5x and -4y to both sides of the objective function:

$$-5x - 4y + f = 0$$

3. (continued)

c) We have the following system of equations:

$$3x + 2y + u = 12$$
$$x + y + v = 5$$
$$4x + 6y + w = 13$$
$$-5x - 4y + f = 0$$

The initial simplex tableau is

x	y	u	v	w	f	1
3	2	1	0	0	0	12
1	1	0	1	0	0	5
4	6	0	0	1	0	13
-5	-4	0	0	0	1	0

d) x and y are nonbasic variables. (We need to set them equal to 0 in order to compute values for the rest of the variables.) There are 2 nonbasic variables.

e) u, v, and w are basic variables. (There is exactly one nonzero number in each of those columns.) There are 3 basic variables.

5. Maximize

$f = -2y + 5z$ ⟵ Objective function

Subject to

$3x - 2y + z \leqslant 8$
$-4x + 3y + 2z \leqslant 4$ ⟵ Structural constraints
$3x + y - 6z \leqslant 6$

$x, y, z \geqslant 0$ ⟵ Nonnegative constraints

a) Add a different slack variable to each structural constraint.

$$3x - 2y + z + u = 8 \quad (u \geqslant 0)$$
$$-4x + 3y + 2z + v = 4 \quad (v \geqslant 0)$$
$$3x + y - 6z + w = 6 \quad (w \geqslant 0)$$

b) Add 2y and -5z to both sides of the objective function:

$$2y - 5z + f = 0$$

c) We have the following system of equations:

$$3x - 2y + z + u = 8$$
$$-4x + 3y + 2z + v = 4$$
$$3x + y - 6z + w = 6$$
$$2y - 5z + f = 0$$

The initial simplex tableau is

x	y	z	u	v	w	f	1
3	-2	1	1	0	0	0	8
-4	3	2	0	1	0	0	4
3	1	-6	0	0	1	0	6
0	2	-5	0	0	0	1	0

d) There are 3 nonbasic variables, x, y, and z.

e) There are 3 basic variables, u, v, and w.

Exercise Set 4.2

<u>1</u>. Maximize

$$F = x + 2y$$

Subject to

$$x + y \leqslant 6$$
$$y \leqslant 5$$
$$x, y \geqslant 0$$

Convert the structural constraints to equations by adding slack variables, and convert the objective function to a form with 0 on one side.

$$x + y + u \qquad = 6$$
$$y \qquad + v \qquad = 5$$
$$-x - 2y \qquad + F = 0$$

Set up the initial simplex tableau:

x	y	u	v	F	1
1	1	1	0	0	6
0	1	0	1	0	5
-1	-2	0	0	1	0

Carry out the pivoting. In general, our pivoting strategy will be first to multiply the row being changed by a positive integer (if necessary) and then to multiply the pivot row by a positive or negative integer (if necessary) and add the two rows. Follow the steps given in "The Simplex Algorithm for Maximum-Type Programs" in the text.

x	y	u	v	F	1	q
1	1	1	0	0	6	6/1 = 6
0	1*	0	1	0	5	5/1 = 5 ← Min
-1	⟨-2⟩	0	0	1	0	F = 0
1*	0	1	-1	0	1	1/1 = 1 ← Min
0	1	0	1	0	5	—
⟨-1⟩	0	0	2	1	10	F = 10/1 = 10
1	0	1	-1	0	1	
0	1	0	1	0	5	
0	0	1	1	1	11	F = 11/1 = 11

All of the indicators on the bottom row are nonnegative, so there are no more pivot columns and the algorithm terminates. The nonbasic variables are u and v. Setting u and v equal to 0 we find that x = 1/1, or 1, from the first row, y = 5/1, or 5, from the second row, and F = 11/1, or 11, from the third row. The solution is Max = 11 and x = 1, y = 5.

<u>3</u>. Maximize

$$f = 5x + 4y$$

Subject to

$$3x + 2y \leqslant 12$$
$$x + y \leqslant 5$$
$$x, y \geqslant 0$$

Convert the structural constraints to equations by adding slack variables, and convert the objective function to a form with 0 on one side.

$$3x + 2y + u \qquad = 12$$
$$x + y \qquad + v \qquad = 5$$
$$-5x - 4y \qquad + f = 0$$

3. (continued)

Set up the initial simplex tableau.

x	y	u	v	f	1
3	2	1	0	0	12
1	1	0	1	0	5
-5	-4	0	0	1	0

Carry out the pivoting. Follow the steps given in "The Simplex Algorithm for Maximum-Type Programs" in the text.

x	y	u	v	f	1	q
3*	2	1	0	0	12	12/3 = 4 ← Min
1	1	0	1	0	5	5/1 = 5
⊝5	-4	0	0	1	0	f = 0
3	2	1	0	0	12	12/2 = 6
0	1*	-1	3	0	3	3/1 = 3 ← Min
0	⊝2	5	0	3	60	f = 60/3 = 20
3	0	3	-6	0	6	
0	1	-1	3	0	3	
0	0	3	6	3	66	f = 66/3 = 22

All of the indicators on the bottom row are nonnegative, so there are no more pivot columns and the algorithm terminates. The nonbasic variables are u and v. Setting them equal to 0 we find that x = 6/3, or 2, from the first row, y = 3/1, or 3, from the second row, and f = 66/3, or 22, from the third row. The solution is Max = 22 at x = 2, y = 3.

5. Maximize

$$G = 3x + 4y$$

Subject to

$$x + 2y \leqslant 14$$
$$4x + 3y \leqslant 26$$
$$2x + y \leqslant 12$$
$$x, y \geqslant 0$$

Convert the structural constraints to equations by adding slack variables, and convert the objective function to a form with 0 on one side.

$$x + 2y + u \qquad\qquad = 14$$
$$4x + 3y \qquad + v \qquad = 26$$
$$2x + y \qquad\qquad + w \quad = 12$$
$$-3x - 4y \qquad\qquad\qquad + G = 0$$

Set up the initial simplex tableau, and carry out the pivoting.

<u>5</u>. (continued)

x	y	u	v	w	G		1	q
1	2*	1	0	0	0		14	14/2 = 7 ← Min
4	3	0	1	0	0		26	$26/3 = 8\frac{2}{3}$
2	1	0	0	1	0		12	12/1 = 12
-3	④	0	0	0	1		0	G = 0
1	2	1	0	0	0		14	14/1 = 14
5*	0	-3	2	0	0		10	10/5 = 2 ← Min
3	0	-1	0	2	0		10	$10/3 = 3\frac{1}{3}$
①	0	2	0	0	1		28	G = 28/1 = 28
0	10	8	-2	0	0		60	
5	0	-3	2	0	0		10	
0	0	4	-6	10	0		20	
0	0	7	2	0	5		150	G = 150/5 = 30

All of the indicators on the bottom row are nonnegative, so the algorithm
terminates. The nonbasic variables are u and v. Setting them equal to zero
we find:

   y = 60/10 = 6 from row 1,

   x = 10/5 = 2 from row 2,

   w = 20/10 = 2 from row 3,

   G = 150/5 = 30 from row 4

The solution is Max = 30 at x = 2, y = 6.

<u>7</u>. Maximize

   f = 120x + 75y

Subject to

   x + 2y ⩽ 60

   4x + 3y ⩽ 120

      x, y ⩾ 0

Convert the structural constraints to equations by adding slack variables,
and convert the objective function to a form with 0 on one side.

   x + 2y + u   = 60

    4x + 3y   + v   = 120

   -120x - 75y   + f =   0

Set up the initial simplex tableau, and carry out the pivoting.

x	y	u	v	f		1	q
1	2	1	0	0		60	60/1 = 60
4*	3	0	1	0		120	120/4 = 30 ← Min
⓪-120	-75	0	0	1		0	f = 0
0	5	4	-1	0		120	
4	3	0	1	0		120	
0	60	0	120	4		14,400	f = 14,400/4 = 3600

7. (continued)

All the indicators in the bottom row are nonnegative, so the algorithm terminates. We have the following result:

   y (nonbasic) = 0,

   v (nonbasic) = 0,

   u = 120/4 = 30 from row 1,

   x = 120/4 = 30 from row 2,

   f = 14,400/4 = 3600 from row 3

The solution is Max = 3600 at x = 30, y = 0. The check is left to the reader.

9. Maximize

   $f = 3x + 2y$

Subject to

   $3x + 6y \leqslant 90$

   $5x + 3y \leqslant 160$

   $x + y \leqslant 44$

   $x, y \geqslant 0$

Convert to a system of equations containing slack variables.

   $3x + 6y + u \qquad\qquad = 90$

   $5x + 3y \quad + v \qquad = 160$

   $x + y \qquad\quad + w \quad = 44$

   $-3x - 2y \qquad\qquad + f = 0$

Set up the initial simplex tableau, and carry out the pivoting.

x	y	u	v	w	f	1	q
3*	6	1	0	0	0	90	90/3 = 30 ← Min
5	3	0	1	0	0	160	160/5 = 32
1	1	0	0	1	0	44	44/1 = 44
⊖3	-2	0	0	0	1	0	f = 0
3	6	1	0	0	0	90	
0	-21	-5	3	0	0	30	
0	-3	-1	0	3	0	42	
0	12	3	0	0	3	270	f = 270/3 = 90

All the indicators in the bottom row are nonnegative, so the algorithm terminates. We have the following result:

   y (nonbasic) = 0,

   u (nonbasic) = 0,

   x = 90/3 = 30 from row 1,

   v = 30/3 = 10 from row 2,

   w = 42/3 = 14 from row 3,

   f = 270/3 = 90 from row 4

The solution is Max = 90 at x = 30, y = 0. The check is left to the reader.

11. Maximize

$$f = 2x + y$$

Subject to

$$5x + 6y \leqslant 60$$
$$x + y \leqslant 11$$
$$3x + y \leqslant 27$$
$$x, y \geqslant 0$$

Convert to a system of equations containing slack variables.

$$5x + 6y + u \qquad\qquad = 60$$
$$x + y \quad + v \qquad\quad = 11$$
$$3x + y \qquad\quad + w \quad = 27$$
$$-2x - y \qquad\qquad + f = 0$$

Set up the initial simplex tableau, and carry out the pivoting.

x	y	u	v	w	f		1	q
5	6	1	0	0	0		60	60/5 = 12
1	1	0	1	0	0		11	11/1 = 11
3*	1	0	0	1	0		27	27/3 = 9 ← Min
⊘-2	-1	0	0	0	1		0	f = 0
0	13	3	0	-5	0		45	$45/13 = 3\frac{6}{13}$
0	2*	0	3	-1	0		6	6/2 = 3 ← Min
3	1	0	0	1	0		27	27/1 = 27
0	⊘-1	0	0	2	3		54	f = 54/3 = 18
0	0	6	-39	3	0		12	
0	2	0	3	-1	0		6	
6	0	0	-3	3	0		48	
0	0	0	3	3	6		114	f = 114/6 = 19

All the indicators in the bottom row are nonnegative, so the algorithm terminates. We have the following result:

v (nonbasic) = 0,

w (nonbasic) = 0,

u = 12/6 = 2 from row 1,

y = 6/2 = 3 from row 2,

x = 48/6 = 8 from row 3,

f = 114/6 = 19 from row 4

The solution is Max = 19 at x = 8, y = 3. The check is left to the reader.

13. Maximize

$$f = 7x + 4y$$

Subject to

$$3x + 4y \leqslant 48$$
$$x + y \leqslant 13$$
$$2x + y \leqslant 22$$
$$x, y \geqslant 0$$

Convert to a system of equations containing slack variables.

$$3x + 4y + u \qquad\qquad = 48$$
$$x + y \quad + v \qquad\quad = 13$$
$$2x + y \qquad\quad + w \quad = 22$$
$$-7x - 4y \qquad\qquad + f = 0$$

13. (continued)

Set up the initial simplex tableau and carry out the pivoting.

x	y	u	v	w	f		1	q
3	4	1	0	0	0		48	48/3 = 16
1	1	0	1	0	0		13	13/1 = 13
2*	1	0	0	1	0		22	22/2 = 11 ← Min
(-7)	-4	0	0	0	1		0	f = 0
0	5	2	0	-3	0		30	30/5 = 6
0	1*	0	2	-1	0		4	4/1 = 4 ← Min
2	1	0	0	1	0		22	22/1 = 22
0	(-1)	0	0	7	2		154	f = 154/2 = 77
0	0	2	-10	2	0		10	
0	1	0	2	-1	0		4	
2	0	0	-2	2	0		18	
0	0	0	2	6	2		158	f = 158/2 = 79

All the indicators in the bottom row are nonnegative, so the algorithm terminates. We have the following result:

v (nonbasic) = 0,

w (nonbasic) = 0,

u = 10/2 = 5 from row 1,

y = 4/1 = 4 from row 2,

x = 18/2 = 9 from row 3,

f = 158/2 = 79 from row 4

The solution is Max = 79 at x = 9, y = 4. The check is left to the reader.

15. Maximize

$$f = -2y + 5z$$

Subject to

$$3x - 2y + z \leqslant 8$$
$$-4x + 3y + 2z \leqslant 4$$
$$3x + y - 6z \leqslant 6$$
$$x, y, z \geqslant 0$$

Convert to a system of equations containing slack variables.

$$3x - 2y + z + u \qquad\qquad = 8$$
$$-4x + 3y + 2z \qquad + v \qquad\quad = 4$$
$$3x + y - 6z \qquad\qquad + w \quad = 6$$
$$2y - 5z \qquad\qquad\quad + f = 0$$

Set up the initial simplex tableau, and carry out the pivoting.

15. (continued)

x	y	z	u	v	w	f		1	q
3	-2	1	1	0	0	0		8	8/1 = 8
-4	3	2*	0	1	0	0		4	4/2 = 2 ← Min
3	1	-6	0	0	1	0		6	—
0	2	⊙-5	0	0	0	1		0	f = 0
10*	-7	0	2	-1	0	0		12	12/10 = 1.2 ← Min
-4	3	2	0	1	0	0		4	—
-9	10	0	0	3	1	0		18	—
⊙-20	19	0	0	5	0	2		20	f = 20/2 = 10
10	-7	0	2	-1	0	0		12	
0	2	20	8	6	0	0		88	
0	37	0	18	21	10	0		288	
0	5	0	4	3	0	2		44	f = 44/2 = 22

(A "—" in the "q" column of the tableaus indicates that that quotient is negative and can be ignored in determining the minimum positive quotient.) All the indicators in the bottom row are nonnegative, so the algorithm terminates. We have the following result:

   y (nonbasic) = 0,

   u (nonbasic) = 0,

   v (nonbasic) = 0,

   x = 12/10 = 6/5 from row 1,

   z = 88/20 = 22/5 from row 2,

   w = 288/10 = 144/5 from row 3,

   f = 44/2 = 22 from row 4

The solution is Max = 22 at $x = \frac{6}{5}$, y = 0, $z = \frac{22}{5}$.

17. Maximize

   f = x + 2y + 3z

Subject to

   $2x + y + 4z \leqslant 24$

   $x + y + z \leqslant 7$

   $2x - y + 3z \leqslant 12$

   $x, y, z \geqslant 0$

Convert to a system of equations containing slack variables.

   2x + y + 4z + u   = 24

   x + y + z   + v   = 7

   2x - y + 3z   + w = 12

   -x - 2y - 3z   + f = 0

Set up the initial simplex tableau, and carry out the pivoting.

<u>17</u>. (continued)

x	y	z	u	v	w	f		1	q
2	1	4	1	0	0	0		24	24/4 = 6
1	1	1	0	1	0	0		7	7/1 = 7
2	-1	3*	0	0	1	0		12	12/3 = 4 ← Min
-1	-2	(-3)	0	0	0	1		0	f = 0
-2	7	0	3	0	-4	0		24	$24/7 = 3\frac{3}{7}$
1	4*	0	0	3	-1	0		9	$9/4 = 2\frac{1}{4}$ ← Min
2	-1	3	0	0	1	0		12	—
1	(-3)	0	0	0	1	1		12	f = 12/1 = 12
-15	0	0	12	-21	-9	0		33	
1	4	0	0	3	-1	0		9	
9	0	12	0	3	3	0		57	
7	0	0	0	9	1	4		75	f = 75/4

All the indicators in the bottom row are nonnegative, so the algorithm terminates. We have the following result:

   x (nonbasic) = 0,

   v (nonbasic) = 0,

   w (nonbasic) = 0,

   u = 33/12 = 11/4 from row 1,

   y = 9/4 from row 2,

   z = 57/12 = 19/4 from row 3,

   f = 75/4 from row 4

The solution is Max = $\frac{75}{4}$ at x = 0, y = $\frac{9}{4}$, z = $\frac{19}{4}$.

<u>19</u>. Familiarize: Let

   x = the number of product A produced,

   y = the number of product B produced,

   z = the number of product C produced,

and P = total profit from the sale of A, B, and C.

Organize the given information in a table.

	Composition			Number of Units of Supply
	Product A	Product B	Product C	
Number of Units	x	y	z	
Machining	1	3	2	16
Plating	1	1	3	12
Polishing	1	2	2	6
Profit per Unit ($)	$10	$20	$30	Objective Function Maximize P

19. (continued)

Translate: We obtain the translation from the table.

Maximize $P = 10x + 20y + 30z$

Subject to

$$x + 3y + 2z \leqslant 16$$
$$x + y + 3z \leqslant 12$$
$$x + 2y + 2z \leqslant 6$$
$$x, y, z \geqslant 0$$

Carry out: Use the simplex method. First, convert to a system of equations containing slack variables.

$$
\begin{aligned}
x + 3y + 2z + u & = 16 \\
x + y + 3z \quad + v & = 12 \\
x + 2y + 2z \qquad + w & = 6 \\
-10x - 20y - 30z \qquad + P & = 0
\end{aligned}
$$

Set up the initial simplex tableau, and carry out the pivoting.

x	y	z	u	v	w	P	1	q
1	3	2	1	0	0	0	16	16/2 = 8
1	1	3	0	1	0	0	12	12/3 = 4
1	2	2*	0	0	1	0	6	6/2 = 3 ← Min
-10	-20	-30	0	0	0	1	0	P = 0
0	1	0	1	0	-1	0	10	
-1	-4	0	0	2	-3	0	6	
1	2	2	0	0	1	0	6	
5	10	0	0	0	15	1	90	P = 90/1 = 90

All the indicators in the bottom row are nonnegative, so the algorithm terminates. We have the following result:

x (nonbasic) = 0,

y (nonbasic) = 0,

w (nonbasic) = 0,

$u = 10/1 = 10$ from row 1,

$v = 6/2 = 3$ from row 2,

$z = 6/2 = 3$ from row 3,

$P = 90/1 = 90$ from row 4

State: Profit is maximized when 0 of A, 0 of B, and 3 of C are produced. (The maximum profit is $90.)

21. Familiarize: Let

x = the number of inexpensive speaker assemblies made,

y = the number of expensive speaker assemblies made,

z = the number of medium-priced speaker assemblies made, and

I = the total income from the sale of the speakers.

Organize the given information in a table.

21. (continued)

	Composition			Number of Units of Supply
	Inexpensive	Expensive	Medium	
Number of Units	x	y	z	
Woofers	0	1	1	44
Midrange Speakers	1	1	1	60
Tweeters	1	2	1	90
Profit per Unit ($)	$175	$450	$280	Objective Function Maximize I

Translate: We obtain the translation from the table.

Maximize $I = 175x + 450y + 280z$

Subject to

$$y + z \leqslant 44$$
$$x + y + z \leqslant 60$$
$$x + 2y + z \leqslant 90$$
$$x, y, z \geqslant 0$$

Carry out: Use the simplex method. First, convert to a system of equations containing slack variables.

$$y + z + u = 44$$
$$x + y + z + v = 60$$
$$x + 2y + z + w = 90$$
$$-175x - 450y - 280z + I = 0$$

Then set up the initial simplex tableau, and carry out the pivoting.

x	y	z	u	v	w	I	1	q
0	1*	1	1	0	0	0	44	44/1 = 44 ← Min
1	1	1	0	1	0	0	60	60/1 = 60
1	2	1	0	0	1	0	90	90/2 = 45
-175	(-450)	-280	0	0	0	1	0	I = 0
0	1	1	1	0	0	0	44	—
1	0	0	-1	1	0	0	16	16/1 = 16
1*	0	-1	-2	0	1	0	2	2/1 = 2 ← Min
(-175)	0	170	450	0	0	1	19,800	I = 19,800/1 = 19,800
0	1	1	1	0	0	0	44	44/1 = 44
0	0	1*	1	1	-1	0	14	14/1 = 14 ← Min
1	0	-1	-2	0	1	0	2	—
0	0	(-5)	100	0	175	1	20,150	I = 20,150/1 = 20,150
0	1	0	0	-1	1	0	30	
0	0	1	1	1	-1	0	14	
1	0	0	-1	1	0	0	16	
0	0	0	105	5	170	1	20,220	I = 20,220/1 = 20,220

21. (continued)

All of the indicators in the bottom row are nonnegative, so the algorithm terminates. We have the following result:

u (nonbasic) = 0,

v (nonbasic) = 0,

w (nonbasic) = 0,

y = 30/1 = 30 from row 1,

z = 14/1 = 14 from row 2,

x = 16/1 = 16 from row 3,

I = 20,220/1 = 20,220 from row 4

State: The manufacturer should make 16 inexpensive, 30 expensive, and 14 medium-priced speakers assemblies to maximize income. (The maximum income is $20,220.)

23. Familiarize: Let

x = number of lb of Mixture A to be blended,

y = number of lb of Mixture B to be blended,

z = number of lb of Mixture C to be blended,

and I = total income from the sale of A, B, and C.

We organize the information in a table. Express the percents in decimal notation.

	Composition			Number of Units of Supply
	Mixture A	Mixture B	Mixture C	
Number of Units	x	y	z	
Black tea	0.5	0	0.3	400
Pekoe tea	0.3	0.5	0.4	300
Orange Pekoe tea	0.2	0.5	0.3	240
Profit per lb ($)	$1.50	$4.00	$2.60	Objective Function Maximize I

Translate: We obtain the translation from the table.

Maximize $I = 1.50x + 4.00y + 2.60z$

Subject to

$$0.5x + 0.3z \leqslant 400$$
$$0.3x + 0.5y + 0.4z \leqslant 300$$
$$0.2x + 0.5y + 0.3z \leqslant 240$$
$$x, y, z \geqslant 0$$

Multiplying the structural constraints by 10 to clear of decimals we have

Maximize $I = 1.50x + 4.00y + 2.60z$

Subject to

$$5x + 3z \leqslant 4000$$
$$3x + 5y + 4z \leqslant 3000$$
$$2x + 5y + 3z \leqslant 2400$$
$$x, y, z \geqslant 0$$

Carry out: Use the simplex method. First, convert to a system of equations containing slack variables.

$$5x \qquad + 3z + u \qquad\qquad = 4000$$
$$3x + 5y + 4z \quad + v \qquad\quad = 3000$$
$$2x + 5y + 3z \qquad + w \quad = 2400$$
$$-1.50x - 4.00y - 2.60z \qquad\qquad + I = \quad 0$$

23. (continued)

Set up the initial simplex tableau, and carry out the pivoting.

x	y	z	u	v	w	I		1	
5	0	3	1	0	0	0		4000	—
3	5	4	0	1	0	0		3000	3000/5 = 600
2	5*	3	0	0	1	0		2400	2400/5 = 480 ← Min
-1.5	(-4)	-2.6	0	0	0	1		0	I = 0
5	0	3	1	0	0	0		4000	4000/3 = $1333\frac{1}{3}$
1	0	1*	0	1	-1	0		600	600/1 = 600 ← Min
2	5	3	0	0	1	0		2400	2400/3 = 800
0.5	0	(-1)	0	0	4	5		9600	I = 9600/5 = 1920
2	0	0	1	-3	3	0		2200	
1	0	1	0	1	-1	0		600	
-1	5	0	0	-3	4	0		600	
1.5	0	0	0	1	3	5		10,200	I = 10,200/5 = 2040

All the indicators in the bottom row are nonnegative so the algorithm terminates. We have the following result:

x (nonbasic) = 0,

v (nonbasic) = 0,

w (nonbasic) = 0,

u = 2200/1 = 2200 from row 1,

z = 600/1 = 600 from row 2,

y = 600/5 = 120 from row 3,

I = 10,200/5 = 2040 from row 4

State: The merchant should blend 0 lb of Mixture A, 120 lb of B, and 600 lb of C to maximize income. (The maximum income is $2040.)

25. Familiarize: Let

x = number of A1 that can be supported,

y = number of A2 that can be supported,

z = number of A3 that can be supported,

and T = total value of the animals.

Organize the information in a table.

	Composition			Number of Units
	A1	A2	A3	of Supply
Number of Units	x	y	z	
F1	1	1.2	3	600
F2	2	1.8	0.8	960
F3	2	0.6	1	720
Value per Animal	$150	$180	$130	Objective Function Maximize T

25. (continued)

Translate: We obtain the translation from the table.

Maximize T = 150x + 180y + 130z

Subject to

$$x + 1.2y + 3z \leqslant 600$$
$$2x + 1.8y + 0.8z \leqslant 960$$
$$2x + 0.6y + z \leqslant 720$$
$$x, y, z \geqslant 0$$

Multiplying the structural constraints by 10 to clear of decimals we have

Maximize T = 150x + 180y + 130z

Subject to

$$10x + 12y + 30z \leqslant 6000$$
$$20x + 18y + 8z \leqslant 9600$$
$$20x + 6y + 10z \leqslant 7200$$
$$x, y, z \geqslant 0$$

Carry out: Use the simplex method. First, convert to a system of equations containing slack variables.

$$10x + 12y + 30z + u \qquad\qquad = 6000$$
$$20x + 18y + 8z \qquad + v \qquad\quad = 9600$$
$$20x + 6y + 10z \qquad\qquad + w \quad = 7200$$
$$-150x - 180y - 130z \qquad\qquad\quad + T = 0$$

Set up the initial simplex tableau, and carry out the pivoting.

x	y	z	u	v	w	T	1	
10	12*	30	1	0	0	0	6000	6000/12 = 500 ← Min
20	18	8	0	1	0	0	9600	9600/18 = 533⅓
20	6	10	0	0	1	0	7200	7200/6 = 1200
-150	(-180)	-130	0	0	0	1	0	T = 0
10	12	30	1	0	0	0	6000	
10	0	-74	-3	2	0	0	1200	
30	0	-10	-1	0	2	0	8400	
0	0	320	15	0	0	1	90,000	T = 90,000/1 = 90,000

All the indicators in the bottom row are nonnegative, so the algorithm terminates. We have the following result:

x (nonbasic) = 0,

z (nonbasic) = 0,

u (nonbasic) = 0,

y = 6000/12 = 500 from row 1,

v = 1200/2 = 600 from row 2,

w = 8400/2 = 4200 from row 3,

T = 90,000/1 = 90,000 from row 4

State: Populating the forest with 0 of A1, 500 of A2, and 0 of A3 will maximize the value of the animals. (The maximum value is $90,000.)

27. Familiarize: Let

x = amount invested in home loans,

y = amount invested in car loans,

z = amount invested in securities,

and T = total income from the investments.

27. (continued)

The financial institution can place part or all of its $800 million in these investments, so

$x + y + z \leqslant 800.$

The total amount in loans should be no more than $600 million, so

$x + y \leqslant 600.$

No more than $700 million should be in securities, so

$z \leqslant 700.$

We also have nonnegativity constraints,

$x, y, z \geqslant 0.$

The total income can be found using the formula for simple interest, $I = Prt$. We have

$T = 0.1x + 0.11y + 0.08z.$

<u>Translate</u>: Putting all the information together, we can write a linear program.

Maximize $T = 0.1x + 0.11y + 0.08z$

Subject to

$$x + y + z \leqslant 800$$
$$x + y \qquad \leqslant 600$$
$$z \leqslant 700$$
$$x, y, z \geqslant 0$$

<u>Carry out</u>: Use the simplex method. First, convert to a system of equations containing slack variables.

$$x + \quad y + \quad z + u \qquad\qquad = 800$$
$$x + \quad y \qquad\qquad + v \qquad = 600$$
$$z \qquad + w \quad = 700$$
$$-0.1x - 0.11y - 0.08z \qquad\qquad + T = \quad 0$$

Set up the initial simplex tableau, and carry out the pivoting.

x	y	z	u	v	w	T	1	
1	1	1	1	0	0	0	800	800/1 = 800
1	1*	0	0	1	0	0	600	600/1 = 600 ← Min
0	0	1	0	0	1	0	700	—
-0.1	(-0.11)	-0.08	0	0	0	1	0	T = 0
0	0	1*	1	-1	0	0	200	200/1 = 200 ← Min
1	1	0	0	1	0	0	600	—
0	0	1	0	0	1	0	700	700/1 = 700
0.01	0	(-0.08)	0	0.11	0	1	66	T = 66/1 = 66
0	0	1	1	-1	0	0	200	
1	1	0	0	1	0	0	600	
0	0	0	-1	1	1	0	500	
0.01	0	0	0.08	0.03	0	1	82	T = 82/1 = 82

<u>27</u>. (continued)

All the indicators in the bottom row are nonnegative, so the algorithm terminates. We have the following result:

  x (nonbasic) = 0,

  u (nonbasic) = 0,

  v (nonbasic) = 0,

  z = 200/1 = 200 from row 1,

  y = 600/1 = 600 from row 2,

  w = 500/1 = 500 from row 3,

  T = 82/1 = 82 from row 4

<u>State</u>: Income will be maximized by investing $0 in home loans, $600 million in car loans, and $200 million in securities. (The maximum income is $82 million.)

<u>29</u>. <u>Familiarize</u>: Let

  x = number of lamp A produced,

  y = number of lamp B produced,

  z = number of lamp C produced,

and P = total profit from the sale of the lamps.

Organize the information in a table.

	Composition			Number of Units of Supply
	A	B	C	
Number of Units	x	y	z	
Hours in Dept. I	2	3	1	400
Hours in Dept. II	4	2	3	600
Profit per Lamp	$5	$4	$3	Objective Function Maximize P

<u>Translate</u>: We obtain the translation from the table.

Maximize $P = 5x + 4y + 3z$

Subject to

  $2x + 3y + z \leqslant 400$

  $4x + 2y + 3z \leqslant 600$

   $x, y, z \geqslant 0$

<u>Carry out</u>: Use the simplex method. First, set up a system of equations containing slack variables.

  $2x + 3y + \;\, z + u \qquad\quad = 400$

  $4x + 2y + 3z \qquad + v \quad = 600$

  $-5x - 4y - 3z \qquad\quad + P = \quad 0$

Set up the initial simplex tableau, and carry out the pivoting.

29. (continued)

x	y	z	u	v	P		1	
2	3	1	1	0	0		400	$400/2 = 200$
4*	2	3	0	1	0		600	$600/4 = 150 \leftarrow$ Min
$\boxed{-5}$	-4	-3	0	0	1		0	$P = 0$
0	4*	-1	2	-1	0		200	$200/4 = 50 \leftarrow$ Min
4	2	3	0	1	0		600	$600/2 = 300$
0	$\boxed{-6}$	3	0	5	4		3000	$P = 3000/4 = 750$
0	4	-1	2	-1	0		200	
8	0	7	-2	3	0		1000	
0	0	3	6	7	8		6600	$P = 6600/8 = 825$

All the indicators in the bottom row are nonnegative, so the algorithm terminates. We have the following result:

z (nonbasic) = 0,

u (nonbasic) = 0,

v (nonbasic) = 0,

$y = 200/4 = 50$ from row 1,

$x = 1000/8 = 125$ from row 2,

$P = 6600/8 = 825$ from row 3

State: Maximum profit = \$825 by producing 125 of type A, 50 of type B, and 0 of type C.

Exercise Set 4.3

---

1. Rewrite the exercise in the following form to facilitate writing the dual:

$$3y_1 + y_2 \geqslant 9,$$
$$y_1 + y_2 \geqslant 7,$$
$$y_1 + 2y_2 \geqslant 8,$$

Min $y_0 = 3y_1 + 4y_2,$

$$y_1, y_2 \geqslant 0.$$

First form the dual.

$$3x_1 + x_2 + x_3 \leqslant 3,$$
$$x_1 + x_2 + 2x_3 \leqslant 4,$$

Max $x_0 = 9x_1 + 7x_2 + 8x_3,$

$$x_1, x_2, x_3 \geqslant 0.$$

Add slack variables to the dual, using the structural variables of the primal.

$$3x_1 + x_2 + x_3 + y_1 \qquad = 3$$
$$x_1 + x_2 + 2x_3 \qquad + y_2 \qquad = 4$$
$$-9x_1 - 7x_2 - 8x_3 \qquad\qquad + x_0 = 0$$

Set up the initial simplex tableau, and solve using the simplex method.

<u>1.</u> (continued)

x_1	x_2	x_3	y_1	y_2	x_0		1	q
3*	1	1	1	0	0		3	3/3 = 1 ← Min
1	1	2	0	1	0		4	4/1 = 4
-9	-7	-8	0	0	1		0	$x_0 = 0$
3*	1	1	1	0	0		3	3/1 = 3
0	2	5*	-1	3	0		9	9/5 = 1.8 ← Min
0	-4	-5	3	0	1		9	$x_0 = 9/1 = 9$
15	3*	0	6	-3	0		6	6/3 = 2 ← Min
0	2	5	-1	3	0		9	9/2 = 4.5
0	-2	0	2	3	1		18	$x_0 = 18/1 = 18$
15	3	0	6	-3	0		6	
-30	0	15	-15	15	0		15	
30	0	0	18	3	3		66	$x_0 = 66/3 = 22$

All the entries in the bottom row are nonnegative, so the algorithm
terminates. To find the minimum solution divide each entry in the bottom row
by the basic variable that is the coefficient of x_0. In this case we divide
by 3. The minimum solution is

$x_1 = 30/3 = 10$, $x_2 = 0$, $x_3 = 0$, $y_1 = 18/3 = 6$, $y_2 = 3/3 = 1$,

$x_0 = 66/3 = 22$.

The solution of the original problem is

Min $y_0 = 22$ at $y_1 = 6$, $y_2 = 1$.

<u>3.</u> To facilitate forming the dual rewrite the exercise in the following form:

$2y_1 + y_2 \geqslant 9$,

$4y_1 + 3y_2 \geqslant 23$,

$y_1 + 3y_2 \geqslant 8$,

Min $y_0 = 2y_1 + 5y_2$,

$y_1, y_2 \geqslant 0$.

Then form the dual.

$2x_1 + 4x_2 + x_3 \leqslant 2$,

$x_1 + 3x_2 + 3x_3 \leqslant 5$,

Max $x_0 = 9x_1 + 23x_2 + 8x_3$,

$x_1, x_2, x_3 \geqslant 0$.

Add slack variables to the dual, using the structural variables of the primal.

$2x_1 + 4x_2 + x_3 + y_1 \qquad = 2$

$x_1 + 3x_2 + 3x_3 \qquad + y_2 \qquad = 5$

$-9x_1 - 23x_2 - 8x_3 \qquad + x_0 = 0$

Set up the initial simplex tableau, and solve using the simplex method.

<u>3</u>. (continued)

x_1	x_2	x_3	y_1	y_2	x_0	1	q
2	4*	1	1	0	0	2	$2/4 = \frac{1}{2} \leftarrow$ Min
1	3	3	0	1	0	5	$5/3 = 1\frac{2}{3}$
-9	(-23)	-8	0	0	1	0	$x_0 = 0$
2	4	1	1	0	0	2	$2/1 = 2$
-2	0	9*	-3	4	0	14	$14/9 = 1\frac{5}{9} \leftarrow$ Min
10	0	(-9)	23	0	4	46	$x_0 = 46/4 = 11\frac{1}{2}$
20	36	0	12	-4	0	4	
-2	0	9	-3	4	0	14	
8	0	0	20	4	4	60	$x_0 = 60/4 = 15$

All the entries in the bottom row are nonnegative, so the algorithm terminates. To find the minimum solution, divide each entry in the bottom row by the coefficient of x_0, 4. The minimum solution is

$x_1 = 8/4 = 2$, $x_2 = 0$, $x_3 = 0$, $y_1 = 20/4 = 5$,

$y_2 = 4/4 = 1$, $y_0 = 60/4 = 15$.

The solution of the original problem is

Min $y_0 = 15$ at $y_1 = 5$, $y_2 = 1$.

<u>5</u>. To facilitate forming the dual rewrite the exercise in the following form.

$$4y_1 + y_2 \geqslant 9,$$
$$3y_1 + 2y_2 \geqslant 13,$$
$$2y_1 + 5y_2 \geqslant 16,$$
Min $y_0 = 3y_1 + 5y_2$,
$$y_1, y_2 \geqslant 0.$$

Then form the dual.

$$4x_1 + 3x_2 + 2x_3 \leqslant 3,$$
$$x_1 + 2x_2 + 5x_3 \leqslant 5,$$
Max $x_0 = 9x_1 + 13x_2 + 16x_3$,
$$x_1, x_2, x_3 \geqslant 0.$$

Add slack variables to the dual, using the structural variables of the primal.

$$4x_1 + 3x_2 + 2x_3 + y_1 \qquad = 3$$
$$x_1 + 2x_2 + 5x_3 \qquad + y_2 \qquad = 5$$
$$-9x_1 - 13x_2 - 16x_3 \qquad + x_0 = 0$$

Set up the initial simplex tableau, and solve using the simplex method.

x_1	x_2	x_3	y_1	y_2	x_0	1	
4	3	2	1	0	0	3	$3/2 = 1.5$
1	2	5*	0	1	0	5	$5/5 = 1 \leftarrow$ Min
-9	-13	(-16)	0	0	1	0	$x_0 = 0$
18	11*	0	5	-2	0	5	$5/11 \leftarrow$ Min
1	2	5	0	1	0	5	$5/2$
-29	(-33)	0	0	16	5	80	$x_0 = 80/5 = 16$
18	11	0	5	-2	0	5	
-25	0	55	-10	15	0	45	
25	0	0	15	10	5	95	$x_0 = 95/5 = 19$

<u>5.</u> (continued)

All the entries in the bottom row are nonnegative, so that algorithm terminates. To find the minimum solution, divide each entry in the bottom row by the coefficient of x_0, 5. The minimum solution is

$x_1 = 25/5 = 5$, $x_2 = 0$, $x_3 = 0$, $y_1 = 15/5 = 3$, $y_2 = 10/5 = 2$,

$y_0 = 95/5 = 19$.

The solution of the original problem is

Min $y_0 = 19$ at $y_1 = 3$, $y_2 = 2$.

<u>7.</u>
$$3y_1 + y_2 \geqslant 14,$$
$$4y_1 + 3y_2 \geqslant 34,$$
$$3y_1 + 4y_2 \geqslant 36,$$

Min $y_0 = 7y_1 + 8y_2$,

$$y_1, y_2 \geqslant 0.$$

First form the dual.

$$3x_1 + 4x_2 + 3x_3 \leqslant 7,$$
$$x_1 + 3x_2 + 4x_3 \leqslant 8,$$

Max $x_0 = 14x_1 + 34x_2 + 36x_3$,

$$x_1, x_2, x_3 \geqslant 0.$$

Add as slack variables the structural variables of the primal.

$$3x_1 + 4x_2 + 3x_3 + y_1 \qquad\qquad = 7$$
$$x_1 + 3x_2 + 4x_3 \qquad + y_2 \qquad = 8$$
$$-14x_1 - 34x_2 - 36x_3 \qquad\qquad + x_0 = 0$$

Set up the initial simplex tableau, and solve using the simplex method.

x_1	x_2	x_3	y_1	y_2	x_0	1	
3	4	3	1	0	0	7	$7/3 = 2\frac{1}{3}$
1	3	4*	0	1	0	8	$8/4 = 2 \leftarrow$ Min
-14	-34	(-36)	0	0	1	0	$x_0 = 0$
9	7*	0	4	-3	0	4	$4/7 \leftarrow$ Min
1	3	4	0	1	0	8	$8/3 = 2\frac{2}{3}$
-5	(-7)	0	0	9	1	72	$x_0 = 72/1 = 72$
9	7	0	4	-3	0	4	
-20	0	28	-12	16	0	44	
4	0	0	4	6	1	76	$x_0 = 76/1 = 76$

All the entries in the bottom row are nonnegative, so the algorithm terminates. To find the minimum solution, divide each entry in the bottom row by the coefficient of x_0, 1. The minimum solution is

$x_1 = 4$, $x_2 = 0$, $x_3 = 0$, $y_1 = 4$, $y_2 = 6$, $y_0 = 76$.

<u>Check</u>: In order to check your solution, you must also find the dual solution as you did in Section 4.2. The dual solution is $x_1 = 0$, $x_2 = 4/7$, $x_3 = 44/28 = 11/7$, $y_1 = 0$, $y_2 = 0$. Use the primal and dual solutions as indicated in Example 5, Section 4.3, of the text to check your solution.

<u>9.</u>
$$3y_1 + 2y_2 \geqslant 29,$$
$$4y_1 + 5y_2 \geqslant 55,$$
$$y_1 + 2y_2 \geqslant 18,$$

Min $y_0 = 5y_1 + 4y_2$,

$$y_1, y_2 \geqslant 0.$$

9. (continued)

First form the dual.

$$3x_1 + 4x_2 + x_3 \leqslant 5,$$
$$2x_1 + 5x_2 + 2x_3 \leqslant 4,$$

Max $x_0 = 29x_1 + 55x_2 + 18x_3$,

$$x_1, x_2, x_3 \geqslant 0.$$

Add as slack variables the structural variables of the primal.

$$3x_1 + 4x_2 + x_3 + y_1 \qquad\qquad = 5$$
$$2x_1 + 5x_2 + 2x_3 \qquad + y_2 \qquad = 4$$
$$-29x_1 - 55x_2 - 18x_3 \qquad\qquad + x_0 = 0$$

Set up the initial simplex tableau, and solve using the simplex method.

x_1	x_2	x_3	y_1	y_2	x_0	1	
3	4	1	1	0	0	5	5/4 = 1.25
2	5*	2	0	1	0	4	4/5 = 0.8 ← Min
-29	(-55)	-18	0	0	1	0	$x_0 = 0$
7*	0	-3	5	-4	0	9	$9/7 = 1\frac{2}{7}$ ← Min
2	5	2	0	1	0	4	4/2 = 2
(-7)	0	4	0	11	1	44	$x_0 = 44/1 = 44$
7	0	-3	5	-4	0	9	
0	35	20	-10	15	0	10	
0	0	1	5	7	1	53	$x_0 = 53/1 = 53$

All the entries in the bottom row are nonnegative, so the algorithm terminates. To find the minimum solution, divide each entry in the bottom row by the coefficient of x_0, 1. The minimum solution is

$$x_1 = 0, \quad x_2 = 0, \quad x_3 = 1, \quad y_1 = 5, \quad y_2 = 7, \quad y_0 = 53.$$

Check: Find the dual solution as in Section 4.2:

$$x_1 = 9/7, \quad x_2 = 10/35 = 2/7, \quad x_3 = y_1 = y_2 = 0$$

Use the primal and dual solutions as indicated in Example 5, Section 4.3, of the text to check your solution.

11.
$$2y_1 + 9y_2 + 4y_3 \geqslant 6,$$
$$3y_1 + 6y_2 - 2y_3 \geqslant 8,$$
$$y_1 + y_2 + y_3 \geqslant 3,$$

Min $y_0 = 5y_1 + 2y_2 + y_3$,

$$y_1, y_2, y_3 \geqslant 0.$$

First find the dual.

$$2x_1 + 3x_2 + x_3 \leqslant 5,$$
$$9x_1 + 6x_2 + x_3 \leqslant 2,$$
$$4x_1 - 2x_2 + x_3 \leqslant 1,$$

Max $x_0 = 6x_1 + 8x_2 + 3x_3$,

$$x_1, x_2, x_3 \geqslant 0.$$

Add as slack variables the structural var..._ _s of the primal.

$$2x_1 + 3x_2 + x_3 + y_1 \qquad\qquad = 5$$
$$9x_1 + 6x_2 + x_3 \qquad + y_2 \qquad = 2$$
$$4x_1 - 2x_2 + x_3 \qquad\qquad + y_3 \qquad = 1$$
$$-6x_1 - 8x_2 - 3x_3 \qquad\qquad\qquad + x_0 = 0$$

11. (continued)

Set up the initial simplex tableau, and solve using the simplex method.

x_1	x_2	x_3	y_1	y_2	y_3	x_0		1	
2	3	1	1	0	0	0		5	$5/3 = 1\frac{2}{3}$
9	6*	1	0	1	0	0		2	$2/6 = \frac{1}{3} \leftarrow$ Min
4	-2	1	0	0	1	0		1	—
-6	⑧	-3	0	0	0	1		0	$x_0 = 0$
-5	0	1	2	-1	0	0		8	$8/1 = 8$
9	6	1	0	1	0	0		2	$2/1 = 2$
21	0	4*	0	1	3	0		5	$5/4 = 1\frac{1}{4} \leftarrow$ Min
18	0	⑤	0	4	0	3		8	$x_0 = 8/3 = 2\frac{2}{3}$
41	0	0	-8	5	3	0		-27	
-15	-24	0	0	-3	3	0		-3	
21	0	4	0	1	3	0		5	
177	0	0	0	21	15	12		57	$x_0 = 57/12 = 19/4$

All the entries in the bottom row are nonnegative, so the algorithm terminates. To find the minimum solution, divide each entry in the bottom row by the coefficient of x_0, 12. The minimum solution is

$x_1 = 177/12 = 59/4$, $x_2 = x_3 = y_1 = 0$, $y_2 = 21/12 = 7/4$,

$y_3 = 15/12 = 5/4$, $y_0 = 57/12 = 19/4$.

Check: Find the dual solution as in Section 4.2:

$x_1 = 0$, $x_2 = -3/-24 = 1/8$, $x_3 = 5/4$, $y_1 = -27/-8 = 27/8$, $y_2 = y_3 = 0$.

Use the primal and dual solutions as indicated in Example 5, Section 4.3, to check your solution.

13.
$$2y_1 + 3y_2 + y_3 \geqslant 2,$$
$$5y_1 + 2y_2 - 3y_3 \geqslant 4,$$
$$7y_1 + 6y_2 + 4y_3 \geqslant 5,$$
Min $y_0 = 8y_1 + 9y_2 + 5y_3$,
$$y_1, y_2, y_3 \geqslant 0.$$
First form the dual.
$$2x_1 + 5x_2 + 7x_3 \leqslant 8,$$
$$3x_1 + 2x_2 + 6x_3 \leqslant 9,$$
$$x_1 - 3x_2 + 4x_3 \leqslant 5,$$
Max $x_0 = 2x_1 + 4x_2 + 5x_3$,
$$x_1, x_2, x_3 \geqslant 0.$$

Add as slack variables the structural variables of the primal.
$$2x_1 + 5x_2 + 7x_3 + y_1 \qquad\qquad = 8$$
$$3x_1 + 2x_2 + 6x_3 \qquad + y_2 \qquad = 9$$
$$x_1 - 3x_2 + 4x_3 \qquad\qquad + y_3 \qquad = 5$$
$$-2x_1 - 4x_2 - 5x_3 \qquad\qquad\qquad + x_0 = 0$$
Set up the initial simplex tableau, and solve using the simplex method.

<u>13.</u> (continued)

x_1	x_2	x_3	y_1	y_2	y_3	x_0	1	
2	5	7*	1	0	0	0	8	$8/7 = 1\frac{1}{7}$ ← Min
3	2	6	0	1	0	0	9	$9/6 = 1\frac{1}{2}$
1	-3	4	0	0	1	0	5	$5/4 = 1\frac{1}{4}$
-2	-4	⊘-5	0	0	0	1	0	$x_0 = 0$
2	5	7	1	0	0	0	8	$8/2 = 4$
9*	-16	0	-6	7	0	0	15	$15/9 = 1\frac{2}{3}$ ← Min
-1	-41	0	-4	0	7	0	3	—
⊘-4	-3	0	5	0	0	7	40	$x_0 = 40/7$
0	77*	63	21	-14	0	0	42	$42/77 = 6/11$ ← Min
9	-16	0	-6	7	0	0	15	—
0	-385	0	-42	7	63	0	42	—
0	⊘-91	0	21	28	0	63	420	$x_0 = 420/63 = 6\frac{2}{3}$
0	77	63	21	-14	0	0	42	
693	0	1008	-126	315	0	0	1827	
0	0	315	63	-63	63	0	252	
0	0	819	504	126	0	693	5166	$x_0 = 5166/693 = 82/11$

All the entries in the bottom row are nonnegative so the algorithm terminates.
To find the minimum solution, divide each entry in the bottom row by the
coefficient of x_0, 693. The minimum solution is

$$x_1 = x_2 = 0, \quad x_3 = 819/693 = 13/11, \quad y_1 = 504/693 = 8/11,$$
$$y_2 = 126/693 = 2/11, \quad y_3 = 0, \quad y_0 = 5166/693 = 82/11.$$

<u>Check</u>: Find the dual solution as in Section 4.2:

$$x_1 = 1827/693 = 29/11, \quad x_2 = 42/77 = 6/11, \quad x_3 = y_1 = y_2 = 0,$$
$$y_3 = 252/63 = 4$$

Use the primal and dual solutions as indicated in Example 5, Section 4.3, to
check your solution.

$$y_1 - y_2 \geqslant 2,$$
$$-y_1 + y_2 \geqslant 5,$$
Min $y_0 = 4y_1 + 8y,$
$$y_1, y_2 \geqslant 0.$$

First form the dual.

$$x_1 - x_2 \leqslant 4,$$
$$-x_1 + x_2 \leqslant 8,$$
Max $x_0 = 2x_1 + 5x_2,$
$$x_1, x_2 \geqslant 0.$$

Add as slack variables the structural variables of the primal.

$$x_1 - x_2 + y_1 \qquad = 4$$
$$-x_1 + x_2 \qquad + y_2 \qquad = 8$$
$$-2x_1 - 5x_2 \qquad + x_0 = 0$$

Set up the initial simplex tableau, and solve using the simplex method.

<u>15.</u> (continued)

x_1	x_2	y_1	y_2	x_0	1	q
1	-1	1	0	0	4	—
-1	1*	0	1	0	8	8/1 = 8 ← Min
-2	(-5)	0	0	1	0	
0	0	1	1	0	12	
-1	1	0	1	0	8	
-7	0	0	5	1	40	

└ Negative indicator, and the rest of the numbers in the column are either
   0's or negative

We have a column with a negative indicator in the bottom row, and the rest of
the column has either negative or 0 entries. Therefore, the maximum program
is unbounded, and the minimum program has no solution.

<u>17.</u> <u>Familiarize</u>: Let

y_1 = number of sacks of soybean meal used,

y_2 = number of sacks of oats used, and

y_0 = total cost of the feed.

Organize the information in a table.

	Composition		Pounds Required
	Soybean Meal	Oats	
Number of 100-lb sacks	y_1	y_2	
x_1: Protein	50	15	120
x_2: Fat	8	5	24
x_3: Mineral Ash	5	1	10
Cost per 100-lb sack	$15	$5	Objective Function Minimize y_0

<u>Translate</u>: Use the table to write a linear program.

Minimize $y_0 = 15y_1 + 5y_2$

Subject to

$$50y_1 + 15y_2 \geqslant 120$$
$$8y_1 + 5y_2 \geqslant 24$$
$$5y_1 + y_2 \geqslant 10$$
$$y_1, y_2 \geqslant 0$$

<u>Carry out</u>: First form the dual.

$$50x_1 + 8x_2 + 5x_3 \leqslant 15,$$
$$15x_1 + 5x_2 + x_3 \leqslant 5,$$

Max $x_0 = 120x_1 + 24x_2 + 10x_3,$

$$x_1, x_2, x_3 \geqslant 0.$$

Add as slack variables the structural variables of the primal.

$$50x_1 + 8x_2 + 5x_3 + y_1 \qquad = 15$$
$$15x_1 + 5x_2 + x_3 \qquad + y_2 \qquad = 5$$
$$-120x_1 - 24x_2 - 10x_3 \qquad\qquad + x_0 = 0$$

<u>17.</u> (continued)

Solve using the simplex method.

x_1	x_2	x_3	y_1	y_2	x_0	1	
50*	8	5	1	0	0	15	15/50 = 3/10 ← Min
15	5	1	0	1	0	5	5/15 = 1/3
(-120)	-24	-10	0	0	1	0	$x_0 = 0$
50	8	5	1	0	0	15	$15/8 = 1\frac{7}{8}$
0	26*	-5	-3	10	0	5	$\frac{5}{26}$ ← Min
0	(-24)	10	12	0	5	180	$x_0 = 180/5 = 36$
650	0	85	25	-40	0	175	
0	26	-5	-3	10	0	5	
0	0	70	120	120	65	2400	$x_0 = 2400/65$

All the entries in the bottom row are nonnegative, so the algorithm terminates. To find the minimum solution, divide each entry in the bottom row by the coefficient of x_0, 65. The minimum solution is

$$x_1 = x_2 = 0, \quad x_3 = 70/65 = 1\frac{1}{13}, \quad y_1 = 120/65 = 1\frac{11}{13},$$
$$y_2 = 120/65 = 1\frac{11}{13}, \quad y_0 = 2400/65 = 36\frac{12}{13}.$$

<u>State:</u> $1\frac{11}{13}$ sacks of soybean meal and $1\frac{11}{13}$ sacks of oats should be used. (The minimum cost is $\$36\frac{12}{13}$.)

<u>19.</u> Familiarize: Let

y_1 = number of sacks of soybean meal used,

y_2 = number of sacks of alfalfa used, and

y_0 = total cost of the feed.

Organize the information in a table.

	Composition		Pounds Required
	Soybean Meal	Alfalfa	
Number of 100-lb sacks	y_1	y_2	
x_1: Protein	50	20	120
x_2: Fat	8	6	24
x_3: Mineral Ash	5	8	20
Cost per 100-lb sack	$\$15$	$\$8$	Objective Function Minimize y_0

<u>Translate:</u> Use the table to write a linear program.

Minimize $y_0 = 15y_1 + 8y_2$

Subject to

$$50y_1 + 20y_2 \geq 120$$
$$8y_1 + 6y_2 \geq 24$$
$$5y_1 + 8y_2 \geq 20$$
$$y_1, y_2 \geq 0$$

19. (continued)

Carry out: First form the dual.

$$50x_1 + 8x_2 + 5x_3 \leq 15,$$
$$20x_1 + 6x_2 + 8x_3 \leq 8,$$

Max $x_0 = 120x_1 + 24x_2 + 20x_3$,

$$x_1, x_2, x_3 \geq 0.$$

Add as slack variables the structural variables of the primal.

$$50x_1 + 8x_2 + 5x_3 + y_1 \qquad = 15$$
$$20x_1 + 6x_2 + 8x_3 \qquad + y_2 \quad = 8$$
$$-120x_1 - 24x_2 - 20x_3 \qquad\qquad + x_0 = 0$$

Solve using the simplex method.

x_1	x_2	x_3	y_1	y_2	x_0		1	
50*	8	5	1	0	0		15	15/50 = 3/10 ← Min
20	6	8	0	1	0		8	8/20 = 2/5
(−120)	−24	−20	0	0	1		0	x_0 = 0
50	8	5	1	0	0		15	15/5 = 3
0	14	30*	−2	5	0		10	10/30 = 1/3 ← Min
0	−24	(−40)	12	0	5		180	x_0 = 180/5 = 36
300	34	0	8	−5	0		80	$80/34 = 2\frac{6}{17}$
0	14*	30	−2	5	0		10	10/14 = 5/7 ← Min
0	(−16)	0	28	20	15		580	$x_0 = 580/15 = 38\frac{2}{3}$
2100	0	−510	90	−120	0		390	
0	14	30	−2	5	0		10	
0	0	240	180	180	105		4140	x_0 = 4140/105

All the entries in the bottom row are nonnegative, so the algorithm terminates. To find the minimum solution, divide each entry on the bottom row by the coefficient of x_0, 105. The minimum solution is

$$x_1 = x_2 = 0, \quad x_3 = 240/105 = 2\frac{2}{7}, \quad y_1 = 180/105 = 1\frac{5}{7},$$

$$y_2 = 180/105 = 1\frac{5}{7}, \quad y_0 = 4140/105 = 39\frac{3}{7}.$$

State: $1\frac{5}{7}$ sacks of soybean meal and $1\frac{5}{7}$ sacks of alfalfa should be used. (The minimum cost is $\$39\frac{3}{7}$.)

21. Familiarize: Let

y_1 = number of type P1 used,

y_2 = number of type P2 used, and

y_0 = total operating cost (in thousands of dollars).

Organize the information in a table.

21. (continued)

	Composition		Number of Passengers
	P1	P2	
Number of Airplanes	y_1	y_2	
x_1: First-class	40	80	2000
x_2: Tourist	40	30	1500
x_3: Economy	120	40	2400
Operating Cost per Airplane (thousands of \$)	\$12	\$10	Objective Function Minimize y_0

Translate: Use the table to write a linear program.

Minimize $y_0 = 12y_1 + 10y_2$

Subject to $40y_1 + 80y_2 \geqslant 2000$

$40y_1 + 30y_2 \geqslant 1500$

$120y_1 + 40y_2 \geqslant 2400$

$y_1, y_2 \geqslant 0$

Carry out: First form the dual.

$40x_1 + 40x_2 + 120x_3 \leqslant 12,$

$80x_1 + 30x_2 + 40x_3 \leqslant 10,$

Max $x_0 = 2000x_1 + 1500x_2 + 2400x_3,$

$x_1, x_2, x_3 \geqslant 0.$

Add as slack variables the structural variables of the primal.

$40x_1 +\ \ \ 40x_2 +\ \ 120x_3 + y_1 \qquad\qquad = 12$

$80x_1 +\ \ \ 30x_2 +\ \ \ 40x_3 \qquad + y_2 \qquad = 10$

$-2000x_1 - 1500x_2 - 2400x_3 \qquad\qquad + x_0 = 0$

Solve using the simplex method.

x_1	x_1	x_3	y_1	y_2	x_0	1	
40	40	120*	1	0	0	12	12/120 = 1/10 ← Min
80	30	40	0	1	0	10	10/40 = 1/4
-2000	-1500	(-2400)	0	0	1	0	$x_0 = 0$
40	40	120	1	0	0	12	12/40 = 3/10
200*	50	0	-1	3	0	18	18/200 = 9/100 ← Min
(-1200)	-700	0	20	0	1	240	$x_0 = 240/1 = 240$
0	150*	600	6	-3	0	42	42/150 = 7/25 ← Min
200	50	0	-1	3	0	18	18/50 = 9/25
0	(-400)	0	14	18	1	348	$x_0 = 348/1 = 348$
0	150	600	6	-3	0	42	
600	0	-600	-9	12	0	12	
0	0	4800	90	30	3	1380	$x_0 = 1380/3 = 460$

All the entries in the bottom row are nonnegative, so the algorithm terminates. To find the minimum solution, divide each entry in the bottom row by the coefficient of x_0, 3. The minimum solution is

$x_1 = x_2 = 0,\ \ x_3 = 4800/3 = 1600,\ \ y_1 = 90/3 = 30,\ \ y_2 = 30/3 = 10,$

$y_0 = 1380/3 = 460.$

<u>21</u>. (continued)

   <u>State</u>: The airline should use 30 airplanes of type P1 and 10 of type P2.
   (The minimum operating cost is $460 thousand.)

<u>23</u>. <u>Familiarize</u>: Let

   y_1 = number of type P1 used,

   y_2 = number of type P3 used, and

   y_0 = total operating cost (in thousands of dollars).

Organize the information in a table.

	Composition		Number of Passengers
	P1	P3	
Number of Airplanes	y_1	y_2	
x_1: First-class	40	40	2000
x_2: Tourist	40	80	1500
x_3: Economy	120	80	2400
Operating Cost per Airplane (thousands of $)	$12	$15	Objective Function Minimize y_0

<u>Translate</u>: Use the table to write a linear program.

Minimize $y_0 = 12y_1 + 15y_2$

Subject to

$$40y_1 + 40y_2 \geqslant 2000$$
$$40y_1 + 80y_2 \geqslant 1500$$
$$120y_1 + 80y_2 \geqslant 2400$$
$$y_1,\ y_2 \geqslant 0$$

<u>Carry out</u>: First form the dual.

$$40x_1 + 40x_2 + 120x_3 \leqslant 12,$$
$$40x_1 + 80x_2 + 80x_3 \leqslant 15,$$

Max $x_0 = 2000x_1 + 1500x_2 + 2400x_3,$

$$x_1,\ x_2,\ x_3 \geqslant 0.$$

Add as slack variables the structural variables of the primal.

$$40x_1 + \quad 40x_2 + 120x_3 + y_1 \qquad\qquad = 12$$
$$40x_1 + \quad 80x_2 + \ 80x_3 \qquad + y_2 \qquad = 15$$
$$-2000x_1 - 1500x_2 - 2400x_3 \qquad\qquad + x_0 = 0$$

Solve using the simplex method.

x_1	x_2	x_3	y_1	y_2	x_0	1	
40	40	120*	1	0	0	12	12/120 = 1/10 ← Min
40	80	80	0	1	0	15	15/80 = 3/16
-2000	-1500	(-2400)	0	0	1	0	$x_0 = 0$
40*	40	120	1	0	0	12	12/40 = 3/10 ← Min
40	160	0	-2	3	0	21	21/40
(-1200)	-700	0	20	0	1	240	$x_0 = 240/1 = 240$
40	40	120	1	0	0	12	
0	120	-120	-3	3	0	9	
0	500	3600	50	0	1	600	$x_0 = 600/1 = 600$

<u>23.</u> (continued)

All the entries in the bottom row are nonnegative, so the algorithm terminates. To find the minimum solution, divide each entry in the bottom row by the coefficient of x_0, 1. The minimum solution is

$$x_1 = 0, \quad x_2 = 500, \quad x_3 = 3600, \quad y_1 = 50, \quad y_2 = 0, \quad y_0 = 600.$$

<u>State:</u> The airline should use 50 airplanes of type P1 and 0 of type P3. (The minimum operating cost is \$600 thousand.)

<u>25.</u> <u>Familiarize:</u> Define the variables as in Exercise 21, and organize the information in a table.

	Composition		Number of Passengers
	P1	P2	
Number of Airplanes	y_1	y_2	
x_1: First-class	40	80	1600
x_2: Tourist	40	30	2100
x_3: Economy	120	40	2400
Operating Cost per Airplane (thousands of \$)	\$12	\$10	Objective Function Minimize y_0

<u>Translate:</u> Use the table to write a linear program.

Minimize $y_0 = 12y_1 + 10y_2$

Subject to

$$40y_1 + 80y_2 \geqslant 1600$$
$$40y_1 + 30y_2 \geqslant 2100$$
$$120y_1 + 40y_2 \geqslant 2400$$
$$y_1, y_2 \geqslant 0$$

<u>Carry out:</u> First form the dual.

$$40x_1 + 40x_2 + 120x_3 \leqslant 12,$$
$$80x_1 + 30x_2 + 40x_3 \leqslant 10,$$

Max $x_0 = 1600x_1 + 2100x_2 + 2400x_3,$

$$x_1, x_2, x_3 \geqslant 0.$$

Add as slack variables the structural variables of the primal.

$$40x_1 + 40x_2 + 120x_3 + y_1 \qquad = 12$$
$$80x_1 + 30x_2 + 40x_3 \qquad + y_2 \qquad = 10$$
$$-1600x_1 - 2100x_2 - 2400x_3 \qquad\qquad + x_0 = 0$$

Solve using the simplex method.

x_1	x_2	x_3	y_1	y_2	x_0	1	
40	40	120*	1	0	0	12	12/120 = 1/10 ← Min
80	30	40	0	1	0	10	10/40 = 1/4
-1600	-2100	(-2400)	0	0	1	0	$x_0 = 0$
40	40*	120	1	0	0	12	12/40 = 0.3 ← Min
200	50	0	-1	3	0	18	18/50 = 0.36
-800	(-1300)	0	20	0	1	240	$x_0 = 240/1 = 240$
40	40	120	1	0	0	12	
600	0	-600	-9	12	0	12	
1000	0	7800	105	0	2	1260	$x_0 = 1260/2 = 630$

<u>25</u>. (continued)

All the entries in the bottom row are nonnegative, so the algorithm
terminates. To find the minimum solution, divide each entry on the bottom row
by the coefficient of x_0, 2. The minimum solution is

$$x_1 = 1000/2 = 500, \quad x_2 = 0, \quad x_3 = 7800/2 = 3900,$$

$$y_1 = 105/2 = 52.5, \quad y_2 = 0, \quad y_0 = 1260/2 = 630.$$

<u>State</u>: The airline should use 52.5 airplanes of type P1 and 0 of type P2.
(The minimum cost is $630 thousand.)

<u>27</u>. <u>Familiarize</u>: Define the variables as in Exercise 23, and organize the
information in a table.

	Composition		Number of Passengers
	P1	P3	
Number of Airplanes	y_1	y_2	
x_1: First-class	40	40	1600
x_2: Tourist	40	80	2100
x_3: Economy	120	80	2400
Operating Cost per Airplane (thousands of $)	$12	$15	Objective Function Minimize y_0

<u>Translate</u>: Use the table to write a linear program.

Minimize $y_0 = 12y_1 + 15y_2$

Subject to

$$40y_1 + 40y_2 \geqslant 1600$$
$$40y_1 + 80y_2 \geqslant 2100$$
$$120y_1 + 80y_2 \geqslant 2400$$
$$y_1, y_2 \geqslant 0$$

<u>Carry out</u>: First form the dual.

$$40x_1 + 40x_2 + 120x_3 \leqslant 12,$$
$$40x_1 + 80x_2 + 80x_3 \leqslant 15,$$

Max $x_0 = 1600x_1 + 2100x_2 + 2400x_3,$

$$x_1, x_2, x_3 \geqslant 0.$$

Add as slack variables the structural variables of the primal.

$$40x_1 + 40x_2 + 120x_3 + y_1 \qquad = 12$$
$$40x_1 + 80x_2 + 80x_3 \qquad + y_2 \quad = 15$$
$$-1600x_1 - 2100x_2 - 2400x_3 \qquad + x_0 = 0$$

27. (continued)

Solve using the simplex method.

x_1	x_2	x_3	y_1	y_2	x_0	1	
40	40	120*	1	0	0	12	$12/120 = 1/10 \leftarrow$ Min
40	80	80	0	1	0	15	$15/80 = 3/16$
-1600	-2100	⟨-2400⟩	0	0	1	0	$x_0 = 0$
40	40	120	1	0	0	12	$12/40 = 3/10$
40	160*	0	-2	3	0	21	$21/160 \leftarrow$ Min
-800	⟨-1300⟩	0	20	0	1	240	$x_0 = 240/1 = 240$
120*	0	480	6	-3	0	27	$27/120 = 9/40 \leftarrow$ Min
40	160	0	-2	3	0	21	$21/40$
⟨-3800⟩	0	0	30	195	8	3285	$x_0 = 3285/8 = 410.625$
120	0	480	6	-3	0	27	
0	480	-480	-12	12	0	36	
0	0	45,600	660	300	24	12,420	$x_0 = 12,420/24 = 517.5$

All the entries in the bottom row are nonnegative, so the algorithm terminates. To find the minimum solution, divide each entry in the bottom row by the coefficient of x_0, 24. The minimum solution is

$x_1 = x_2 = 0$, $x_3 = 45,600/24 = 1900$, $y_1 = 660/24 = 27.5$,

$y_2 = 300/24 = 12.5$, $y_0 = 12,420/24 = 517.5$.

State: The airline should use 27.5 airplanes of type P1 and 12.5 of type P3. (The minimum cost is $517.5 thousand.)

29. Familiarize: Let

y_1 = number of days Refinery JR should be operated,

y_2 = number of days Refinery Bobby should be operated, and

y_0 = total product cost.

Organize the information in a table.

	Composition		Number of Barrels
	Refinery JR	Refinery Bobby	
Number of Days	y_1	y_2	
x_1: Low-grade oil	6000	3000	30,000
x_2: High-grade oil	2500	5000	20,000
Operating Cost per Day	$20,000	$15,000	Objective Function Minimize y_0

Translate: Use the table to write a linear program.

Minimize $y_0 = 20,000y_1 + 15,000y_2$

Subject to

$6000y_1 + 3000y_2 \geqslant 30,000$

$2500y_1 + 5000y_2 \geqslant 20,000$

$y_1, y_2 \geqslant 0$

29. (continued)

    <u>Carry out</u>: First form the dual.

$$6000x_1 + 2500x_2 \leqslant 20{,}000,$$
$$3000x_1 + 5000x_2 \leqslant 15{,}000,$$

Max $x_0 = 30{,}000x_1 + 20{,}000x_2$,

$$x_1, x_2 \geqslant 0.$$

Add as slack variables the structural variables of the primal.

$$6000x_1 + 2500x_2 + y_1 \qquad\qquad = 20{,}000$$
$$3000x_1 + 5000x_2 \qquad + y_2 \qquad = 15{,}000$$
$$-30{,}000x_1 - 20{,}000x_2 \qquad\qquad + x_0 = \qquad 0$$

Solve using the simplex method.

x_1	x_2	y_1	y_2	x_0	1	
6000*	2500	1	0	0	20,000	$20{,}000/6000 = 3\frac{1}{3}$ ← Min
3000	5000	0	1	0	15,000	$15{,}000/3000 = 5$
⟨-30,000⟩	-20,000	0	0	1	0	$x_0 = 0$
6000	2500	1	0	0	20,000	$20{,}000/2500 = 8$
0	7500*	-1	2	0	10,000	$10{,}000/7500 = 1\frac{1}{3}$ ← Min
0	⟨-7500⟩	5	0	1	100,000	$x_0 = 100{,}000/1 = 100{,}000$
18,000	0	4	-2	0	50,000	
0	7500	-1	2	0	10,000	
0	0	4	2	1	110,000	$x_0 = 110{,}000/1 = 110{,}000$

All the entries in the bottom row are nonnegative, so the algorithm terminates. To find the minimum solution, divide each entry in the bottom row by the coefficient of x_0, 1. The minimum solution is

$$x_1 = x_2 = 0, \quad y_1 = 4, \quad y_2 = 2, \quad y_0 = 110{,}000.$$

<u>State</u>: Refinery JR should be operated 4 days, and Refinery Bobby should be operated 2 days.

31. x_1 = price per pound of protein, x_2 = price per pound of fat, and x_3 = price per pound of mineral ash. The values of these variables in the solution of the dual are the prices at which income is maximized while the animal's minimum nutritional requirements are satisfied.

33. x_1 = cost per barrel of low-grade oil, and x_2 = cost per barrel of high-grade oil.

35. <u>Familiarize</u>: Let

    y_1 = amount of eggs (in hundreds of grams),

    y_2 = amount of beef (in hundreds of grams),

    y_3 = amount of cheese (in hundreds of grams),

    y_4 = amount of soybeans (in hundreds of grams),

and y_0 = total cost of diet ($).

Organize the information in a table.

35. (continued)

	Composition				Minimum Daily Requirement
	Eggs	Beef	Cheese	Soybeans	
Amount in Diet	y_1	y_2	y_3	y_4	
Calcium	54	7	570	260	750
Iron	2.7	6.6	0.7	10.0	10
Protein	12.8	18.6	20.5	34.9	50
Cost per 100 g ($)	0.22	0.36	0.32	0.10	Objective Function Minimize y_0

Translate: Use the table to write a linear program.

Minimize $y_0 = 0.22y_1 + 0.36y_2 + 0.32y_3 + 0.10y_4$

Subject to

$$54y_1 + 7y_2 + 570y_3 + 260y_4 \geqslant 750$$
$$2.7y_1 + 6.6y_2 + 0.7y_3 + 10.0y_4 \geqslant 10$$
$$12.8y_1 + 18.6y_2 + 20.5y_3 + 34.9y_4 \geqslant 50$$
$$y_1, y_2, y_3, y_4 \geqslant 0$$

Carry out: First form the dual.

$$54x_1 + 2.7x_2 + 12.8x_3 \leqslant 0.22,$$
$$7x_1 + 6.6x_2 + 18.6x_3 \leqslant 0.36,$$
$$570x_1 + 0.7x_2 + 20.5x_3 \leqslant 0.32,$$
$$260x_1 + 10.0x_2 + 34.9x_3 \leqslant 0.10,$$

Max $x_0 = 750x_1 + 10x_2 + 50x_3,$

$$x_1, x_2, x_3 \geqslant 0.$$

Add as slack variables the structural constraints of the primal.

$$54x_1 + 2.7x_2 + 12.8x_3 + y_1 = 0.22$$
$$7x_1 + 6.6x_2 + 18.6x_3 + y_2 = 0.36$$
$$570x_1 + 0.7x_2 + 20.5x_3 + y_3 = 0.32$$
$$260x_1 + 10.0x_2 + 34.9x_3 + y_4 = 0.10$$
$$-750x_1 - 10x_2 - 50x_3 + x_0 = 0$$

Solve using the simplex method.

x_1	x_2	x_3	y_1	y_2	y_3	y_4	x_0	1	
54	2.7	12.8	1	0	0	0	0	0.22	$0.22/54 \approx 0.0041$
7	6.6	18.6	0	1	0	0	0	0.36	$0.36/7 \approx 0.0514$
570	0.7	20.5	0	0	1	0	0	0.32	$0.32/570 \approx 0.0006$
260*	10.0	34.9	0	0	0	1	0	0.10	$0.10/260 \approx 0.0004 \leftarrow$ Min
(-750)	-10	-50	0	0	0	0	1	0	$x_0 = 0$
0	81	721.7	130	0	0	-27	0	25.9	
0	1646	4591.7	0	260	0	-7	0	92.9	
0	-551.8	1456.3	0	0	26	-57	0	2.62	
260	10.0	34.9	0	0	0	1	0	0.10	
0	490	1317.5	0	0	0	75	26	7.5	$x_0 = 7.5/26$

35. (continued)

All the entries in the bottom row are nonnegative, so the algorithm terminates. To find the minimum solution, divide each entry in the bottom row by the coefficient of x_0, 26. The minimum solution is

$x_1 = 0$, $x_2 \approx 18.8$, $x_3 \approx 50.7$, $y_1 = y_2 = y_3 = 0$, $y_4 \approx 2.88$,
$y_0 \approx 0.288$.

State: The minimum cost diet consists of 0 g of eggs, beef, and cheese and about 288 g (2.88×100 g) of soybeans and costs about $0.29.

37. Familiarize: Let

y_1 = amount of chicken (in hundreds of grams),

y_2 = amount of filberts (in hundreds of grams),

y_3 = amount of cashews (in hundreds of grams),

and y_0 = total cost of diet ($).

Organize the information in a table.

	Composition			Minimum Daily Requirement
	Chicken	Filberts	Cashews	
Amount in Diet	y_1	y_2	y_3	
Calcium	14	209	38	750
Iron	1.5	3.4	3.8	10
Protein	20.2	12.6	17.2	50
Cost per 100 g ($)	0.16	0.40	0.48	Objective Function Minimize y_0

Translate: Use the table to write a linear program.

Minimize $y_0 = 0.16y_1 + 0.40y_2 + 0.48y_3$

Subject to

$$14y_1 + 209y_2 + 38y_3 \geqslant 750$$
$$1.5y_1 + 3.4y_2 + 3.8y_3 \geqslant 10$$
$$20.2y_1 + 12.6y_2 + 17.2y_3 \geqslant 50$$
$$y_1, y_2, y_3 \geqslant 0$$

Carry out: First form the dual.

$$14x_1 + 1.5x_2 + 20.2x_3 \leqslant 0.16,$$
$$209x_1 + 3.4x_2 + 12.6x_3 \leqslant 0.40,$$
$$38x_1 + 3.8x_2 + 17.2x_3 \leqslant 0.48,$$

Max $x_0 = 750x_1 + 10x_2 + 50x_3$,

$$x_1, x_2, x_3 \geqslant 0.$$

Add as slack variables the structural variables of the primal.

$$14x_1 + 1.5x_2 + 20.2x_3 + y_1 \qquad\qquad = 0.16$$
$$209x_1 + 3.4x_2 + 12.6x_3 \qquad + y_2 \qquad = 0.40$$
$$38x_1 + 3.8x_2 + 17.2x_3 \qquad\qquad + y_3 \qquad = 0.48$$
$$-750x_1 - 10x_2 - 50x_3 \qquad\qquad\qquad + x_0 = \quad 0$$

Solve using the simplex method. Because of space limitations, we show the quotient column below the tableau.

<u>37</u>. (continued)

x_1	x_2	x_3	y_1	y_2	y_3	x_0	1
14	1.5	20.2	1	0	0	0	0.16
209*	3.4	12.6	0	1	0	0	0.40
38	3.8	17.2	0	0	1	0	0.48
(−750)	−10	−50	0	0	0	1	0
0	265.9	4045.4*	209	−14	0	0	27.84
209	3.4	12.6	0	1	0	0	0.40
0	665	3116	0	−38	209	0	85.12
0	460	(−1000)	0	750	0	209	300
0	265.9	4045.4	209	−14	0	0	27.84
845,488.6	10,404.02	0	−2633.4	4221.8	0	0	1267.376
0	1,861,646.6	0	−651,244	−110,101.2	845,488.6	0	257,595.008
0	2,126,784	0	209,000	3,020,050	0	845,488.6	1,241,460

<u>First tableau, q</u>

$0.16/14 \approx 0.0114$

$0.40/209 \approx 0.0019 \leftarrow$ Min

$0.48/38 \approx 0.0126$

<u>Second tableau, q</u>

$27.84/4045.4 \approx 0.0069 \leftarrow$ Min

$0.40/12.6 \approx 0.0317$

$85.12/3116 \approx 0.02732$

All the entries in the bottom row are nonnegative, so the algorithm terminates. To find the minimum solution, divide each of the entries on the bottom row by the coefficient of x_0, 845,488.6. The minimum solution is

$x_1 = 0, \quad x_2 \approx 2.5, \quad x_3 = 0, \quad y_1 \approx 0.25, \quad y_2 \approx 3.6, \quad y_3 = 0, \quad y_0 \approx 1.47.$

<u>State</u>: The minimum cost diet consists of approximately 25 g (0.25×100 g) of chicken, 360 g (3.6×100 g) of filberts, and 0 g of cashews and costs approximately \$1.47.

<u>39</u>. <u>Familiarize</u>: Let

y_1 = amount of whole wheat bread (in hundreds of grams),

y_2 = amount of sesame seeds (in hundreds of grams),

y_3 = amount of almonds (in hundreds of grams),

y_4 = amount of chicken (in hundreds of grams),

and y_0 = total cost of diet (\$).

Organize the information in a table.

	Composition				Minimum Daily Requirement
	Bread	Sesame Seeds	Almonds	Chicken	
Amount in Diet	y_1	y_2	y_3	y_4	
Calcium	96	72	234	14	750
Iron	2.2	7.7	4.7	1.5	10
Protein	9.3	23.4	18.6	20.2	50
Cost per 100 g (\$)	0.10	0.18	0.48	0.16	Objective Function Minimize y_0

39. (continued)

Translate: Use the table to write a linear program.

Minimize $y_0 = 0.10y_1 + 0.18y_2 + 0.48y_3 + 0.16y_4$

Subject to

$$96y_1 + 72y_2 + 234y_3 + 14y_4 \geqslant 750$$
$$2.2y_1 + 7.7y_2 + 4.7y_3 + 1.5y_4 \geqslant 10$$
$$9.3y_1 + 23.4y_2 + 18.6y_3 + 20.2y_4 \geqslant 50$$
$$y_1, \; y_2, \; y_3, \; y_4 \geqslant 0$$

Carry out: First form the dual.

$$96x_1 + 2.2x_2 + 9.3x_3 \leqslant 0.10,$$
$$72x_1 + 7.7x_2 + 23.4x_3 \leqslant 0.18,$$
$$234x_1 + 4.7x_2 + 18.6x_3 \leqslant 0.48,$$
$$14x_1 + 1.5x_2 + 20.2x_3 \leqslant 0.16,$$
$$\text{Max } x_0 = 750x_1 + 10x_2 + 50x_3,$$
$$x_1, \; x_2, \; x_3 \geqslant 0.$$

Add as slack variables the structural variables of the primal.

$$96x_1 + 2.2x_2 + 9.3x_3 + y_1 \qquad\qquad\qquad = 0.10$$
$$72x_1 + 7.7x_2 + 23.4x_3 \qquad + y_2 \qquad\qquad = 0.18$$
$$234x_1 + 4.7x_2 + 18.6x_3 \qquad\qquad + y_3 \qquad = 0.48$$
$$14x_1 + 1.5x_2 + 20.2x_3 \qquad\qquad\qquad + y_4 \qquad = 0.16$$
$$-750x_1 - 10x_2 - 50x_3 \qquad\qquad\qquad\qquad + x_0 = 0$$

Solve using the simplex method.

x_1	x_2	x_3	y_1	y_2	y_3	y_4	x_0	1	
96*	2.2	9.3	1	0	0	0	0	0.10	$0.10/96 \approx 0.0010 \leftarrow$ Min
72	7.7	23.4	0	1	0	0	0	0.18	$0.18/72 = 0.0025$
234	4.7	18.6	0	0	1	0	0	0.48	$0.48/234 \approx 0.0021$
14	1.5	20.2	0	0	0	1	0	0.16	$0.16/14 \approx 0.0114$
(−750)	−10	−50	0	0	0	0	1	0	$x_0 = 0$
96	2.2	9.3	1	0	0	0	0	0.10	
0	24.2	65.7	−3	4	0	0	0	0.42	
0	−10.6	−65.1	−39	0	16	0	0	3.78	
0	56.6	904.5	−7	0	0	48	0	6.98	
0	115	362.5	125	0	0	0	16	12.5	$x_0 = 12.5/16$

All the entries in the bottom row are nonnegative, so the algorithm
terminates. To find the minimum solution, divide each entry in the bottom row
by the coefficient of x_0, 16. The minimum solution is

$$x_1 = 0, \quad x_2 = 7.1875, \quad x_3 = 22.65625, \quad y_1 = 7.8125,$$
$$y_2 = y_3 = y_4 = 0, \quad y_0 = 0.78125.$$

State: The minimum cost diet consists of approximately 781 g (7.81 × 100 g)
of whole-wheat bread and 0 g of sesame seeds, almonds, and chicken and costs
approximately \$0.78.

Exercise Set **4.4**

<u>1.</u> Minimize Z = 3u + 4v

Subject to

$$3u + v \geqslant 9$$
$$u + v \geqslant 7$$
$$u + 2v \geqslant 8$$
$$u, v \geqslant 0$$

We must first convert the program to a standard maximum-type program. The nonnegativity constraints are satisfied. Multiplying the objective function and the structural constraints by -1 we obtain

Maximize -Z = -3u - 4v

Subject to

$$-3u - v \leqslant -9$$
$$-u - v \leqslant -7$$
$$-u - 2v \leqslant -8$$
$$u, v \geqslant 0.$$

Add nonnegative slack variables.

$$-3u - v + w \qquad\qquad = -9,$$
$$-u - v \qquad + x \qquad\quad = -7,$$
$$-u - 2v \qquad\quad + y \quad = -8,$$
$$3u + 4v \qquad\qquad - Z = 0.$$

Set up the initial simplex tableau.

u	v	w	x	y	-Z		1	q
⊘-3	-1	1	0	0	0		⊘-9	-9/-3 = 3
-1	-1	0	1	0	0		-7	-7/-1 = 7
-1	-2	0	0	1	0		-8	-8/-1 = 8 ← Max
3	4	0	0	0	1		0	

The uppermost negative entry in the right-hand column is -9 in the first row. The first row becomes the "pseudo"-objective function row. The most negative entry in the first row is -3 in the first column, so the first column becomes the pivot column. There are no positive/positive quotients, so we seek the maximum of the negative/negative quotients. It is 8 in the third row, so the pivot element is -1. Since this pivot is negative we multiply the pivot row by -1 before pivoting. We obtain the second tableau. Pivoting we obtain the third tableau.

u	v	w	x	y	-Z		1	q
-3	-1	1	0	0	0		-9	
-1	-1	0	1	0	0		-7	
1*	2	0	0	-1	0		8	
3	4	0	0	0	1		0	
0	5	1	0	-3	0		15	15/5 = 3
0	1*	0	1	-1	0		1	1/1 = 1 ← Min
1	2	0	0	-1	0		8	8/2 = 4
0	⊘-2	0	0	3	1		-24	

There are no more negative entries in the right-hand column. (The bottom objective function entry is not considered.) We move on using the simplex algorithm.

<u>1</u>. (continued)

u	v	w	x	y	-Z		1
0	0	1	-5	2	0		10
0	1	0	1	-1	0		1
1	0	0	-2	1	0		6
0	0	0	2	1	1		-22

The algorithm terminates. We get $x = y = 0$, $w = 10/1 = 10$, $v = 1/1 = 1$, $u = 6/1 = 6$, $-Z = -22/1 = -22$. To solve the original problem we change "Max $-Z = -22$" to "Min $Z = 22$" by multiplying by -1. Thus, the solution to the original problem is

   Min $= 22$ at $u = 6$, $v = 1$.

<u>3</u>. Minimize $A = 2a + 5b$

Subject to

$$2a + b \geqslant 9$$
$$4a + 3b \geqslant 23$$
$$a + 3b \geqslant 8$$
$$a, b \geqslant 0$$

Convert the program to a standard maximum-type program. The nonnegativity constraints are satisfied. Multiplying the objective function and the structural constraints by -1 we obtain

Maximize $-A = -2a - 5b$

Subject to

$$-2a - b \leqslant -9$$
$$-4a - 3b \leqslant -23$$
$$-a - 3b \leqslant -8$$
$$a, b \geqslant 0.$$

Add nonnegative slack variables.

$$-2a - b + u \qquad\qquad = -9,$$
$$-4a - 3b \qquad + v \qquad = -23,$$
$$-a - 3b \qquad\qquad + w \quad = -8,$$
$$2a + 5b \qquad\qquad\qquad - A = 0.$$

Set up the initial simplex tableau.

a	b	u	v	w	-A		1		q
⊘-2	-1	1	0	0	0		⊘-9		-9/-2 = 4.5
-4	-3	0	1	0	0		-23		-23/-4 = 5.75
-1	-3	0	0	1	0		-8		-8/-1 = 8 ← Max
2	5	0	0	0	1		0		

The uppermost negative entry in the right-hand column is -9 in the first row, so that row becomes the "pseudo"-objective function row. The most negative entry in the first row is -2 in the first column, so the first column becomes the pivot column. Since there are no positive-positive quotients we seek the maximum negative/negative quotient. It is 8 in the third row, so the pivot element is -1. Since this pivot is negative we multiply the pivot row by -1 before pivoting. We obtain the second tableau. Pivoting we obtain the third tableau.

<u>3</u>. (continued)

a	b	u	v	w	-A		1	q
-2	-1	1	0	0	0		-9	
-4	-3	0	1	0	0		-23	
1*	3	0	0	-1	0		8	
2	5	0	0	0	1		0	
0	5	1	0	-2	0		7	$7/5 = 1\frac{2}{5}$
0	9*	0	1	-4	0		9	$9/9 = 1 \leftarrow$ Min
1	3	0	0	-1	0		8	$8/3 = 2\frac{2}{3}$
0	(-1)	0	0	2	1		-16	

Since there are no more negative entries in the right-hand column (the bottom objective function entry is not considered), we move on using the simplex algorithm.

a	b	u	v	w	-A		1
0	0	9	-5	2	0		18
0	9	0	1	-4	0		9
3	0	0	-1	1	0		15
0	0	0	1	14	9		-135

The algorithm terminates. We have v = w = 0, u = 18/9 = 2, b = 9/9 = 1, a = 15/3 = 5, -A = -135/9 = -15. To solve the original problem we multiply by -1 to change "Max -A = -15" to "Min A = 15." The solution is

Min A = 15 at a = 5, b = 1.

<u>5</u>. Minimize W = 3u + 5v

Subject to

$$4u + v \geqslant 9$$
$$3u + 2v \geqslant 13$$
$$2u + 5v \geqslant 16$$
$$u, v \geqslant 0$$

Convert to a standard maximum-type program:

Maximize -W = -3u - 5v

Subject to

$$-4u - v \leqslant -9$$
$$-3u - 2v \leqslant -13$$
$$-2u - 5v \leqslant -16$$
$$u, v \geqslant 0$$

Add nonnegative slack variables.

$$-4u - v + w \qquad\qquad = -9,$$
$$-3u - 2v \qquad + x \qquad\quad = -13,$$
$$-2u - 5v \qquad\qquad + y \quad = -16,$$
$$3u + 5v \qquad\qquad\quad - W = \quad 0.$$

Set up the initial simplex tableau.

u	v	w	x	y	-W		1	q
(-4)	-1	1	0	0	0		(-9)	$-9/-4 = 2\frac{1}{4}$
-3	-2	0	1	0	0		-13	$-13/-3 = 4\frac{1}{3}$
-2	-5	0	0	1	0		-16	$-16/-2 = 8 \leftarrow$ Max
3	5	0	0	0	1		0	

<u>5</u>. (continued)

Row 1 is the "pseudo"-objective function row, and the first column is the
pivot column. There are no positive/positive quotients, so we seek the
maximum negative/negative quotient. The pivot element is -2. We multiply the
pivot row by -1 before pivoting.

u	v	w	x	y	-W		1
-4	-1	1	0	0	0		-9
-3	-2	0	1	0	0		-13
2*	5	0	0	-1	0		16
3	5	0	0	0	1		0

u	v	w	x	y	-W		1	
0	9	1	0	-2	0		23	$23/9 = 2\frac{5}{9}$
0	11*	0	2	-3	0		22	$22/11 = 2 \leftarrow$ Min
2	5	0	0	-1	0		16	$16/5 = 3\frac{1}{5}$
0	(-5)	0	0	3	2		-48	

There are no more negative entries in the right-hand column (we do not
consider the bottom objective function entry), so we continue using the
simplex algorithm.

u	v	w	x	y	-W		1
0	0	11	-18	5	0		55
0	11	0	2	-3	0		22
22	0	0	-10	4	0		66
0	0	0	10	18	22		-418

The algorithm terminates. We have $x = y = 0$, $w = 55/11 = 5$, $v = 22/11 = 2$,
$u = 66/22 = 3$, $-W = -418/22 = -19$. Multiply by -1 to change "Max $-W = -19$" to
"Min $W = 19$." Then the solution is

Min $W = 19$ at $u = 3$, $v = 2$.

<u>7</u>. Minimize $G = 14x + 6y$

Subject to

$3x + 2y \leqslant 12$

$7x + 5y \leqslant 29$

$x, y \geqslant 0$

Convert to a standard maximum-type program by multiplying the objective
function by -1.

Maximize $-G = -14x - 6y$

Subject to

$3x + 2y \leqslant 12$

$7x + 5y \leqslant 29$

$x, y \geqslant 0$

Add nonnegative slack variables.

$3x + 2y + u \qquad\qquad = 12,$

$7x + 5y \qquad + v \qquad = 29,$

$14x + 6y \qquad\qquad - G = 0.$

Set up the initial simplex tableau.

7. (continued)

x	y	u	v	-G		1
3	2	1	0	0		12
7	5	0	1	0		29
14	6	0	0	1		0

There are no negative entries in the right-hand column or in the bottom row. Therefore, the algorithm terminates, and we have $x = y = 0$, $u = 12/1 = 12$, $v = 29/1 = 29$, $-G = 0$. Multiply by -1 to change "Max -G = 0" to "Min G = 0." The solution is

   Min G = 0 at $x = 0$, $y = 0$.

9. Minimize $F = x + y$

   Subject to

   $y - x \leqslant 1$

   $y - x \leqslant 3$

   $ x \leqslant 5$

   $ x \geqslant 2$

   $x, y \geqslant 0$

Change to a standard maximum-type program by multiplying the objective function and the inequality $x \geqslant 2$ by -1. We will also write the first two structural constraints with the x-term first.

Maximize $-F = -x - y$

   Subject to

   $-x + y \leqslant 1$

   $-x + y \leqslant 3$

   $x \leqslant 5$

   $-x \leqslant -2$

   $x, y \geqslant 0$

Add nonnegative slack variables.

$$-x + y + u = 1,$$
$$-x + y + v = 3,$$
$$x + w = 5,$$
$$-x + z = -2,$$
$$x + y - F = 0.$$

Set up the initial simplex tableau.

x	y	u	v	w	z	-F		1
-1	1	1	0	0	0	0		1
-1	1	0	1	0	0	0		3
1*	0	0	0	1	0	0		5
(-1)	0	0	0	0	1	0		(-2)
1	1	0	0	0	0	1		0

The fourth row is the "pseudo"-objective function row, and the pivot column is the first column. The only positive/positive quotient is $5/1 = 5$ in the third row, so the pivot element is 1. Carry out the pivoting.

9. (continued)

x	y	u	v	w	z	-F		1	q
0	1	1	0	1	0	0		6	6/1 = 6
0	1	0	1	1	0	0		8	8/1 = 8
1	0	0	0	1	0	0		5	5/1 = 5
0	0	0	0	1*	1	0		3	3/1 = 3 ← Min
0	1	0	0	-1	0	1		-5	

Continue using the simplex algorithm.

x	y	u	v	w	z	-F		1
0	1	1	0	0	-1	0		3
0	1	0	1	0	-1	0		5
1	0	0	0	0	-1	0		2
0	0	0	0	1	1	0		3
0	1	0	0	0	1	1		-2

The algorithm terminates. We have $y = z = 0$, $u = 3$, $v = 5$, $x = 2$, $w = 3$, $-F = -2$. Multiply by -1 to change "Max $-F = -2$" to "Min $F = 2$." Then the solution is

Min $F = 2$ at $x = 2$, $y = 0$.

11.
$$x + 2y \leqslant 14,$$
$$4x + 3y \leqslant 26,$$
$$2x + y \leqslant 12,$$
$$3x + 4y \geqslant 12,$$
Max $f = 3x + 4y$; $x, y \geqslant 0$.

Change to a standard maximum-type program by multiplying the inequality $3x + 4y \geqslant 12$ by -1. It becomes $-3x - 4y \leqslant -12$. Add nonnegative slack variables.

$$x + 2y + u \qquad\qquad = 14,$$
$$4x + 3y \quad + v \qquad\quad = 26,$$
$$2x + y \qquad + w \qquad = 12,$$
$$-3x - 4y \qquad\quad + z \quad = -12,$$
$$-3x - 4y \qquad\qquad + f = 0,$$
$$x, y, u, v, w, z \geqslant 0.$$

Set up the initial simplex tableau.

x	y	u	v	w	z	f		1	q
1	2*	1	0	0	0	0		14	14/2 = 7 ← Min
4	3	0	1	0	0	0		26	26/3 = $8\frac{2}{3}$
2	1	0	0	1	0	0		12	12/1 = 12
-9	-4	0	0	0	1	0		-12	
-3	-4	0	0	0	0	1		0	

The fourth row is the "pseudo"-objective function row, and the second column is the pivot column. The minimum positive-positive quotient occurs in the first row, so the pivot element is 2. Carry out the pivoting.

11. (continued)

x	y	u	v	w	z	f		1	q
1	2	1	0	0	0	0		14	14/1 = 14
5*	0	-3	2	0	0	0		10	10/5 = 2 ← Min
3	0	-1	0	2	0	0		10	10/3 = $3\frac{1}{3}$
-1	0	2	0	0	1	0		16	—
⊝1	0	2	0	0	0	1		28	

Continue using the simplex algorithm.

x	y	u	v	w	z	f		1
0	10	8	-2	0	0	0		60
5	0	-3	2	0	0	0		10
0	0	4	-6	10	0	0		20
0	0	7	2	0	5	0		90
0	0	7	2	0	0	5		150

The algorithm terminates. We have $u = v = 0$, $y = 60/10 = 6$, $x = 10/5 = 2$, $w = 20/10 = 2$, $z = 90/5 = 18$, $f = 150/5 = 30$. The solution is

Max $f = 30$ at $x = 2$, $y = 6$.

13. $5x + 6y \leqslant 60$,

         $x + y \leqslant 11$,

        $3x + y \leqslant 27$,

        $5x + 2y \geqslant 10$,

        $3x + 2y \geqslant 6$,

Max $f = 2x + y$; x, $y \geqslant 0$.

Change to a standard minimum-type program by multiplying the fourth and fifth inequalities by -1. They become $-5x - 2y \leqslant -10$ and $-3x - 2y \leqslant -6$. Add nonnegative slack variables

      $5x + 6y + s$ $= 60$,

       $x + y$ $+ t$ $= 11$,

      $3x + y$ $+ u$ $= 27$,

     $-5x - 2y$ $+ v$ $= -10$,

     $-3x - 2y$ $+ w$ $= -6$,

     $-2x - y$ $+ f =$ 0,

              x, y, s, t, u, v, $w \geqslant 0$.

Set up the initial simplex tableau, determine the pivot element, and pivot.

13. (continued)

x	y	s	t	u	v	w	f	1	q
5	6	1	0	0	0	0	0	60	60/5 = 12
1	1	0	1	0	0	0	0	11	11/1 = 11
3*	1	0	0	1	0	0	0	27	27/3 = 9 ← Min
(-5)	-2	0	0	0	1	0	0	(-10)	—
-3	-2	0	0	0	0	1	0	-6	—
-2	-1	0	0	0	0	0	1	0	
0	13	3	0	-5	0	0	0	45	$45/13 = 3\frac{6}{13}$
0	2*	0	3	-1	0	0	0	6	6/2 = 3 ← Min
3	1	0	0	1	0	0	0	27	27/1 = 27
0	-1	0	0	5	3	0	0	105	—
0	-1	0	0	1	0	1	0	21	—
0	(-1)	0	0	2	0	0	3	54	

Continue using the simplex method.

x	y	s	t	u	v	w	f	1
0	0	6	-39	3	0	0	0	12
0	2	0	3	-1	0	0	0	6
6	0	0	-3	3	0	0	0	48
0	0	0	3	9	6	0	0	216
0	0	0	3	1	0	2	0	48
0	0	0	3	3	0	0	6	114

The algorithm terminates. We have t = u = 0, s = 12/6 = 2, y = 6/2 = 3,
x = 48/6 = 8, v = 216/6 = 36, w = 48/2 = 24, f = 114/6 = 19. The solution is

   Max f = 19 at x = 8, y = 3.

15. $3x + 4y \leqslant 48,$
              $x + y \leqslant 13,$
            $2x + y \leqslant 22,$
            $5x + 3y \geqslant 30,$
            $2x + 7y = 28,$
Max f = 7x + 4y; x, y $\geqslant$ 0.

Convert to a standard maximum-type solution by first multiplying the fourth
inequality by -1 to get -5x - 3y $\leqslant$ -30. Then replace the equality constraint
2x + 7y = 28 by two inequality constraints 2x + 7y $\leqslant$ 28 and 2x + 7y $\geqslant$ 28.
Finally, multiply the latter by -1 to get -2x - 7y $\leqslant$ -28. We have

        $3x + 4y \leqslant 48,$
          $x + y \leqslant 13,$
         $2x + y \leqslant 22,$
       $-5x - 3y \leqslant -30,$
        $2x + 7y \leqslant 28,$
       $-2x - 7y \leqslant -28,$
Max f = 7x + 4y; x, y $\geqslant$ 0.

15. (continued)

Add nonnegative slack variables.

$$3x + 4y + r = 48,$$
$$x + y + s = 13,$$
$$2x + y + t = 22,$$
$$-5x - 3y + u = -30,$$
$$2x + 7y + v = 28,$$
$$-2x - 7y + w = -28,$$
$$-7x - 4y + f = 0,$$
$$x, y, r, s, t, u, v, w \geqslant 0.$$

Set up the initial simplex tableau, determine the pivot element and pivot.

x	y	r	s	t	u	v	w	f	1	q
3	4	1	0	0	0	0	0	0	48	48/3 = 16
1	1	0	1	0	0	0	0	0	13	13/1 = 13
2*	1	0	0	1	0	0	0	0	22	22/2 = 11 ← Min
(-5)	-3	0	0	0	1	0	0	0	(-30)	—
2	7	0	0	0	0	1	0	0	28	28/2 = 14
-2	-7	0	0	0	0	0	1	0	-28	—
-7	-4	0	0	0	0	0	0	1	0	
0	5	2	0	-3	0	0	0	0	30	30/5 = 6
0	1	0	2	-1	0	0	0	0	4	4/1 = 4
2	1	0	0	1	0	0	0	0	22	22/1 = 22
0	-1	0	0	5	2	0	0	0	50	—
0	6*	0	0	-1	0	1	0	0	6	6/6 = 1 ← Min
0	(-6)	0	0	1	0	0	1	0	(-6)	—
0	-1	0	0	7	0	0	0	2	154	
0	0	12	0	-13	0	-5	0	0	150	
0	0	0	12	-5	0	-1	0	0	18	
12	0	0	0	7	0	-1	0	0	126	
0	0	0	0	29	12	1	0	0	306	
0	6	0	0	-1	0	1	0	0	6	
0	0	0	0	0	0	1	1	0	0	
0	0	0	0	41	0	1	0	12	930	

The algorithm terminates. We have $t = v = 0$, $r = 150/12 = 25/2$, $s = 18/12 = 3/2$, $x = 126/12 = 21/2$, $u = 306/12 = 51/2$, $y = 6/6 = 1$, $w = 0/1 = 0$, $f = 930/12 = 155/2$. The solution is

Max $f = \dfrac{155}{2}$ at $x = \dfrac{21}{2}$, $y = 1$.

17. $$3x - 2y + z \leqslant 8,$$
$$-4x + 3y - z \leqslant 4,$$
$$2x - 3y - 6z \leqslant 6,$$
$$x - y + z \geqslant 1,$$
$$x + y + z = 5,$$

Max $f = -2y + 5z$; $x, y \geqslant 0$.

Convert to a standard maximum-type program by first multiplying the fourth inequality by -1 to get $-x + y - z \leqslant -1$. Then replace the equality constraint $x + y + z = 5$ by two inequality constraints $x + y + z \leqslant 5$ and $x + y + z \geqslant 5$. Finally, multiply the latter by -1 to get $-x - y - z \leqslant -5$.

17. (continued)

We have

$$3x - 2y + z \leq 8,$$
$$-4x + 3y - z \leq 4,$$
$$2x - 3y - 6z \leq 6,$$
$$-x + y - z \leq -1,$$
$$x + y + z \leq 5,$$
$$-x - y - z \leq -5,$$

Max $f = -2y + 5z$; $x, y, z \geq 0$.

Add nonnegative slack variables.

$$3x - 2y + z + r = 8,$$
$$-4x + 3y - z + s = 4,$$
$$2x - 3y - 6z + t = 6,$$
$$-x + y - z + u = -1,$$
$$x + y + z + v = 5,$$
$$-x - y - z + w = -5,$$
$$2y - 5z + f = 0,$$
$$x, y, z, r, s, t, u, v, w, \geq 0.$$

Set up the initial simplex tableau, determine the pivot element, and pivot.

x	y	z	r	s	t	u	v	w	f	1	q
3*	-2	1	1	0	0	0	0	0	0	8	$8/3 = 2\frac{2}{3}$ ← Min
-4	3	-1	0	1	0	0	0	0	0	4	—
2	-3	-6	0	0	1	0	0	0	0	6	$6/2 = 3$
(-1)	1	-1	0	0	0	1	0	0	0	(-1)	—
1	1	1	0	0	0	0	1	0	0	5	$5/1 = 5$
-1	-1	-1	0	0	0	0	0	1	0	-5	
0	2	-5	0	0	0	0	0	0	1	0	
3	-2	1	1	0	0	0	0	0	0	8	—
0	1	1	4	3	0	0	0	0	0	44	$44/1 = 44$
0	-5	-20	-2	0	3	0	0	0	0	2	—
0	1	-2	1	0	0	3	0	0	0	5	$5/1 = 5$
0	5*	2	-1	0	0	0	3	0	0	7	$7/5 = 1.4$ ← Min
0	(-5)	-2	1	0	0	0	0	3	0	(-7)	—
0	2	-5	0	0	0	0	0	0	1	0	
15	0	9	3	0	0	0	6	0	0	54	$54/9 = 6$
0	0	3	21	15	0	0	-3	0	0	213	$213/3 = 71$
0	0	-18	-3	0	3	0	3	0	0	9	—
0	0	-12	6	0	0	15	-3	0	0	18	—
0	5	2*	-1	0	0	0	3	0	0	7	$7/2 = 3.5$ ← Min
0	0	0	0	0	0	0	3	3	0	0	
0	0	(-29)	2	0	0	0	-6	0	5	-14	
30	-45	0	15*	0	0	0	-15	0	0	45	$45/15 = 3$ ← Min
0	-15	0	45	30	0	0	-15	0	0	405	$405/45 = 9$
0	45	0	-12	0	3	0	30	0	0	72	—
0	30	0	0	0	0	15	15	0	0	60	—
0	5	2	-1	0	0	0	3	0	0	7	—
0	0	0	0	0	0	0	3	3	0	0	
0	145	0	(-25)	0	0	0	75	0	10	175	
30	-45	0	15	0	0	0	-15	0	0	45	
-90	120	0	0	30	0	0	30	0	0	270	
120	45	0	0	0	15	0	90	0	0	540	
0	30	0	0	0	0	15	15	0	0	60	
30	30	30	0	0	0	0	30	0	0	150	
0	0	0	0	0	0	0	3	3	0	0	
150	210	0	0	0	0	0	150	0	30	750	

The algorithm terminates. We have $x = y = v = 0$, $r = 45/15 = 3$,
$s = 270/30 = 9$, $t = 540/15 = 36$, $u = 60/15 = 4$, $z = 150/30 = 5$, $w = 0/3 = 0$,
$f = 750/30 = 25$. The solution is

Max $f = 25$ at $x = 0$, $y = 0$, $z = 5$.

19. <u>Familiarize</u>: Let

x = number of Type A questions to be done,

y = number of Type B questions to be done,

z = number of Type C questions to be done,

and S = total test score.

Questions of Type A, B, and C are worth 4, 7, and 8 points, respectively so $S = 4x + 7y + 8z$. You must do at least 5 questions of Type A, but you can do no more than 10, so $5 \leqslant x \leqslant 10$. You must do at least 3 questions of Type B, but you can do no more than 10, so $3 \leqslant y \leqslant 10$. You must do at least 2 questions of Type C, but you can do no more than 5, so $2 \leqslant z \leqslant 5$. You can do no more than 22 questions, so $x + y + z \leqslant 22$.

<u>Translate</u>: Write a linear program using the information above.

$$x \geqslant 5,$$
$$x \leqslant 10,$$
$$y \geqslant 3,$$
$$y \leqslant 10,$$
$$z \geqslant 2,$$
$$z \leqslant 5,$$
$$x + y + z \leqslant 22$$

Max $S = 4x + 7y + 8z$; $x, y, z \geqslant 0$.

<u>Carry out</u>: Convert to a standard maximum-type program by multiplying the first, third, and fifth inequalities by -1. We have

$$-x \leqslant -5,$$
$$x \leqslant 10,$$
$$-y \leqslant -3,$$
$$y \leqslant 10,$$
$$-z \leqslant -2,$$
$$z \leqslant 5,$$
$$x + y + z \leqslant 22,$$

Max $S = 4x + 7y + 8z$; $x, y, z \geqslant 0$.

Add nonnegative slack variables:

$$
\begin{array}{rcl}
-x \quad\quad + q & = & -5 \\
x \quad\quad + r & = & 10 \\
-y \quad\quad + s & = & -3 \\
y \quad\quad + t & = & 10 \\
-z \quad\quad + u & = & -2 \\
z \quad\quad + v & = & 5 \\
x + y + z \quad\quad + w & = & 22 \\
-4x - 7y - 8z \quad\quad + S & = & 0
\end{array}
$$

Set up the initial simplex tableau, determine the pivot element, and pivot.

19. (continued)

x	y	z	q	r	s	t	u	v	w	S		1	q
⊖1	0	0	1	0	0	0	0	0	0	0		⊖5	—
1*	0	0	0	1	0	0	0	0	0	0		10	10/1 = 10 ← Min
0	-1	0	0	0	1	0	0	0	0	0		-3	—
0	1	0	0	0	0	1	0	0	0	0		10	—
0	0	-1	0	0	0	0	1	0	0	0		-2	—
0	0	1	0	0	0	0	0	1	0	0		5	—
1	1	1	0	0	0	0	0	0	1	0		22	22/1 = 22
-4	-7	-8	0	0	0	0	0	0	0	1		0	
0	0	0	1	1	0	0	0	0	0	0		5	—
1	0	0	0	1	0	0	0	0	0	0		10	—
0	⊖1	0	0	0	1	0	0	0	0	0		⊖3	—
0	1*	0	0	0	0	1	0	0	0	0		10	10/1 = 10 ← Min
0	0	-1	0	0	0	0	1	0	0	0		-2	—
0	0	1	0	0	0	0	0	1	0	0		5	—
0	1	1	0	-1	0	0	0	0	1	0		12	12/1 = 12
0	-7	-8	0	4	0	0	0	0	0	1		40	
0	0	0	1	1	0	0	0	0	0	0		5	—
1	0	0	0	1	0	0	0	0	0	0		10	—
0	0	0	0	0	1	1	0	0	0	0		7	—
0	1	0	0	0	0	1	0	0	0	0		10	—
0	0	-1	0	0	0	0	1	0	0	0		-2	—
0	0	1	0	0	0	0	0	1	0	0		5	5/1 = 5
0	0	1*	0	-1	0	-1	0	0	1	0		2	2/1 = 2 ← Min
0	0	⊖8	0	4	0	7	0	0	0	1		110	
0	0	0	1	1	0	0	0	0	0	0		5	5/1 = 5
1	0	0	0	1	0	0	0	0	0	0		10	10/1 = 10
0	0	0	0	0	1	1	0	0	0	0		7	—
0	1	0	0	0	0	1	0	0	0	0		10	—
0	0	0	0	-1	0	-1	1	0	1	0		0	—
0	0	0	0	1*	0	1	0	1	-1	0		3	3/1 = 3 ← Min
0	0	1	0	-1	0	-1	0	0	1	0		2	—
0	0	0	0	⊖4	0	-1	0	0	8	1		126	
0	0	0	1	0	0	-1	0	-1	1	0		2	
1	0	0	0	0	0	-1	0	-1	1	0		7	
0	0	0	0	0	1	1	0	0	0	0		7	
0	1	0	0	0	0	1	0	0	0	0		10	
0	0	0	0	0	0	0	1	1	0	0		3	
0	0	0	0	1	0	1	0	1	-1	0		3	
0	0	1	0	0	0	0	0	1	0	0		5	
0	0	0	0	0	0	3	0	4	4	1		138	

The algorithm terminates. We have t = v = w = 0, q = 2, x = 7, s = 7, y = 10, u = 3, r = 3, z = 5, S = 138.

State: You should do 7 questions of Type A, 10 of Type B, and 5 of Type C. The maximum score is 138.

21. Familiarize: Let

x = amount invested in Bank X,

y = amount invested in Bank Y,

z = amount invested in Bank Z, and

T = total income, in dollars.

The man plans to invest up to $80,000, so x + y + z ≤ 80,000. He wants to invest at least $20,000 but no more than $64,000 in Bank X. We have 20,000 ≤ x ≤ 64,000. He will invest no more than $40,000 in Bank Y, so y ≤ 40,000. He will invest at least $26,000 but no more than $50,000 in Bank Z. This gives us 26,000 ≤ z ≤ 50,000. Using the simple interest formula, I = Prt, we have T = 0.06x + 0.085y + 0.065z.

21. (continued)

Translate: Write a linear program using the information above.

$$x + y + z \leqslant 80{,}000,$$
$$x \geqslant 20{,}000,$$
$$x \leqslant 64{,}000,$$
$$y \leqslant 40{,}000,$$
$$z \geqslant 26{,}000,$$
$$z \leqslant 50{,}000,$$

Max $T = 0.06x + 0.085y + 0.065z$; $x, y, z \geqslant 0$.

Carry out: Convert to a standard maximum-type program by multiplying the second and fifth inequalities by -1. We have

$$x + y + z \leqslant 80{,}000,$$
$$-x \leqslant -20{,}000,$$
$$x \leqslant 64{,}000,$$
$$y \leqslant 40{,}000,$$
$$-z \leqslant -26{,}000,$$
$$z \leqslant 50{,}000,$$

Max $T = 0.06x + 0.085y + 0.065z$; $x, y, z \geqslant 0$.

Add nonnegative slack variables.

$$\begin{aligned}
x + \quad y + \quad z + r \quad\quad\quad\quad\quad\quad\quad\quad &= 80{,}000 \\
-x \quad\quad\quad\quad\quad\quad + s \quad\quad\quad\quad\quad\quad &= -20{,}000 \\
x \quad\quad\quad\quad\quad\quad\quad\quad + t \quad\quad\quad\quad\quad &= 64{,}000 \\
y \quad\quad\quad\quad\quad\quad\quad\quad\quad + u \quad\quad\quad &= 40{,}000 \\
-z \quad\quad\quad\quad\quad\quad\quad\quad\quad + v \quad\quad &= -26{,}000 \\
z \quad\quad\quad\quad\quad\quad\quad\quad\quad\quad + w &= 50{,}000 \\
-0.06x - 0.085y - 0.065z \quad\quad\quad\quad\quad\quad\quad\quad + T &= 0
\end{aligned}$$

Set up the initial simplex tableau, determine the pivot element, and pivot.

x	y	z	r	s	t	u	v	w	T	1	q
1	1	1	1	0	0	0	0	0	0	80,000	80,000/1 = 80,000
(-1)	0	0	0	1	0	0	0	0	0	-20,000	—
1*	0	0	0	0	1	0	0	0	0	64,000	64,000/1 = 64,000
0	1	0	0	0	0	1	0	0	0	40,000	— ⌐ Min
0	0	-1	0	0	0	0	1	0	0	-26,000	—
0	0	1	0	0	0	0	0	1	0	50,000	—
-0.06	-0.085	-0.065	0	0	0	0	0	0	1	0	
0	1	1*	1	0	-1	0	0	0	0	16,000	16,000/1 = 16,000
0	0	0	0	1	1	0	0	0	0	44,000	— ⌐ Min
1	0	0	0	0	1	0	0	0	0	64,000	—
0	1	0	0	0	0	1	0	0	0	40,000	—
0	0	(-1)	0	0	0	0	1	0	0	-26,000	—
0	0	1	0	0	0	0	0	1	0	50,000	50,000/1 = 50,000
0	-0.085	-0.065	0	0	0.06	0	0	0	1	3840	
0	1	1	1	0	-1	0	0	0	0	16,000	—
0	0	0	0	1	1	0	0	0	0	44,000	44,000/1 = 44,000
1	0	0	0	0	1	0	0	0	0	64,000	64,000/1 = 64,000
0	1	0	0	0	0	1	0	0	0	40,000	—
0	1	0	1	0	(-1)	0	1	0	0	-10,000	—
0	-1	0	-1	0	1*	0	0	1	0	34,000	34,000/1 = 34,000
0	-0.02	0	0.065	0	-0.005	0	0	0	1	4880	⌐ Min
0	0	1	0	0	0	0	0	1	0	50,000	— ⌐ Min
0	1*	0	1	1	0	0	0	-1	0	10,000	10,000/1 = 10,000
1	1	0	1	0	0	0	0	-1	0	30,000	30,000/1 = 30,000
0	1	0	0	0	0	1	0	0	0	40,000	40,000/1 = 40,000
0	0	0	0	0	0	0	1	1	0	24,000	—
0	-1	0	-1	0	1	0	0	1	0	34,000	—
0	-0.025	0	0.06	0	0	0	0	0.005	1	5050	

21. (continued)

x	y	z	r	s	t	u	v	w	T		1	
0	0	1	0	0	0	0	0	1	0		50,000	50,000/1 = 50,000
0	1	0	1	1	0	0	0	-1	0		10,000	—
1	0	0	0	-1	0	0	0	0	0		20,000	—
0	0	0	-1	-1	0	1	0	1	0		30,000	30,000/1 = 30,000
0	0	0	0	0	0	0	1	1*	0		24,000	24,000/1 = 24,000
0	0	0	0	1	1	0	0	0	0		44,000	—
0	0	0	0.085	0.025	0	0	0	(-0.02)	1		5300	

$\llcorner$ Min

x	y	z	r	s	t	u	v	w	T		1
0	0	1	0	0	0	0	-1	0	0		26,000
0	1	0	1	1	0	0	1	0	0		34,000
1	0	0	0	-1	0	0	0	0	0		20,000
0	0	0	-1	-1	0	1	-1	0	0		6000
0	0	0	0	0	0	0	1	1	0		24,000
0	0	0	0	1	1	0	0	0	0		44,000
0	0	0	0.085	0.025	0	0	0.02	0	1		5780

The algorithm terminates. We have $r = s = v = 0$, $z = 26,000$, $y = 34,000$, $x = 20,000$, $u = 6000$, $w = 24,000$, $t = 44,000$, $T = 5780$.

State: The man should invest $20,000 in Bank X, $34,000 in Bank Y, and $26,000 in Bank Z. (The maximum income is $5780.)

23. Familiarize: Let

   x = number of servings of potatoes,

   y = number of servings of steak,

   z = number of servings of broccoli, and

   C = total cost of the meal.

Organize the information in a table.

	Composition			Amount Required
	Potatoes	Steak	Broccoli	
Number of Servings	x	y	z	
Vitamin C	20 mg	0 mg	100 mg	220 mg
Protein	5 g	20 g	5 g	55 g
Calories	100	300	50	800
Cost per Serving	$0.23	$1.30	$0.50	Objective Function Minimize C

Translate: Use the table to write a linear program.

Minimize $C = 0.23x + 1.30y + 0.50z$

Subject to

$$20x + 100z \geqslant 220$$
$$5x + 20y + 5z \geqslant 55$$
$$100x + 300y + 50z \leqslant 800$$
$$x, y, z \geqslant 0$$

Carry out: Convert to a standard maximum-type program by multiplying the objective function and the first two inequalities by –1.

$$-20x - 100z \leqslant -220,$$
$$-5x - 20y - 5z \leqslant -55,$$
$$100x + 300y + 50z \leqslant 800,$$

Max $-C = -0.23x - 1.30y - 0.50z$; $x, y, z \geqslant 0$.

<u>23</u>. (continued)

Add nonnegative slack variables.

$$-20x \qquad\quad - 100z + u \qquad\qquad\qquad = -220$$
$$-5x - 20y - 5z \quad + v \qquad\qquad\quad = -55$$
$$100x + 300y + 50z \qquad\quad + w \qquad = 800$$
$$0.23x + 1.30y + 0.50z \qquad\qquad\quad - C = \quad 0$$

Set up the initial simplex tableau, determine the pivot element, and pivot.

x	y	z	u	v	w	-C		1	q
-20	0	(-100)	1	0	0	0		(-220)	—
-5	-20	-5	0	1	0	0		-55	—
100	300	50*	0	0	1	0		800	800/50 = 16 ← Min
0.23	1.30	0.50	0	0	0	1		0	
180	600*	0	1	0	2	0		1380	$1380/600 = 2\frac{3}{10}$ Min
50	100	0	0	10	1	0		250	$250/100 = 2\frac{1}{2}$
100	300	50	0	0	1	0		800	$800/300 = 2\frac{2}{3}$
-0.77	(-1.7)	0	0	0	-0.01	1		-8	
180	600	0	1	0	2	0		1380	$1380/180 = 7\frac{2}{3}$
120*	0	0	-1	60	4	0		120	120/120 = 1 ← Min
20	0	100	-1	0	0	0		220	220/20 = 11
(-156)	0	0	1.7	0	-2.6	600		-2454	
0	1200	0	5	-180	-8	0		2400	
120	0	0	-1	60	4	0		120	
0	0	600	-5	-60	-4	0		1200	
0	0	0	4	780	26	6000		-22,980	

The algorithm terminates. We have u = v = w = 0, y = 2400/1200 = 2, x = 120/120 = 1, z = 1200/600 = 2, -C = -22,980/6000 = -3.83. Multiply "Max -C = -3.83" by -1 to get "Min C = 3.83."

<u>State</u>: The meal should consist of 1 serving of baked potato, 2 servings of steak, and 2 servings of broccoli. (The minimum cost is $3.83.)

<u>25</u>. <u>Familiarize</u>: Let

x = percent of investment in Fund I (expressed as a decimal quantity),

y = percent of investment in Fund II (expressed as a decimal quantity), and

R = rate of return on total investment.

Organize the information in a table.

	Composition		Amount of Total Investment
	Fund I	Fund II	
Percent Invested	x	y	
Stocks	40%	15%	25%
Bonds	35%	70%	70%
Mortgages	25%	15%	
Rate of Return	15%	10%	Objective Function Maximize R

25. (continued)

Translate: Use the table to write a linear program. Remember also that the two investments will add up to 100% of the available funds. Since x and y are expressed as decimal quantities we will express 100% as 1.

$$40x + 15y \leqslant 25,$$
$$60x + 85y \geqslant 70,$$
$$x + y = 1,$$

Max R = 15x + 10y; x, y $\geqslant$ 0.

Carry out: Convert to a standard maximum-type program by first multiplying the second inequality by -1 and by writing the equation x + y = 1 as two inequalities.

$$40x + 15y \leqslant 25,$$
$$-60x - 85y \leqslant -70,$$
$$x + y \leqslant 1,$$
$$x + y \geqslant 1,$$

Max R = 15x + 10y; x, y $\geqslant$ 0.

Then multiply the fourth inequality by -1. Now we have a standard maximum-type linear program.

$$40x + 15y \leqslant 25,$$
$$-60x - 85y \geqslant -70,$$
$$x + y \leqslant 1,$$
$$-x - y \leqslant -1,$$

Max R = 15x + 10y; x, y $\geqslant$ 0.

Add nonnegative slack variables.

$$40x + 15y + t \qquad\qquad\qquad = 25$$
$$-60x - 85y \qquad + u \qquad\qquad = -70$$
$$x + y \qquad\qquad + v \qquad\quad = 1$$
$$-x - y \qquad\qquad\qquad + w \quad = -1$$
$$-15x - 10y \qquad\qquad\qquad\quad + R = 0$$

Set up the initial simplex tableau, determine the pivot element, and pivot.

x	y	t	u	v	w	R		1	q
40	15	1	0	0	0	0		25	$25/15 = 1\frac{2}{3}$
-60	(-85)	0	1	0	0	0		(70)	—
1	1*	0	0	1	0	0		1	$1/1 = 1 \leftarrow$ Min
-1	-1	0	0	0	1	0		-1	—
-15	-10	0	0	0	0	1		0	
25*	0	1	0	-15	0	0		10	$10/25 = 0.4 \leftarrow$ Min
25	0	0	1	85	0	0		15	$15/25 = 0.6$
1	1	0	0	1	0	0		1	$1/1 = 1$
0	0	0	0	1	1	0		0	—
(-5)	0	0	0	10	0	1		10	
25	0	1	0	-15	0	0		10	
0	0	-1	1	100	0	0		5	
0	25	-1	0	40	0	0		15	
0	0	0	0	1	1	0		0	
0	0	1	0	35	0	5		60	

The algorithm terminates. We have t = v = 0, x = 10/25 = 0.4, u = 5/1 = 5, y = 15/25 = 0.6, w = 0/1 = 0, R = 60/5 = 12.

State: 0.4, or 40%, of the investment should be in Fund I, and 0.6, or 60%, should be in Fund II. (The maximum rate of return is 12%.)

Exercise Set 5.1

<u>1</u>. We know that 100 people were surveyed, and 57 of them wore glasses or contacts. Therefore, the probability that a person wears either glasses or contacts is P, where

$$P = \frac{57}{100}, \text{ or } 0.57, \text{ or } 57\%.$$

The number of people who wore neither glasses nor contacts was 100 - 57, or 43. The probability that a person wears neither is P, where

$$P = \frac{43}{100}, \text{ or } 0.43, \text{ or } 43\%.$$

<u>3</u>. We know that 600 of the 16,000 applicants are accepted. Therefore, the probability of a person being accepted for employment by the FBI is P, where

$$P = \frac{600}{16,000} = 0.0375, \text{ or } 3.75\%.$$

16,000 - 600, or 15,400 applicants are not accepted. The probability of a person not being accepted is P, where

$$P = \frac{15,400}{16,000} = 0.9625, \text{ or } 96.25\%.$$

<u>5</u>. The probability that a person is obese is P, where

$$P = \frac{45,000,000}{243,000,000} \approx 0.185, \text{ or } 18.5\%.$$

243,000,000 - 45,000,000, or 198,000,000 people are not obese. Therefore, the probability that a person is not obese is P, where

$$P = \frac{198,000,000}{243,000,000} \approx 0.815, \text{ or } 81.5\%.$$

<u>7</u>. The company can expect 78% of the 15,000 pieces of advertising to be opened and read. Now,

$$78\% \text{ of } 15,000 = 0.78 \times 15,000 = 11,700,$$

so 11,700 would be opened and read.

<u>9</u>. The probability that a baby will die at birth is P, where

$$P = \frac{21}{1000} = 0.021, \text{ or } 2.1\%.$$

The number of babies that will live at birth is 1000 - 21, or 979. The probability that a baby will live at birth is P, where

$$P = \frac{979}{1000} = 0.979, \text{ or } 97.9\%.$$

<u>11</u>. Let D = the number of deer in the perserve. If there are D deer and 318 of them are tagged, then the probability that a deer is tagged is the ratio

$$\frac{318}{D}.$$

Later, the ratio of tagged deer to deer caught is

$$\frac{56}{168}.$$

This can be used as an estimate of the probability that a deer is tagged. Thus we assume the two ratios are the same, and we have a proportion:

$$\frac{318}{D} = \frac{56}{168}$$

<u>11</u>. (continued)

Solve the proportion for D. First multiply on both sides by the least common multiple of the denominators, 168D:

$$168D \cdot \frac{318}{D} = 168D \cdot \frac{56}{168}$$

$$168 \cdot 318 = D \cdot 56$$

$$\frac{168 \cdot 318}{56} = D$$

$$954 = D$$

We estimate that there are about 954 deer in the preserve.

<u>13</u>. The total number of gum drops is 7 + 8 + 9 + 4 + 5 = 33. There are 8 lemon gum drops, so the probability of getting a lemon gum drop is 8/33. There are 9 orange gum drops, so the probability of getting an orange gum drop is 9/33, or 3/11. There are 4 cherry gum drops, so the probability of getting a cherry gum drop is 4/33. There are 5 lime gum drops, so the probability of getting a lime gum drop is 5/33. There are no licorice gum drops, so the probability of getting a licorice gum drop is 0/33, or 0. There are 7 strawberry gum drops, so the probability of getting a strawberry gum drop is 7/33.

<u>15</u>. Add the number of occurences for A, E, I, O, and U to find the number of occurences of a vowel:

$$853 + 1229 + 539 + 705 + 240 = 3566$$

There were 9136 letters in all so the probability of a vowel occurring is

$$\frac{3566}{9136} \approx 0.39, \text{ or } 39\%.$$

<u>17</u>. Reading from the completed table (see Exercise 14), we see that the five consonants that occur with the greatest probability are T(8.6%), S(8.2%), R(6.7%), N(6.5%), and L(4.6%).

<u>19</u>. Since R, S, T, L, N, and E are the five consonants and vowel with the greatest probability of occurring, the results of Exercises 17 and 18 support their choice.

<u>21</u>. Using the table preceding Exercise 14, add the number of occurences of A, S, D, F, G, H, J, K, and L (the letters on the home row on Shole's keyboard):

$$853 + 745 + 286 + 173 + 190 + 399 + 21 + 57 + 417 = 3141$$

Divide by 9136 to find the probability or finding a letter you wish to type on the home row of Shole's keyboard.

$$\frac{3141}{9136} \approx 0.344, \text{ or } 34.4\%$$

Add the number of occurences of A, O, E, U, I, D, H, T, N, and S (the letters on the home row of Dvorak's keyboard):

$$853 + 705 + 1229 + 240 + 539 + 286 + 399 + 789 + 597 + 745 = 6382$$

Divide by 9136 to find the probability of finding a letter you wish to type on the home row of Dvorak's keyboard.

$$\frac{6382}{9136} \approx 0.699, \text{ or } 69.9\%$$

Exercise Set 5.2

1. $5! = 5 \cdot 4 \cdot 3 \cdot 2 \cdot 1 = 120$

3. $0!$ is defined to be 1.

5. $\dfrac{6!}{4!} = \dfrac{6 \cdot 5 \cdot 4!}{4!} = 6 \cdot 5$, or 30

7. $\dfrac{9!}{5!} = \dfrac{9 \cdot 8 \cdot 7 \cdot 6 \cdot 5!}{5!} = 9 \cdot 8 \cdot 7 \cdot 6$, or 3024

9. $(8 - 5)! = 3! = 3 \cdot 2 \cdot 1 = 6$

11. $8! - 5! = (8 \cdot 7 \cdot 6 \cdot 5 \cdot 4 \cdot 3 \cdot 2 \cdot 1) - (5 \cdot 4 \cdot 3 \cdot 2 \cdot 1) = 40{,}320 - 120 = 40{,}200$

13. $\dfrac{10!}{7!3!} = \dfrac{10 \cdot 9 \cdot 8 \cdot 7!}{7!(3 \cdot 2 \cdot 1)} = \dfrac{10 \cdot 9 \cdot 8}{3 \cdot 2 \cdot 1} = \dfrac{720}{6} = 120$

15. $\dfrac{9!}{(9 - 2)!} = \dfrac{9!}{7!} = \dfrac{9 \cdot 8 \cdot 7!}{7!} = 9 \cdot 8$, or 72

17. $_6P_6 = 6! = 6 \cdot 5 \cdot 4 \cdot 3 \cdot 2 \cdot 1 = 720$

19. Using formula (1) of Theorem 3, we have

    The 20 tells where to start.
    $_{20}P_2 = 20 \cdot 19 = 380$.
    The 2 tells how many factors.

    Using formula (2) of Theorem 3, we have
    $_{20}P_2 = \dfrac{20!}{(20 - 2)!} = \dfrac{20!}{18!} = \dfrac{20 \cdot 19 \cdot 18!}{18!} = 20 \cdot 19 = 380$.

21. Using formula (1) of Theorem 3, we have

    The 7 tells where to start.
    $_7P_5 = 7 \cdot 6 \cdot 5 \cdot 4 \cdot 3 = 2520$
    The 5 tells how many factors.

    Using formula (2) of Theorem 3, we have
    $_7P_5 = \dfrac{7!}{(7 - 5)!} = \dfrac{7!}{2!} = \dfrac{7 \cdot 6 \cdot 5 \cdot 4 \cdot 3 \cdot 2!}{2!} = 7 \cdot 6 \cdot 5 \cdot 4 \cdot 3 = 2520$.

23. $_7P_1 = \dfrac{7!}{(7 - 1)!} = \dfrac{7!}{6!} = \dfrac{7 \cdot 6!}{6!} = 7$

25. $_9P_0 = \dfrac{9!}{(9 - 0)!} = \dfrac{9!}{9!} = 1$

27. $_nP_3 = \dfrac{n!}{(n - 3)!} = \dfrac{n(n - 1)(n - 2)(n - 3)!}{(n - 3)!} = n(n - 1)(n - 2)$

29. There are 5 letters, so we have
    $_5P_5 = 5!$, or 120.

31. There are 6 letters, so we have
    $_6P_6 = 6!$, or 720.

33. There are 7 letters, so we have
    $_7P_7 = 7!$, or 5040.

35. There are 9 letters, so we have
    $_9P_9 = 9!$, or 362,880.

37. We have $_7P_4 = \dfrac{7!}{(7 - 4)!} = \dfrac{7!}{3!} = \dfrac{7 \cdot 6 \cdot 5 \cdot 4 \cdot 3!}{3!}$
    $= 7 \cdot 6 \cdot 5 \cdot 4$, or 840.

39. The total number of arrangements is $_4P_4 = 4! = 24$. Other than the one arrangement given there are $24 - 1$, or 23 other ways to arrange the programs.

41. By Theorem 1, without repetition we have $_5P_5 = 5!$, or 120. By Theorem 5, with repetition, we have 5^5, or 3125.

43. By Theorem 1, we have $_6P_6 = 6!$, or 720, arrangements in a straight line. By Theorem 4, we have $(6 - 1)! = 5!$, or 120, arrangements in a circle.

45. Think of a series of seven product blanks:

    __ · __ __ __ __ __ __

    First digit chosen from 1,2,3,4,5,6, 7,8,9

    Last 6 digits chosen from 0 and the eight remaining digits of 1 through 9 after choice of first digit is made

    The first digit is one of 9 numbers. This is equivalent to $_9P_1$. There are 9 choices for the second digit, 8 for the third digit, and so on down to 4 choices for the seventh digit. This is equivalent to $_9P_6$. Filling in the product blanks we have

    9 · 9 · 8 · 7 · 6 · 5 · 4.

    Thus, 544,320 seven-digit phone numbers can be formed.

47. Thirteen people are to be seated in circular arrangements. By Theorem 4, we have $(13 - 1)! = 12!$, or 479,001,600 arrangements.

49. Think of a series of 5 product blanks representing the number of choices of routes between Boston and Washington, Washington and Atlanta, Atlanta and New Orleans, New Orleans and Denver, and Denver and Los Angeles, respectively. Filling in the blanks we have

    6 · 2 · 4 · 5 · 7, or 1680.

51. a) By Theorem 3, we have $_6P_5 = \dfrac{6!}{(6 - 5)!} = \dfrac{6!}{1!} = \dfrac{6 \cdot 5 \cdot 4 \cdot 3 \cdot 2 \cdot 1}{1} = 720$.

    b) By Theorem 5, we have $6^5 = 7776$.

    c) The first letter can be chosen in $_1P_1 = 1!$, or 1, way. The last four letters can be chosen from the 5 remaining letters in $_5P_4 = \dfrac{5!}{(5 - 4)!} = \dfrac{5!}{1!} = \dfrac{5 \cdot 4 \cdot 3 \cdot 2 \cdot 1}{1} = 120$ ways. Using the Fundamental Counting Principle, the number of words that can be formed is $1 \cdot 120$, or 120.

51. (continued)

    d) The first two letters can be chosen from the three letters J, K, L in $_3P_2 = \frac{3!}{(3-2)!} = \frac{3!}{1!} = \frac{3 \cdot 2 \cdot 1}{1} = 6$ ways. The last three letters can each be chosen in $_1P_1 = 1!$, or 1, way. By the Fundamental Counting Principle, the number of words that can be formed is $6 \cdot 1 \cdot 1 \cdot 1$, or 6.

53. By Theorem 5, we have $4^4 = 256$ if repetitions are permitted.

    By Theorem 1, we have $_4P_4 = 4!$, or 24, if repetitions are not permitted.

55. a) By Theorem 5, we have 5^3, or 125.

    b) By Theorem 1, we have $_5P_3 = \frac{5!}{(5-3)!} = \frac{5!}{2!} = \frac{5 \cdot 4 \cdot 3 \cdot 2!}{2!} = 5 \cdot 4 \cdot 3$, or 60.

57. In the first part of the exercise repetition is allowed. We have 7^4, or 2401, ways. If no firm gets more than one contract, we have $_7P_4 = \frac{7!}{(7-4)!} = \frac{7!}{3!} = \frac{7 \cdot 6 \cdot 5 \cdot 4 \cdot 3!}{3!} = 7 \cdot 6 \cdot 5 \cdot 4$, or 840, ways.

59. The firm that will receive the two particular contracts can be chosen in $_7P_1 = \frac{7!}{(7-1)!} = \frac{7!}{6!} = \frac{7 \cdot 6!}{6!}$, or 7 ways. Now consider the two particular contracts to be awarded to the same firm as a single item and the other two contracts as two items. Since there are no further restrictions, then repetition is allowed and we have a total of 7 objects to be taken 3 at a time. This is 7^3, or 343.

61. a) We want to find the number of permutations of 10 objects taken 5 at a time with repetition. This is 10^5, or 100,000.

    b) Since there are 100,000 possible zip-codes, there could be 100,000 post offices.

63. a) We want to find the number of permutations of 10 objects taken 9 at a time with repetition. This is 10^9, or 1,000,000,000.

    b) Since there are more than 243 million social security numbers, each person can have one.

65. In the first part of the exercise there are no restrictions, so we will find the number of permutations of 7 objects taken 7 at a time:
    $$_7P_7 = 7! = 5040$$
    If a member of the host company sits at each end, the two who sit at the ends can be chosen in $_4P_2 = \frac{4!}{2!} = \frac{4 \cdot 3 \cdot 2!}{2!} = 12$ ways. The remaining five seats may be filled in $_5P_5 = 5! = 120$ ways. The total number of arrangements is $12 \cdot 120 = 1440$. If they must alternate host, visitor, and so on, they will sit H V H V H V H. The hosts can be seated in $_4P_4 = 4! = 24$ ways, and the visitors can be seated in $_3P_3 = 3! = 6$ ways. Then the total number of arrangements is $24 \cdot 6 = 144$.

67. $_nP_5 = 7 \cdot {_nP_4}$

    $$\frac{n!}{(n-5)!} = 7 \cdot \frac{n!}{(n-4)!}$$

    $$\frac{n!}{(n-5)!} = 7 \cdot \frac{n!}{(n-4)(n-5)!}$$

    $$(n-4)(n-5)! \cdot \frac{n!}{(n-5)!}$$
    $$= (n-4)(n-5)!7 \cdot \frac{n!}{(n-4)(n-5)!}$$

    Multiplying by $(n-4)(n-5)!$

    $$(n-4)n! = 7 \cdot n!$$

    $$\frac{1}{n!} \cdot (n-4)n! = \frac{1}{n!} \cdot 7 \cdot n! \quad \text{Multiplying by } \frac{1}{n!}$$

    $$n - 4 = 7$$
    $$n = 11$$

69. $_nP_5 = 9 \cdot {_{n-1}P_4}$

    $$\frac{n!}{(n-5)!} = 9 \cdot \frac{(n-1)!}{(n-1-4)!}$$

    $$\frac{n!}{(n-5)!} = 9 \cdot \frac{(n-1)!}{(n-5)!}$$

    $$(n-5)! \frac{n!}{(n-5)!} = (n-5)! 9 \cdot \frac{(n-1)!}{(n-5)!}$$

    Multiplying by $(n-5)!$

    $$n! = 9 \cdot (n-1)!$$

    $$n(n-1)! = 9 \cdot (n-1)!$$

    $$\frac{1}{(n-1)!} \cdot n(n-1)! = \frac{1}{(n-1)!} \cdot 9 \cdot (n-1)!$$

    Multiplying by $\frac{1}{(n-1)!}$

    $$n = 9$$

71. To find the number of ways the 6 people can be seated in the car we find

$$_6P_6 = 6!, \text{ or } 720.$$

If a given couple must sit together, first consider seating them in the front seat with one on the left side and the other in the middle. This can be done $_2P_2 = 2!$, or 2, ways. The other four people can then be seated in $_4P_4 = 4!$, or 24, ways. This gives a total of $2 \cdot 24$, or 48, arrangements. Next consider seating the couple in the front seat with one on the right side and one in the middle. Again we have 48 arrangements possible. If the couple is seated in the back seat with one on the left and the other in the middle, there are another 48 arrangements possible. Finally, if the couple is seated in the back seat with one on the right and one in the middle, there are another 48 arrangements possible. Thus, the people can be seated in $4 \cdot 48$, or 192, ways.

73. In order to complete the tournament there must be exactly one team without a loss. In other words, $n - 1$ teams must each have lost one game. Therefore, $n - 1$ games are required.

75. Ordinarily n objects can be arranged in $(n - 1)!$ distinct circular arrangements, and orientation is important. That is,

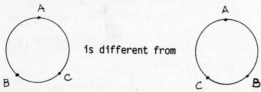

is different from

However, with a key ring the two are not different since one can be obtained from the other by turning the key ring over. Therefore, only half of the $(n - 1)!$ arrangements are distinct, so there are $\dfrac{(n - 1)!}{2}$ arrangements.

Exercise Set 5.3

1. $_{13}C_2 = \dfrac{13 \cdot 12}{2 \cdot 1} = 78$

The 13 tells where to start.

The 2 tells how many factors there are in both numerator and denominator and where to start the denominator.

We used formula (2) from Theorem 6.

3. $\begin{pmatrix} 13 \\ 11 \end{pmatrix} = \dfrac{13!}{11!(13 - 11)!} = \dfrac{13!}{11!2!} = \dfrac{13 \cdot 12 \cdot 11!}{11!2!} = \dfrac{13 \cdot 12}{2 \cdot 1} = 78$

We used formula (1) from Theorem 6. Using Theorem 7 we could also have observed that $\begin{pmatrix} 13 \\ 11 \end{pmatrix} = \begin{pmatrix} 13 \\ 2 \end{pmatrix}$, giving us 78 as in Exercise 1.

5. $\begin{pmatrix} 7 \\ 1 \end{pmatrix} = \dfrac{7}{1} = 7$ Formula (2)

7. $\dfrac{_5P_3}{3!} = {_5}C_3 = \dfrac{5 \cdot 4 \cdot 3}{3 \cdot 2 \cdot 1} = 10$ Formula (2)

9. $\begin{pmatrix} 6 \\ 0 \end{pmatrix} = \dfrac{6!}{0!(6 - 0)!} = \dfrac{6!}{1 \cdot 6!} = 1$ Formula (1)

11. $\begin{pmatrix} 6 \\ 2 \end{pmatrix} = \dfrac{6 \cdot 5}{2 \cdot 1} = 15$ Formula (2)

13. $_{12}C_{11} = {_{12}}C_1$ Theorem 7

$= \dfrac{12}{1}$ Formula (2)

$= 12$

15. $_{12}C_9 = {_{12}}C_3$ Theorem 7

$= \dfrac{12 \cdot 11 \cdot 10}{3 \cdot 2 \cdot 1}$ Formula (2)

$= 220$

17. $\begin{pmatrix} m \\ 2 \end{pmatrix} = \dfrac{m(m - 1)}{2 \cdot 1}$ Formula (2)

$= \dfrac{m(m - 1)}{2}, \text{ or } \dfrac{m^2 - m}{2}$

19. $\begin{pmatrix} p \\ 3 \end{pmatrix} = \dfrac{p(p - 1)(p - 2)}{3 \cdot 2 \cdot 1}$ Formula (2)

$= \dfrac{p(p - 1)(p - 2)}{6}, \text{ or } \dfrac{p^3 - 3p^2 + 2p}{6}$

21. $\begin{pmatrix} 7 \\ 0 \end{pmatrix} + \begin{pmatrix} 7 \\ 1 \end{pmatrix} + \begin{pmatrix} 7 \\ 2 \end{pmatrix} + \begin{pmatrix} 7 \\ 3 \end{pmatrix} + \begin{pmatrix} 7 \\ 4 \end{pmatrix} + \begin{pmatrix} 7 \\ 5 \end{pmatrix} + \begin{pmatrix} 7 \\ 6 \end{pmatrix} + \begin{pmatrix} 7 \\ 7 \end{pmatrix}$

is the total number of subsets of a set of 7 elements. By Theorem 9, this is 2^7, or 128.

23. $_{100}C_0 + {_{100}}C_1 + \cdots + {_{100}}C_{100}$ is the total number of subsets of a set of 100 elements. By Theorem 9, this is 2^{100}.

25. Since the order of choosing is irrelevant, the number of ways the secretaries can be selected is

$$_6C_4 = \dfrac{6 \cdot 5 \cdot 4 \cdot 3}{4 \cdot 3 \cdot 2 \cdot 1} = 15.$$

27. Since the order in which the questions are answered is irrelevant, the student can answer 7 of the first 10 questions in $\begin{pmatrix} 10 \\ 7 \end{pmatrix}$ ways and 5 of the remaining 8 in $\begin{pmatrix} 8 \\ 5 \end{pmatrix}$ ways. Using the Fundamental Counting Principle, the total number of ways the questions can be answered is

$$\begin{pmatrix} 10 \\ 7 \end{pmatrix} \cdot \begin{pmatrix} 8 \\ 5 \end{pmatrix} = \dfrac{10 \cdot 9 \cdot 8 \cdot 7 \cdot 6 \cdot 5 \cdot 4}{7 \cdot 6 \cdot 5 \cdot 4 \cdot 3 \cdot 2 \cdot 1} \cdot \dfrac{8 \cdot 7 \cdot 6 \cdot 5 \cdot 4}{5 \cdot 4 \cdot 3 \cdot 2 \cdot 1} = 6720.$$

29. a) The order of hiring is unimportant. If it does not matter how many are men and how many are women, then we are choosing 5 people from a group of $6 + 8$, or 14. This can be done in

$$\begin{pmatrix} 14 \\ 5 \end{pmatrix} = \dfrac{14 \cdot 13 \cdot 12 \cdot 11 \cdot 10}{5 \cdot 4 \cdot 3 \cdot 2 \cdot 1} = 2002 \text{ ways.}$$

b) We can choose 3 women in $\begin{pmatrix} 8 \\ 3 \end{pmatrix}$ ways. The other 2 must be men and can be chosen in $\begin{pmatrix} 6 \\ 2 \end{pmatrix}$ ways. Using the Fundamental Counting Principle, the total number of choices is

$$\begin{pmatrix} 8 \\ 3 \end{pmatrix} \begin{pmatrix} 6 \\ 2 \end{pmatrix} = \dfrac{8 \cdot 7 \cdot 6}{3 \cdot 2 \cdot 1} \cdot \dfrac{6 \cdot 5}{2 \cdot 1} = 840.$$

29. (continued)

c) We can choose 4 women in $\binom{8}{4}$ ways. The other person must be a man and can be chosen in $\binom{6}{1}$ ways. Using the Fundamental Counting Principle, the total number of choices is

$$\binom{8}{4}\binom{6}{1} = \frac{8\cdot7\cdot6\cdot5}{4\cdot3\cdot2\cdot1} \cdot \frac{6}{1} = 420.$$

d) We can choose 1 woman in $\binom{8}{1}$ ways. The other 4 must be men and can be chosen in $\binom{6}{4}$ ways. Using the Fundamental Counting Principle, the total number of choices is

$$\binom{8}{1}\binom{6}{4} = \frac{8}{1} \cdot \frac{6\cdot5\cdot4\cdot3}{4\cdot3\cdot2\cdot1} = 120.$$

31. The first employee can get 3 tools in $\binom{11}{3}$ ways. The second employee can get 4 of the remaining 11 - 3, or 8, tools in $\binom{8}{4}$ ways. The third can get 2 of the remaining 8 - 4, or 4, tools in $\binom{4}{2}$ ways, and the fourth can get 2 of the remaining 4 - 2, or 2, tools in $\binom{2}{2}$ ways. By the Fundamental Counting Principle, the total number of ways in which the tools can be distributed is

$$\binom{11}{3}\binom{8}{4}\binom{4}{2}\binom{2}{2} =$$

$$\frac{11\cdot10\cdot9}{3\cdot2\cdot1} \cdot \frac{8\cdot7\cdot6\cdot5}{4\cdot3\cdot2\cdot1} \cdot \frac{4\cdot3}{2\cdot1} \cdot \frac{2\cdot1}{2\cdot1} = 69,300.$$

33. We have five letters consisting of one D, two I's, one G, and one T. By Theorem 8, the number of distinguishable words that can be made up is

$$\frac{5!}{1!2!1!1!} = \frac{5\cdot4\cdot3\cdot2\cdot1}{1\cdot2\cdot1\cdot1\cdot1} = 60.$$

35. We have ten letters consisting of 2 C's, 3 I's, 3 N's, 1 A, and 1 T. By Theorem 8, the number of distinguishable words that can be made up is

$$\frac{10!}{2!3!3!1!1!} = \frac{10\cdot9\cdot8\cdot7\cdot6\cdot5\cdot4\cdot3\cdot2\cdot1}{2\cdot1\cdot3\cdot2\cdot1\cdot3\cdot2\cdot1\cdot1\cdot1} = 50,400.$$

37. a) If order is considered and repetition is not allowed, the number of possible cones is

$$_{33}P_3 = 33\cdot32\cdot31 = 32,736.$$

b) If order is considered and repetition is allowed, by Theorem 5 the number of possibilities is

$$33^3 = 35,937.$$

c) If order is not considered and there is no repetition, the number of possibilities is

$$_{33}C_3 = \frac{33\cdot32\cdot31}{3\cdot2\cdot1} = 5456.$$

39. We have a set of 20 objects consisting of 2 A's, 5 B's, 8 C's, 3 D's, and 2 F's. By Theorem 8, the number of distinguishable arrangements is

$$\frac{20!}{2!5!8!3!2!} = 20,951,330,400.$$

41. We have a set of 9 objects consisting of 3 a's, 2 b's, and 4 c's. By Theorem 8, the number of distinguishable arrangements is

$$\frac{9!}{3!2!4!} = 1260.$$

43. a) Since order is not considered the number of ways in which the delegation can be selected is

$$\binom{20}{3} = \frac{20\cdot19\cdot18}{3\cdot2\cdot1} = 1140.$$

b) We can select 3 delegates in $\binom{20}{3}$ ways and we can select one of the three as spokesman in $\binom{3}{1}$ ways. By the Fundamental Counting Principle, the total number of choices is

$$\binom{20}{3}\binom{3}{1} = 1140 \cdot \frac{3}{1} = 3420.$$

45. When order is considered, we have

$$_5P_3 = 5\cdot4\cdot3 = 60.$$

When order is not considered, we have

$$_5C_3 = \frac{5\cdot4\cdot3}{3\cdot2\cdot1} = 10.$$

47. a) We are selecting 6 dances from a group of 10 + 20, or 30, without regard to order. We have

$$_{30}C_6 = \frac{30\cdot29\cdot28\cdot27\cdot26\cdot25}{6\cdot5\cdot4\cdot3\cdot2\cdot1} = 593,775.$$

b) We can select 4 elementary dances in $\binom{10}{4}$ ways and 2 advanced dances in $\binom{20}{2}$ ways. By the Fundamental Counting Principle, the total number of ways in which the dances can be selected is

$$\binom{10}{4}\binom{20}{2} = \frac{10\cdot9\cdot8\cdot7}{4\cdot3\cdot2\cdot1} \cdot \frac{20\cdot19}{2\cdot1} = 39,900.$$

49. The number of plain pizzas (pizzas with no toppings) is $\binom{12}{0}$. The number of pizzas with 1 topping is $\binom{12}{1}$. The number of pizzas with 2 toppings is $\binom{12}{2}$, and so on. The number of pizzas with all 12 toppings is $\binom{12}{12}$. The total number of pizzas is $\binom{12}{0} + \binom{12}{1} + \binom{12}{2} + \cdots + \binom{12}{12} = 2^{12} = 4096.$

51. We want to know the number of ways that 2 people can be selected from a group of 27 without regard to order. This is

$$\binom{27}{2} = \frac{27\cdot26}{2\cdot1} = 351.$$

53. We want to know the number of arrangements of 9 objects taken 4 at a time. (Order is relevant.) This is

$$_9P_4 = 9\cdot8\cdot7\cdot6 = 3024.$$

55. We first want to know the number of ways that 2 teams can be selected from a group of n teams without regard to order. This is

$$\binom{n}{2} = \frac{n(n - 1)}{2 \cdot 1} = \frac{n^2 - n}{2}.$$

If each team plays each other team twice, the number of games will be twice the number found in the first part of the exercise, or $2\binom{n}{2} =$

$2 \cdot \frac{n^2 - n}{2} = n^2 - n.$

57. JACK: We find the number of arrangements of 4 different objects taken 4 at a time:

$$_4P_4 = 4! = 24$$

BUSINESS: We find the number of arrangements of 8 objects taken 8 at a time where we have 3 S's and 1 each of B, U, I, N and E:

$$\frac{8!}{1!1!1!1!1!3!} = \frac{8 \cdot 7 \cdot 6 \cdot 5 \cdot 4 \cdot 3!}{3!} = 6720$$

PHILOSOPHICAL: We find the number of arrangements of 13 objects taken 13 at a time where we have 2 P's, 2 H's, 2 I's, 2 L's, 2 O's, and 1 each of S, C, and A:

$$\frac{13!}{2!2!2!2!2!1!1!1!} = 194,594,400$$

59. We will select 2 of the first 5 lines and 2 of the next 8 lines without regard to order. This can be done in

$$\binom{5}{2}\binom{8}{2} = \frac{5 \cdot 4}{2 \cdot 1} \cdot \frac{8 \cdot 7}{2 \cdot 1} = 280 \text{ ways.}$$

61. We have $_1P_1 \cdot _1P_1 = 1 \cdot 1 = 1$. (That is, a one-symbol permutation consists of either one dot or one dash, so the total number of possibilities is 2.)

63. Each of the three symbols can be chosen in 2 ways (dot or dash). By the Fundamental Counting Principle there are

$$2 \cdot 2 \cdot 2 = 2^3 = 8$$

possible 3-symbol permutations.

65. If the full house consists of 2 kings and 3 sevens, the kings can be chosen in $\binom{4}{2}$ ways, and the sevens in $\binom{4}{3}$ ways. This gives $\binom{4}{2}\binom{4}{3} =$ $\frac{4 \cdot 3}{2 \cdot 1} \cdot \frac{4 \cdot 3 \cdot 2}{3 \cdot 2 \cdot 1} = 24$ ways. If the full house consists of 2 sevens and 3 kings, the sevens can be chosen in $\binom{4}{2}$ ways and the kings in $\binom{4}{3}$ ways. This gives $\binom{4}{2}\binom{4}{3} = 24$ additional ways. The total number of full houses is 24 + 24, or 48.

67. The diagonals are formed by drawing segments between nonadjacent pairs of vertices. To find the number of diagonals we find the number of ways in which pairs of vertices can be formed and subtract the number of pairs of adjacent vertices:

$$_6C_2 - 6 = \frac{6 \cdot 5}{2 \cdot 1} - 6 = 15 - 6 = 9$$

69. We will select 2 of the m lines and 2 of the n lines without regard to order. The number of ways in which this can be done is

$$\binom{m}{2}\binom{n}{2} = \frac{m(m - 1)}{2 \cdot 1} \cdot \frac{n(n - 1)}{2 \cdot 1} =$$
$$\frac{(m^2 - m)(n^2 - n)}{4} = \frac{m^2n^2 - m^2n - mn^2 + mn}{4}.$$

71. $\binom{n + 1}{3} = 2 \cdot \binom{n}{2}$

$$\frac{(n + 1)(n)(n - 1)}{3 \cdot 2 \cdot 1} = 2 \cdot \frac{n(n - 1)}{2 \cdot 1}$$
$$\frac{(n + 1)(n)(n - 1)}{6} = n(n - 1)$$
$$6 \cdot \frac{(n + 1)(n)(n - 1)}{6} = 6 \cdot n(n - 1)$$
$$(n + 1)(n)(n - 1) = 6n(n - 1)$$
$$\frac{1}{n(n - 1)} \cdot (n + 1)(n)(n - 1) = \frac{1}{n(n - 1)} \cdot 6n(n - 1)$$
$$n + 1 = 6$$
$$n = 5$$

73. $\binom{n + 2}{4} = 6 \cdot \binom{n}{2}$

$$\frac{(n + 2)(n + 1)(n)(n - 1)}{4 \cdot 3 \cdot 2 \cdot 1} = 6 \cdot \frac{n(n - 1)}{2 \cdot 1}$$
$$\frac{(n + 2)(n + 1)(n)(n - 1)}{24} = 3n(n - 1)$$
$$24 \cdot \frac{(n + 2)(n + 1)(n)(n - 1)}{24} = 24 \cdot 3n(n - 1)$$
$$(n + 2)(n + 1)(n)(n - 1) = 72n(n - 1)$$
$$\frac{1}{n(n - 1)} \cdot (n + 2)(n + 1)(n)(n - 1) =$$
$$\frac{1}{n(n - 1)} \cdot 72n(n - 1)$$
$$(n + 2)(n + 1) = 72$$
$$n^2 + 3n + 2 = 72$$
$$n^2 + 3n - 70 = 0$$
$$(n + 10)(n - 7) = 0$$
$$n + 10 = 0 \quad \text{or} \quad n - 7 = 0$$
$$n = -10 \quad \text{or} \quad n = 7$$

-10 is not a solution of the original problem, because a set cannot have a negative number of objects. The solution is 7.

75. 7 letter words: There are 2 A's and 1 of each of the other 5 letters. We have

$$\frac{7!}{2!1!1!1!1!1!} = \frac{7 \cdot 6 \cdot 5 \cdot 4 \cdot 3 \cdot 2!}{2!} = 2520.$$

75. (continued)

6 letter words: If the 6 letters consist of 2 A's and 4 other letters, for each such set the number of distinguishable words is $\frac{6!}{2!1!1!1!1!}$. The 4 other letters can be chosen from the letters L, G, E, B, R in $\binom{5}{4}$ ways. Therefore, there are $\binom{5}{4}\frac{6!}{2!1!1!1!1!}$ distinguishable words containing 2 A's. If the 6 letters consist of A, L, G, E, B, R (that is, there is only one A), then the number of distinguishable words is $_6P_6 = 6!$ The total number of distinguishable 6 letter words is
$\binom{5}{4}\frac{6!}{2!1!1!1!1!} + 6! = \frac{5\cdot4\cdot3\cdot2}{4\cdot3\cdot2\cdot1}\cdot\frac{6\cdot5\cdot4\cdot3\cdot2!}{2!} +$
$6\cdot5\cdot4\cdot3\cdot2\cdot1 = 2520.$

5 letter words: If the 5 letters consist of 2 A's and 3 other letters, reasoning similar to that above for 6 letter words containing 2 A's shows that there are $\binom{5}{3}\frac{5!}{2!1!1!1!}$ distinguishable words. If exactly one of the letters is an A, the other 4 letters can be chosen from L, G, E, B, R in $\binom{5}{4}$ ways and each such choice produces $_5P_5 = 5!$ distinguishable words. Therefore, there are $\binom{5}{4}5!$ distinguishable words containing exactly one A. If the letters are L, G, E, B, R (that is, there are no A's), then $_5P_5 = 5!$ words can be formed. The total number of distinguishable 5 letter words is
$\binom{5}{3}\frac{5!}{2!1!1!1!} + \binom{5}{4}5! + 5! =$
$\frac{5\cdot4\cdot3}{3\cdot2}\cdot\frac{5\cdot4\cdot3\cdot2!}{2!} + \frac{5\cdot4\cdot3\cdot2}{4\cdot3\cdot2\cdot1}\cdot5! + 5! =$
$600 + 600 + 120 = 1320.$

77. There are 4 women and 6 men of whom 3 are married couples. The women must stay in the doubles and the men in the triples or one couple must stay in a double, the other 3 women stay in a triple, and the other 5 men stay in a double and a triple. Any other arrangement will result in opposite sexes occupying a room in an unacceptable manner. First consider the number of ways the women can distribute themselves in the doubles and the men in the triples. Choose 2 of the 4 women to stay together, and then choose 2 of the remaining 2 women to stay together. Also, choose 3 of the 6 men to stay together, and then choose 3 of the remaining 3 men to stay together. This gives
$\binom{4}{2}\binom{2}{2}\binom{6}{3}\binom{3}{3} = 120$ ways.

Then consider the other possibility. Choose 1 of the 2 doubles, choose 1 of the 3 couples to occupy it, and choose 2 of the remaining 5 men to occupy the other double. Also, choose 1 of the 2 triples, choose 3 of the 3 remaining women to occupy it, and choose 3 of the remaining 3 men to occupy the other triple. This gives
$\binom{2}{1}\binom{3}{1}\binom{5}{2}\binom{2}{1}\binom{3}{3}\binom{3}{3} = 120$ ways.

77. (continued)

The total number of ways the students can distribute themselves in the rooms is 120 + 120 = 240.

79. The number of subsets of the set {1, 3, 5, 7} that do not contain 5 is the same as the number of all the subsets of {1, 3, 7}, or $2^3 = 8$.

Exercise Set 5.4

1. The outcomes are hitting red (R), hitting white (W), and hitting black (B). Therefore, the sample space is {R, W, B}.

3. The outcomes (possible sums) are 2, 3, 4, 5, 6, 7, 8, 9, 10, 11, and 12. The sample space is {2, 3, 4, 5, 6, 7, 8, 9, 10, 11, 12}.

5. Use a tree diagram to find the outcomes. We first show the results of tossing a coin. Then we show the results of rolling a die.

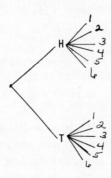

The outcomes are H1, H2, H3, H4, H5, H6, T1, T2, T3, T4, T5, and T6. The sample space is {H1, H2, H3, H4, H5, H6, T1, T2, T3, T4, T5, T6}.

7. Think of dividing each red area into two equal areas, each the same size as white or black. Then there are 6 equally likely outcomes, and there are 4 ways to hit red. By Principle P, m = 4 and n = 6, so $p(R) = \frac{4}{6}$, or $\frac{2}{3}$.

9. Think of dividing each red area as in Exercise 7. Then there are 6 equally likely outcomes, and there is 1 way to hit white. By Principle P, m = 1 and n = 6, so $p(W) = \frac{1}{6}$.

11. There are 16 equally likely outcomes, and there are 6 ways to get a red ball. By Principle P, m = 6 and n = 16, so $p(R) = \frac{6}{16}$, or $\frac{3}{8}$.

13. There are 16 equally likely outcomes, and there are 0 ways to get a chartreuse ball. By Principle P, m = 0 and n = 16, so $p(C) = \frac{0}{16}$, or 0.

15. There are 36 equally likely outcomes (see Section 5.4, Example 12), and there is 1 one way to get a sum of 2 (1 on each die). By Principle P, m = 1 and n = 36, so p(2) = $\frac{1}{36}$.

17. There are 36 equally likely outcomes (see Section 5.4, Example 12), and there are 3 ways to get a sum of 10 (4 + 6, 5 + 5, and 6 + 4). By Principle P, m = 3 and n = 36, so p(10) = $\frac{3}{36}$, or $\frac{1}{12}$.

19. There are 36 equally likely outcomes (see Section 5.4, Example 12), and there are 6 ways to get a sum of 7 (1 + 6, 2 + 5, 3 + 4, 4 + 3, 5 + 2, 6 + 1). By Principle P, m = 6 and n = 36, so p(7) = $\frac{6}{36}$, or $\frac{1}{6}$.

21. There are 36 equally likely outcomes, and there are 4 ways to get a product of 6 (1·6, 2·3, 3·2, 6·1). By Principle P, m = 4 and n = 36, so p(6) = $\frac{4}{36}$, or $\frac{1}{9}$.

23. There are 36 equally likely outcomes, and there is 1 way to get a product of 25 (5·5). By Principle P, m = 1 and n = 36, so p(25) = $\frac{1}{36}$.

25. There are 12 equally likely outcomes, and there are 3 ways to get a head on the coin and an odd number on the die (H1, H3, H5). By Principle P, m = 3 and n = 12, so p (head on the coin, odd number on the die) = $\frac{3}{12}$, or $\frac{1}{4}$.

27. There are 16 equally likely outcomes, and there are 4 ways of selecting a red marble. By Principle P, m = 4 and n = 16, so p(R) = $\frac{4}{16}$, or $\frac{1}{4}$.

29. There are 52 equally likely outcomes, and there are 4 ways to draw a queen. Therefore, p(Q) = $\frac{4}{52}$, or $\frac{1}{13}$.

31. There are 52 equally likely outcomes, and there are 13 ways to get a club. Therefore, p(C) = $\frac{13}{52}$, or $\frac{1}{4}$.

33. There are 52 equally likely outcomes, and there are 4 ways to get an 8. Therefore, p(8) = $\frac{4}{52}$, or $\frac{1}{13}$.

35. There are 52 equally likely outcomes, and there are 26 ways to get a black card. Therefore, p(B) = $\frac{26}{52}$, or $\frac{1}{2}$.

37. There are 52 equally likely outcomes, and there are 8 ways to get a 7 or a jack. Therefore, p(7 or J) = $\frac{8}{52}$, or $\frac{2}{13}$.

39. The number of ways of selecting 4 people from a group of 15 is $\binom{15}{4}$. The number of ways of selecting 2 men from a group of 7 is $\binom{7}{2}$. The number of ways of selecting 2 women from a group of 8 is $\binom{8}{2}$. By the Fundamental Counting Principle, the number of ways of selecting 2 men and 2 women is $\binom{7}{2} \cdot \binom{8}{2}$. Then p(2 men and 2 women) is $\dfrac{\binom{7}{2} \cdot \binom{8}{2}}{\binom{15}{4}} = \dfrac{\frac{7 \cdot 6}{2 \cdot 1} \cdot \frac{8 \cdot 7}{2 \cdot 1}}{\frac{15 \cdot 14 \cdot 13 \cdot 12}{4 \cdot 3 \cdot 2 \cdot 1}} = \frac{28}{65}$.

41. The number of ways to select 7 coins from a group of 25 is $\binom{25}{7}$. The number of ways to select 3 dimes from a group of 7 is $\binom{7}{3}$. The number of ways to select 2 nickels from a group of 8 is $\binom{8}{2}$, and the number of ways to select 2 quarters from a group of 10 is $\binom{10}{2}$. By the Fundamental Counting Principle, the number of ways to select 3 dimes, 2 nickels, and 2 quarters is $\binom{7}{3} \cdot \binom{8}{2} \cdot \binom{10}{2}$. Then p(3 dimes, 2 nickels, and 2 quarters) = $\dfrac{\binom{7}{3} \cdot \binom{8}{2} \cdot \binom{10}{2}}{\binom{25}{7}} = \dfrac{\frac{7 \cdot 6 \cdot 5}{3 \cdot 2 \cdot 1} \cdot \frac{8 \cdot 7}{2 \cdot 1} \cdot \frac{10 \cdot 9}{2 \cdot 1}}{\frac{25 \cdot 24 \cdot 23 \cdot 22 \cdot 21 \cdot 20 \cdot 19}{7 \cdot 6 \cdot 5 \cdot 4 \cdot 3 \cdot 2 \cdot 1}} = \frac{441}{4807}$.

43. The number of ways to deal 5 cards from a deck of 52 is $\binom{52}{5}$. The total number of ways to deal 3 sevens from a group of 4 and 2 kings from a group of 4 is $\binom{4}{3} \cdot \binom{4}{2}$. Then p(3 sevens and 2 kings) = $\dfrac{\binom{4}{3} \cdot \binom{4}{2}}{\binom{52}{5}} = \dfrac{\frac{4 \cdot 3 \cdot 2}{3 \cdot 2 \cdot 1} \cdot \frac{4 \cdot 3}{2 \cdot 1}}{\frac{52 \cdot 51 \cdot 50 \cdot 49 \cdot 48}{5 \cdot 4 \cdot 3 \cdot 2 \cdot 1}} = \frac{1}{108,290}$.

45. The number of ways to deal 5 cards from a deck of 52 is $\binom{52}{5}$. The total number of ways to deal 2 jacks from a group of 4 and 3 aces from a group of 4 is $\binom{4}{2} \cdot \binom{4}{3}$. Then, as in Exercise 43, p(2 jacks and 3 aces) = $\dfrac{\binom{4}{2} \cdot \binom{4}{3}}{\binom{52}{5}} = \frac{1}{108,290}$.

47. The number of ways to deal 5 cards from a deck of 52 is $\binom{52}{5}$. The number of ways to deal 5 aces from a group of 4 is 0. Then p(5 aces) = $\dfrac{0}{\binom{52}{5}} = 0$.

49. The number of ways to deal 5 cards from a deck of 52 is $\binom{52}{5}$. The number of ways to deal 5 spades from a group of 13 is $\binom{13}{5}$. Then p(5 spades) = $\dfrac{\binom{13}{5}}{\binom{52}{5}} = \dfrac{\frac{13\cdot12\cdot11\cdot10\cdot9}{5\cdot4\cdot3\cdot2\cdot1}}{\frac{52\cdot51\cdot50\cdot49\cdot48}{5\cdot4\cdot3\cdot2\cdot1}} = \dfrac{33}{66,640}$.

51. The number of ways to select 4 people from a group of 20 is $\binom{20}{4}$. The total number of ways to select 2 men from a group of 10 and 2 women from a group of 10 is $\binom{10}{2}\cdot\binom{10}{2}$. Then p(2 men and 2 women) = $\dfrac{\binom{10}{2}\cdot\binom{10}{2}}{\binom{20}{4}} = \dfrac{\frac{10\cdot9}{2\cdot1}\cdot\frac{10\cdot9}{2\cdot1}}{\frac{20\cdot19\cdot18\cdot17}{4\cdot3\cdot2\cdot1}} = \dfrac{135}{323}$.

53. The number of ways to select 2 people from a group of 8 is $\binom{8}{2}$. We will use this in each part of this exercise.

   a) The total number of ways to select 1 man from a group of 5 and 1 woman from a group of 3 is $\binom{5}{1}\cdot\binom{3}{1}$. Then p(1 man and 1 woman) = $\dfrac{\binom{5}{1}\cdot\binom{3}{1}}{\binom{8}{2}} = \dfrac{\frac{5}{1}\cdot\frac{3}{1}}{\frac{8\cdot7}{2\cdot1}} = \dfrac{15}{28}$.

   b) The number of ways to select 2 men from a group of 5 is $\binom{5}{2}$. Then p(both are men) = $\dfrac{\binom{5}{2}}{\binom{8}{2}} = \dfrac{\frac{5\cdot4}{2\cdot1}}{\frac{8\cdot7}{2\cdot1}} = \dfrac{5}{14}$.

   c) The number of ways to select 2 women from a group of 3 is $\binom{3}{2}$. Then p(both are women) = $\dfrac{\binom{3}{2}}{\binom{8}{2}} = \dfrac{\frac{3\cdot2}{2\cdot1}}{\frac{8\cdot7}{2\cdot1}} = \dfrac{3}{28}$.

   d) p(both are men or both are women) = p(both are men) + p(both are women) = $\dfrac{5}{14} + \dfrac{3}{28} = \dfrac{10}{28} + \dfrac{3}{28} = \dfrac{13}{28}$.

55. a) $p(E') = 1 - p(E) = 1 - \dfrac{17}{45} = \dfrac{28}{45}$

   b) The odds for the event E to occur are p: 1 − p = 17:28. The odds against the event E are 1 − p: p = 28:17.

57. a) The odds for a person to reach the age of 100 are p: 1 − p = 1047: 98,953. Then the odds against a person reaching the age of 100 are 1 − p: p = 98,953: 1047.

   b) If the odds for an event E are m to n, then $p(E) = \dfrac{m}{m + n}$. Here m = 1047 and n = 98,953, so p(reaching the age of 100 in Alabama) = $\dfrac{1047}{100,000}$.

57. (continued)

   c) If the odds for an event E are m to n, then $p(E') = \dfrac{n}{m + n}$. Here m = 1047 and n = 98,953, so p(not reaching the age of 100 in Alabama) = $\dfrac{98,953}{100,000}$.

59. $p = \dfrac{1}{490,000}$, so $1 - p = 1 - \dfrac{1}{490,000} = \dfrac{489,999}{490,000}$. The odds for getting a ticket are the ratio $\dfrac{p}{1-p} = \dfrac{\frac{1}{490,000}}{\frac{489,999}{490,000}} = \dfrac{1}{489,999}$, or 1 to 489,999. The odds against getting a ticket are $\dfrac{\frac{489,999}{490,000}}{\frac{1}{490,000}} = \dfrac{489,999}{1}$, or 489,999 to 1.

61. a) Since the odds against winning a $20 prize on one ticket are 200:1, then the odds for winning a $20 prize on one ticket are 1:200. Therefore, p(winning $20 on one lottery ticket) = $\dfrac{1}{1 + 200} = \dfrac{1}{201}$.

   b) Since the odds against winning $500 on one ticket are 250,000:3, the odds for this event are 3:250,000. Then p(winning $500 on one lottery ticket) = $\dfrac{3}{3 + 250,000} = \dfrac{3}{250,003}$.

   c) Since the odds against winning $1000 on one ticket are 500,000:3, the odds for this event are 3:500,000. Then p(winning $1000 one one lottery ticket) = $\dfrac{3}{3 + 500,000} = \dfrac{3}{500,003}$.

   d) p(winning $20 on two lottery tickets) = p(winning $20 on one lottery ticket) + p(winning $20 on one lottery ticket) = $\dfrac{1}{201} + \dfrac{1}{201} = \dfrac{2}{201}$.

63. a) There are 38 equally likely outcomes, and there are 18 ways to get a black slot. Therefore, p = p(ball falls in a black slot) $= \dfrac{18}{38} = \dfrac{9}{19}$.

   b) p(ball does not fall in a black slot) = 1 − p = $1 - \dfrac{9}{19} = \dfrac{10}{19}$. Then the odds for the ball to fall in a black slot are $\dfrac{\frac{9}{19}}{\frac{10}{19}} = \dfrac{9}{10}$, or 9:10. The odds against the ball's falling in a black slot are $\dfrac{\frac{10}{19}}{\frac{9}{19}} = \dfrac{10}{9}$, or 10:9.

65. a) There are 38 equally likely outcomes. There are 18 ways to get a red slot and 18 ways to get a black slot, so the total number of ways to get a red or a black slot is 18 + 18, or 36. Then p = p(ball falls in a red or black slot) $= \dfrac{36}{38} = \dfrac{18}{19}$.

65. (continued)

b) p(ball does not fall in a red or black slot) = $1 - p = 1 - \frac{18}{19} = \frac{1}{19}$. Then the odds $\underline{for}$ the ball

to fall in a red or black slot are $\frac{\frac{18}{19}}{\frac{1}{19}} = \frac{18}{1}$,

or 18:1. The odds $\underline{against}$ the ball's falling

in a red or black slot are $\frac{\frac{1}{19}}{\frac{18}{19}} = \frac{1}{18}$, or 1:18.

67. a) There are 38 equally likely outcomes, and there is 1 way to get the 0 slot. Therefore, p = p(ball falls in the 0 slot) = $\frac{1}{38}$.

b) p(ball does not fall in the 0 slot) = 1 - p = $1 - \frac{1}{38} = \frac{37}{38}$. Then the odds $\underline{for}$ the ball to fall

in the 0 slot are $\frac{\frac{1}{38}}{\frac{37}{38}} = \frac{1}{37}$, or 1:37. The odds

against the ball's falling in the 0 slot are $\frac{\frac{37}{38}}{\frac{1}{38}} = \frac{37}{1}$, or 37:1.

69. a), b) The odd-numbered slots are the black slots, so the solution of this exercise is the same as the solution of Exercise 63.

71. Think of dividing the dartboard into 18 equal regions as shown:

R	Y	R
B	Y	R
G	B	R
Y	B	G
Y	R	G
Y	R	G

Then there are 18 equally likely outcomes when a dart thrown randomly at the board always hits the board.

There are 6 ways to hit R, so p(R) = $\frac{6}{18} = \frac{1}{3}$.

There are 3 ways to hit B, so p(B) = $\frac{3}{18} = \frac{1}{6}$.

There are 4 ways to hit G, so p(G) = $\frac{4}{18} = \frac{2}{9}$.

There are 5 ways to hit Y, so p(Y) = $\frac{5}{18}$.

73. The number of ways in which 5 cards can be selected from a group of 52 cards is $_{52}C_5 = \frac{52\cdot51\cdot50\cdot49\cdot48}{5\cdot4\cdot3\cdot2\cdot1} = 2,598,960$.

75. a) In each suit we have the following 9 possibilities for straight flushes:

K-Q-J-10-9, Q-J-10-9-8, J-10-9-8-7, 10-9-8-7-6, 9-8-7-6-5, 8-7-6-5-4, 7-6-5-4-3, 6-5-4-3-2, 5-4-3-2-A

Then the total number of straight flushes is 4·9 = 36.

b) There are 2,598,960 equally likely outcomes (see Exercise 73), and 36 are straight flushes. Then p = p(straight flush) = $\frac{36}{2,598,960} \approx$ 0.0000139.

c) Since p = $\frac{36}{2,598,960}$, then 1 - p = $1 - \frac{36}{2,598,960} = \frac{2,598,924}{2,598,960}$. The odds for getting

a straight flush are $\frac{\frac{36}{2,598,960}}{\frac{2,598,924}{2,598,960}} = \frac{36}{2,598,924} = \frac{3}{216,577}$. This can be expressed as 36 to

2,598,924 or 3 to 216,577. The odds against getting a straight flush are $\frac{\frac{2,598,924}{2,598,960}}{\frac{36}{2,598,960}} =$

$\frac{2,598,924}{36} = \frac{216,577}{3}$. This can be expressed as 2,598,924 to 36 or 216,577 to 3.

77. a) The number of ways we can choose a pair from a group of 4 $\underline{and}$ 3 of a kind from a group of 4 is $\binom{4}{2}\cdot\binom{4}{3}$. The number of ways we can choose one denomination for the pair and another denomination for the three of a kind is $_{13}P_2$. Then the number of full houses is $\binom{4}{2}\binom{4}{3}{}_{13}P_2 = \frac{4\cdot3}{2\cdot1}\cdot\frac{4\cdot3\cdot2}{3\cdot2\cdot1}\cdot13\cdot12 = 3744$.

b) There are 2,598,960 equally likely outcomes, and 3744 are full houses. Then p = p(full house) = $\frac{3744}{2,598,960} \approx 0.00144$.

c) Since p = $\frac{3744}{2,598,960}$, then 1 - p = $1 - \frac{3744}{2,598,960} = \frac{2,595,216}{2,598,960}$. The odds for getting

a full house are $\frac{\frac{3744}{2,598,960}}{\frac{2,595,216}{2,598,960}} = \frac{3744}{2,595,216} = \frac{78}{54,067}$.

This can expressed as 3744 to 2,595,216 or 78 to 54,067. The odds against getting a full

house are $\frac{\frac{2,595,216}{2,598,960}}{\frac{3744}{2,598,960}} = \frac{2,595,216}{3744} = \frac{54,067}{78}$.

This can be expressed as 2,595,216 to 3744 or 54,067 to 78.

79. a) The number of ways 3 cards can be selected from a group of 4 is $\binom{4}{3}$. Two other cards can be selected from the 48 cards of denominations different from the three-of-a-kind in $\binom{48}{2}$ ways. Any of the 13 denominations can be chosen for the three-of-a-kind. The total number of hands consisting of three-of-a-kind and 2 other cards with denominations different from the three-of-a-kind is $13\binom{4}{3}\binom{48}{2}$. However, some of these hands will be full houses. That is, the two other cards selected to go with the three-of-a-kind will have the same denominations. We exclude the 3744 full houses (see Exercise 77) in finding the total number of three-of-a-kind hands:

$$13\binom{4}{3}\binom{48}{2} - 3744 = 13 \cdot \frac{4\cdot3\cdot2}{3\cdot2\cdot1} \cdot \frac{48\cdot47}{2\cdot1} - 3744 = 54,912.$$

b) There are 2,598,960 equally likely outcomes, and 54,912 are three-of-a-kind. Then p = p(three-of-a-kind) = $\frac{54,912}{2,598,960}$ ≈ 0.0211.

c) Since p = $\frac{54,912}{2,598,960}$, then 1 - p = $1 - \frac{54,912}{2,598,960} = \frac{2,544,048}{2,598,960}$. The odds for getting three-of-a-kind are $\frac{\frac{54,912}{2,598,960}}{\frac{2,544,048}{2,598,960}} = \frac{54,912}{2,544,048} = \frac{1144}{53,001}$. The odds can be expressed as 54,912 to 2,544,048 or 1144 to 53,001. The odds against getting three-of-a-kind are $\frac{\frac{2,544,048}{2,598,960}}{\frac{54,912}{2,598,960}} = \frac{2,544,048}{54,912} = \frac{53,001}{1144}$. The odds can be expressed as 2,544,048 to 54,912 or 53,001 to 1144.

81. a) We can choose 2 cards from one group of 4, 2 cards from another group of 4, and 1 card from the 44 cards with denominations different from those already chosen in $\binom{4}{2}\binom{4}{2}\binom{44}{1}$ ways. We can choose the 2 denominations of the pairs from the 13 denominations in $\binom{13}{2}$ ways. The number of two pairs hands is then

$$\binom{13}{2}\binom{4}{2}\binom{4}{2}\binom{44}{1} = \frac{13\cdot12}{2\cdot1} \cdot \frac{4\cdot3}{2\cdot1} \cdot \frac{4\cdot3}{2\cdot1} \cdot \frac{44}{1} = 123,552.$$

b) There are 2,598,960 equally likely outcomes and 123,552 are two pairs hands. Then p = p(two pairs) $= \frac{123,552}{2,598,960}$ ≈ 0.0475.

81. (continued)

c) Since p = $\frac{123,552}{2,598,960}$, then 1 - p = $1 - \frac{123,552}{2,598,960} = \frac{2,475,408}{2,598,960}$. The odds for two pairs are $\frac{\frac{123,552}{2,598,960}}{\frac{2,475,408}{2,598,960}} = \frac{123,552}{2,475,408} = \frac{2574}{51,571}$. These odds can be expressed as 123,552 to 2,475,408 or 2574 to 51,571. The odds against two pairs are $\frac{\frac{2,475,408}{2,598,960}}{\frac{123,552}{2,598,960}} = \frac{2,475,408}{123,552} = \frac{51,571}{2574}$. These odds can be expressed as 2,475,408 to 123,552 or 51,571 to 2574.

83. There are 27 equally likely outcomes for filling each position. (They are the 26 letters of the alphabet and a space.) The probability of filling each position with the one choice shown is $\frac{1}{27}$. The passage contains 50 positions to be filled, each with p = $\frac{1}{27}$. Therefore, the probability this passage could have been written by a monkey is $\left(\frac{1}{27}\right)^{50}$, or $\frac{1}{27^{50}}$.

Exercise Set 5.5

1. p(E ∪ F) = p(E) + p(F) - p(E ∩ F)
   = 0.73 + 0.24 - 0.20 = 0.77
   p[(E ∪ F)'] = 1 - p(E ∪ F) = 1 - 0.77 = 0.23

3. Since E and F are mutually exclusive, p(E ∪ F) = p(E) + p(F) = 37.5% + 33.3% = 70.8%.
   p[(E ∪ F')] = 1 - p(E ∪ F) = 1 - 70.8% = 29.2%

5. There are 31 equally likely outcomes.
   a) One outcome is August 9th, so p(August 9) = $\frac{1}{31}$.
   b) One outcome is August 31st, so p(August 31) = $\frac{1}{31}$.
   c) There are 11 outcomes after August 20th, each with probability $\frac{1}{31}$. These outcomes or events are mutually exclusive (the date chosen cannot be two dates), so p(after August 20) is the sum of the eleven probabilities of $\frac{1}{31}$, or $\frac{11}{31}$.
   d) There are 14 outcomes before August 15th, each with probability $\frac{1}{31}$. These outcomes or events are mutually exclusive so p(before August 15) is the sum of these fourteen probabilities, or $\frac{14}{31}$.
   e) It is impossible to choose a date that is after August 20th and before August 15, so p = 0.

5. (continued)

f) p(after August 20 ∪ before August 15) =

p(after August 20) + p(before August 15) -

p(after August 20 ∩ before August 15) =

$\frac{11}{31} + \frac{14}{31} - 0 = \frac{25}{31}$.

g) Since p(after August 20 ∩ before August 15) = 0, the events are mutually exclusive.

7. Let R = event of getting a red marble,

G = event of getting a green marble, and

B = event of getting a black marble.

There are 22 equally likely outcomes.

a) Seven of the outcomes are red, so p(R) = $\frac{7}{22}$.

b) Four of the outcomes are green, so p(G) = $\frac{4}{22}$ = $\frac{2}{11}$.

c) Eleven of the outcomes are black, so p(B) = $\frac{11}{22} = \frac{1}{2}$.

d) R ∪ G = event of getting red or green. The events R and G are mutually exclusive (in one draw and we cannot get both a red and a green), so p(R ∪ G) = p(R) + p(G) = $\frac{7}{22} + \frac{2}{11} = \frac{7}{22} + \frac{4}{22} = \frac{11}{22} = \frac{1}{2}$.

e) G ∪ B = event of getting green or black. The events G and B are mutually exclusive, so p(G ∪ B) = $\frac{2}{11} + \frac{1}{2} = \frac{4}{22} + \frac{11}{22} = \frac{15}{22}$.

f) R ∪ B = event of getting red or black. The events R and B are mutually exclusive, so p(R ∪ B) = $\frac{7}{22} + \frac{1}{2} = \frac{7}{22} + \frac{11}{22} = \frac{18}{22} = \frac{9}{11}$.

g) R ∪ G ∪ B = event of getting red or green or black. These events are mutually exclusive, so p(R ∪ G ∪ B) = $\frac{7}{22} + \frac{2}{11} + \frac{1}{2} = \frac{7}{22} + \frac{4}{22} + \frac{11}{22} = \frac{22}{22} = 1$.

9. There are 52 equally likely outcomes.

a) Let B = event of drawing a black card and

S = event of drawing a spade.

Now p(B) = $\frac{26}{52}$ since half the cards are black,

p(S) = $\frac{13}{52}$ since there are 13 spades, and

p(B ∩ S) = p(the card is a black spade) = $\frac{13}{52}$ since all the spades are black. Then

p(B ∪ S) = p(B) + p(S) - p(B ∩ S) =

$\frac{26}{52} + \frac{13}{52} - \frac{13}{52} = \frac{26}{52} = \frac{1}{2}$.

9. (continued)

b) Let R = event of drawing a red card and

T = event of drawing a ten.

Now p(R) = $\frac{26}{52}$ since half the cards are red,

p(T) = $\frac{4}{52}$ since there are 4 tens, and

p(R ∩ T) = p(the card is a red ten) = $\frac{2}{52}$

since there are 2 red tens. Then p(R ∪ T) =

p(R) + p(T) - p(R ∩ T) = $\frac{26}{52} + \frac{4}{52} - \frac{2}{52} = \frac{28}{52} = \frac{7}{13}$.

c) Let S = event of drawing a spade and

D = event of drawing a diamond.

These events are mutually exclusive since it is impossible to draw a spade and a diamond at the same time. Now p(S) = $\frac{13}{52}$ since there are 13 spades and p(D) = $\frac{13}{52}$ since there are 13 diamonds. Then p(S ∪ D) = p(S) + p(D) = $\frac{13}{52} + \frac{13}{52} = \frac{26}{52} = \frac{1}{2}$.

d) Define S and D as in part c), and let H = event of drawing a heart. Now p(H) = $\frac{13}{52}$ since there are 13 hearts. The events S, D, and H are mutually exclusive, so p(S ∪ D ∪ H) = p(S) + p(D) + p(H) = $\frac{13}{52} + \frac{13}{52} + \frac{13}{52} = \frac{39}{52} = \frac{3}{4}$.

e) Define S, D, and H as in part d), and let C = event of drawing a club. p(C) = $\frac{13}{52}$. Then

p(S ∪ D ∪ H ∪ C) = $\frac{13}{52} + \frac{13}{52} + \frac{13}{52} + \frac{13}{52} = \frac{52}{52} = 1$.

11. There are 10 + 21 + 18 + 36 = 85 equally likely outcomes. The events 4 games, 5 games, 6 games, and 7 games are mutually exclusive.

a) 10 of the outcomes are 4 games, so p(4) = $\frac{10}{85}$, or $\frac{2}{17}$.

b) 21 of the outcomes are 5 games, so p(5) = $\frac{21}{85}$.

c) 18 of the outcomes are 6 games, so p(6) = $\frac{18}{85}$.

d) 36 of the outcomes are 7 games, so p(7) = $\frac{36}{85}$.

e) p(more than 4 games) = p(5 ∪ 6 ∪ 7) = p(5) + p(6) + p(7) = $\frac{21}{85} + \frac{18}{85} + \frac{36}{85} = \frac{75}{85}$, or $\frac{15}{17}$.

f) p(more than 5 games) = p(6 ∪ 7) = p(6) + p(7) = $\frac{18}{85} + \frac{36}{85} = \frac{54}{85}$.

g) p(more than 6 games) = p(7) = $\frac{36}{85}$.

h) p(4 games or more) = p(4 ∪ 5 ∪ 6 ∪ 7) = $\frac{10}{85} + \frac{21}{85} + \frac{18}{85} + \frac{36}{85} = \frac{85}{85} = 1$.

13. Let R = the event of buying rock music,

    P = the event of buying pop music,

    S = the event of buying soul music,

    C = the event of buying country music, and

    L = the event of buying classical music.

    These events are mutually exclusive.

    a) p(rock or pop) = p(R ∪ P) = p(R) + p(P) =

    43% + 17% = 60%

    b) p(country or soul) = p(C ∪ S) = p(C) + p(S) =

    10% + 10% = 20%

    c) p(classical or rock) = p(L ∪ R) = p(L) + p(R) =

    5% + 43% = 48%

    d) p(rock or pop or soul or country) =

    p(R ∪ P ∪ S ∪ C) = 43% + 17% + 10% + 10% = 80%

15. Check the sample space for this experiment in Example 12 of Section 5.4. p(sum is 7) = p(first is 1 and second is 6 ∪ first is 2 and second is 5 ∪ first is 3 and second is 4 ∪ first is 4 and second is 3 ∪ first is 5 and second is 2 ∪ first is 6 and second is 1) = p(first is 1 and second is 6) + p(first is 2 and second is 5) + p(first is 3 and second is 4) + p(first is 4 and second is 3) + p(first is 5 and second is 2) + p(first is 6 and second is 1) =

$\frac{1}{36} + \frac{1}{36} + \frac{1}{36} + \frac{1}{36} + \frac{1}{36} + \frac{1}{36} = \frac{6}{36} = \frac{1}{6}$

17. Let E_i be the event of drawing exactly i defective chips and $p_i = p(E_i)$. Getting different numbers of defective chips are mutually exclusive events, so $p_0 + p_1 + p_2 = 1$.

    a) The number of ways 2 parts can be drawn from a group of 20 is $\binom{20}{2}$, and the number of ways 2 nondefective chips can be drawn from the 20 − 3, or 17, nondefective chips is $\binom{17}{2}$. Then

    $$p_0 = \frac{\binom{17}{2}}{\binom{20}{2}} = \frac{\frac{17\cdot16}{2\cdot1}}{\frac{20\cdot19}{2\cdot1}} = \frac{17\cdot16}{20\cdot19} = \frac{68}{95}.$$

    b) p(at least one is defective) = p(exactly one is defective ∪ two are defective) = $p_1 + p_2$. The number of ways 1 nondefective chip can be drawn from 17 chips and 1 defective chip can be drawn from 3 chips is $\binom{17}{1}\binom{3}{1}$, so $p_1 = \frac{\binom{17}{1}\binom{3}{1}}{\binom{20}{2}}$.

    The number of ways 2 defective chips can be drawn from 3 chips is $\binom{3}{2}$, so $p_2 = \frac{\binom{3}{2}}{\binom{20}{2}}$. Then,

    $$p_1 + p_2 = \frac{\binom{17}{1}\binom{3}{1}}{\binom{20}{2}} + \frac{\binom{3}{2}}{\binom{20}{2}} = \frac{\frac{17}{1}\cdot\frac{3}{1}}{\frac{20\cdot19}{2\cdot1}} + \frac{\frac{3\cdot2}{2\cdot1}}{\frac{20\cdot19}{2\cdot1}} =$$

    $\frac{51}{190} + \frac{3}{190} = \frac{54}{190} = \frac{27}{95}$. We could find this result more simply as follows:

    p(at least one is defective) = $p_1 \geqslant 1$ =

    $1 - p_0 = 1 - \frac{68}{95} = \frac{27}{95}$.

17. (continued)

    c) p(only one is defective) = $p_1 = \frac{51}{190}$ (See part b).)

    d) p(both are defective) = $p_2 = \frac{3}{190}$ (See part b).)

19. The number of ways 2 people can be chosen from a group of 10 is $\binom{10}{2} = \frac{10\cdot9}{2\cdot1} = 45$.

    a) The number of ways to choose 1 man from a group of 5 is $\binom{5}{1}$, and the number of ways to choose 1 woman from a group of 5 is $\binom{5}{1}$. Then, the number of ways to choose 1 man and 1 woman is $\binom{5}{1}\binom{5}{1}$, and p(1 man and 1 woman) = $\frac{\binom{5}{1}\binom{5}{1}}{45} = \frac{\frac{5}{1}\cdot\frac{5}{1}}{45} = \frac{25}{45} = \frac{5}{9}$.

    b) The number of ways to choose 2 men from a group of 5 is $\binom{5}{2}$, so p(2 men) = $\frac{\binom{5}{2}}{45} = \frac{\frac{5\cdot4}{2\cdot1}}{45} = \frac{10}{45} = \frac{2}{9}$. The number of ways to choose 2 women from a group of 5 is also $\binom{5}{2}$, so p(2 women) = $\frac{\binom{5}{2}}{45} = \frac{2}{9}$. The events 2 men and 2 women are mutually exclusive, so p(2 men or 2 women) = p(2 men) + p(2 women) = $\frac{2}{9} + \frac{2}{9} = \frac{4}{9}$.

    c) The number of ways to choose 2 married people from a group of 10 is $\binom{10}{2}$, so p(they are married) = $\frac{\binom{10}{2}}{45} = \frac{45}{45} = 1$.

    d) The number of ways to choose 1 married couple from a group of 5 couples is $\binom{5}{1}$, so p(they are married to each other) = $\frac{\binom{5}{1}}{45} = \frac{\frac{5}{1}}{45} = \frac{5}{45} = \frac{1}{9}$.

21. The number of ways 2 cards can be dealt from a deck of 52 cards is $\binom{52}{2} = \frac{52\cdot51}{2\cdot1} = 1326$.

    a) The number of ways 2 aces can be dealt from a group of 4 aces is $\binom{4}{2}$, so p(both are aces) = $\frac{\binom{4}{2}}{1326} = \frac{\frac{4\cdot3}{2\cdot1}}{1326} = \frac{6}{1326} = \frac{1}{221}$.

    b) In a particular denomination, the number of ways 2 cards (a pair) can be dealt from a group of 4 cards is $\binom{4}{2}$. There are 13 denominations, so the total number of ways to deal a pair is $13\cdot\binom{4}{2}$. Then p(they are a pair) = $\frac{13\cdot\binom{4}{2}}{1326} = \frac{13\cdot\frac{4\cdot3}{2\cdot1}}{1326} = \frac{78}{1326} = \frac{1}{17}$.

21. (continued)

c) The number of ways 2 cards can be chosen in a particular suit is $\binom{13}{2}$. There are 4 suits, so the total number of ways to deal 2 cards from the same suit is $4 \cdot \binom{13}{2}$. Then p(both are the same suit) $= \dfrac{4 \cdot \binom{13}{2}}{1326} = \dfrac{4 \cdot \frac{13 \cdot 12}{2 \cdot 1}}{1326} = \dfrac{312}{1326} = \dfrac{4}{17}$.

d) The events P = pair and S = same suit are mutually exclusive. Then p(they are neither a pair nor the same suit) $= 1 - p(P \cup S) = 1 - [p(P) + p(S)] = 1 - \left[\dfrac{1}{17} + \dfrac{4}{17}\right] = 1 - \dfrac{5}{17} = \dfrac{12}{17}$.

23. The number of ways 4 seats can be chosen from a group of 6 seats is $_6C_4$. For each set of 4 seats, the 4 people can be arranged in $_4P_4$ ways. Therefore, the total number of seating arrangements is $_6C_4 \cdot {}_4P_4$. In a group of 6 seats there are 5 pairs of adjacent seats in which the given couple can be seated. For each such pair there are $_2P_2$ arrangements of the couple. The number of ways 2 seats can be chosen from the remaining 4 seats for the other couple is $_4C_2$, and for each set of 2 seats the couple can be arranged in $_2P_2$ ways. Therefore, the total number of ways a given couple can be seated together is $5 \cdot {}_2P_2 \cdot {}_4C_2 \cdot {}_2P_2$. Then p(a given couple will sit together) $= \dfrac{5 \cdot {}_2P_2 \cdot {}_4C_2 \cdot {}_2P_2}{{}_6C_4 \cdot {}_4P_4} = \dfrac{5 \cdot 2 \cdot 1 \cdot \frac{4 \cdot 3}{2 \cdot 1} \cdot 2 \cdot 1}{\frac{6 \cdot 5 \cdot 4 \cdot 3}{4 \cdot 3 \cdot 2 \cdot 1} \cdot 4 \cdot 3 \cdot 2 \cdot 1} = \dfrac{120}{360} = \dfrac{1}{3}$.

25. Let S = the event a person smoked and

D = the event a person drank.

a) $p(S \cup D) = p(S) + p(D) - p(S \cap D) = 43\% + 67\% - 24\% = 86\%$

b) The events "smoked or drank" and "neither smoked nor drank" are mutually exclusive, and they partition the sample space. Therefore, p(neither smoked nor drank) $= 1 - p(S \cup D) = 1 - 86\% = 14\%$.

27. The events A = being Type A, B = being Type B, AB = being Type AB, and O = being Type O are mutually exclusive, and they partition the sample space. Then $p(A) + p(B) + p(AB) + p(O) = 1$, so $p(O) = 1 - p(A) - p(B) - p(AB) = 1 - \dfrac{53}{140} - \dfrac{47}{140} - \dfrac{24}{140} = \dfrac{16}{140}$, or $\dfrac{4}{35}$.

29. Since the events form a partition of the sample space, $p(E_1) + p(E_2) + p(E_3) + p(E_4) + p(E_5) = 1$. Then $p(E_2) = 1 - p(E_1) - p(E_3) - p(E_4) - p(E_5) = 1 - \dfrac{1}{10} - \dfrac{1}{4} - \dfrac{1}{5} - \dfrac{3}{20} = \dfrac{20}{20} - \dfrac{2}{20} - \dfrac{5}{20} - \dfrac{4}{20} - \dfrac{3}{20} = \dfrac{6}{20} = \dfrac{3}{10}$.

31. The number of ways 4 balls can be chosen from a group of 16 is $_{16}C_4 = \dfrac{16 \cdot 15 \cdot 14 \cdot 13}{4 \cdot 3 \cdot 2 \cdot 1} = 1820$.

Let W = the event of drawing a white ball,

B = the event of drawing a blue ball, and

R = the event of drawing a red ball.

Each color can be represented in a sample of 4 balls in 3 ways: W, W, B, R; W, B, B, R; W, B, R, R

$p(W, W, B, R) = \dfrac{_5C_2 \cdot {}_4C_1 \cdot {}_7C_1}{1820} = \dfrac{\frac{5 \cdot 4}{2 \cdot 1} \cdot \frac{4}{1} \cdot \frac{7}{1}}{1820} = \dfrac{280}{1820}$

$p(W, B, B, R) = \dfrac{_5C_1 \cdot {}_4C_2 \cdot {}_7C_1}{1820} = \dfrac{\frac{5}{1} \cdot \frac{4 \cdot 3}{2 \cdot 1} \cdot \frac{7}{1}}{1820} = \dfrac{210}{1820}$

$p(W, B, R, R) = \dfrac{_5C_1 \cdot {}_4C_1 \cdot {}_7C_2}{1820} = \dfrac{\frac{5}{1} \cdot \frac{4}{1} \cdot \frac{7 \cdot 6}{2 \cdot 1}}{1820} = \dfrac{420}{1820}$

The events W, W, B, R; W, B, B, R; and W, B, R, R are mutually exclusive so p(each color is represented in a sample of 4 balls) $= \dfrac{280}{1820} + \dfrac{210}{1820} + \dfrac{420}{1820} = \dfrac{910}{1820} = \dfrac{1}{2}$. The correct answer is Choice E.

Exercise Set 6.1

1. $p(E) = \frac{3}{4}$, $p(F) = \frac{5}{7}$, and $p(E \cap F) = \frac{2}{3}$.

$$p(E|F) = \frac{p(F \cap E)}{p(F)} = \frac{\frac{2}{3}}{\frac{5}{7}} = \frac{2}{3} \cdot \frac{7}{5} = \frac{14}{15}$$

$$p(F|E) = \frac{p(E \cap F)}{p(E)} = \frac{\frac{2}{3}}{\frac{3}{4}} = \frac{2}{3} \cdot \frac{4}{3} = \frac{8}{9}$$

$$p(E|F') = \frac{p(F' \cap E)}{p(F')} = \frac{p(E) - p(E \cap F)}{1 - p(F)} = \frac{\frac{3}{4} - \frac{2}{3}}{1 - \frac{5}{7}} =$$

$$\frac{\frac{1}{12}}{\frac{2}{7}} = \frac{1}{12} \cdot \frac{7}{2} = \frac{7}{24}$$

$$p(F|E') = \frac{p(E' \cap F)}{p(E')} = \frac{p(F) - p(F \cap E)}{1 - p(E)} = \frac{\frac{5}{7} - \frac{2}{3}}{1 - \frac{3}{4}} =$$

$$\frac{\frac{1}{21}}{\frac{1}{4}} = \frac{1}{21} \cdot \frac{4}{1} = \frac{4}{21}$$

3. $p(E) = 0.3$, $p(F) = 0.5$, and $p(E \cap F) = 0.15$

$$p(E|F) = \frac{p(F \cap E)}{p(F)} = \frac{0.15}{0.5} = \frac{15}{50} = 0.3$$

$$p(F|E) = \frac{p(E \cap F)}{p(E)} = \frac{0.15}{0.3} = \frac{15}{30} = 0.5$$

$$p(E|F') = \frac{p(F' \cap E)}{p(F')} = \frac{p(E) - p(E \cap F)}{1 - p(F)} =$$

$$\frac{0.3 - 0.15}{1 - 0.5} = \frac{0.15}{0.5} = \frac{15}{50} = 0.3$$

$$p(F|E') = \frac{p(E' \cap F)}{p(E')} = \frac{p(F) - p(F \cap E)}{1 - p(E)} =$$

$$\frac{0.5 - 0.15}{1 - 0.3} = \frac{0.35}{0.7} = \frac{35}{70} = 0.5$$

5. $p(E) = 0.3$ and $p(F|E) = 0.4$

$p(E \cap F) = p(E)p(F|E)$ (Multiplication Theorem)

$\quad\quad\quad = 0.3(0.4)$

$\quad\quad\quad = 0.12$

7. $p(F) = \frac{2}{3}$ and $p(E|F) = \frac{2}{3}$

$p(E \cap F) = p(F)p(E|F)$ (Multiplication Theorem)

$\quad\quad\quad = \frac{2}{3} \cdot \frac{2}{3}$

$\quad\quad\quad = \frac{4}{9}$

9. The sample space is {BB, BG, GB, GG}.

a) There are 4 equally likely outcomes of which 1 is favorable. Then p(two girls) = $\frac{1}{4}$.

9. (continued)

b) The reduced sample space is {BG, GB, GG}. There are 3 equally like outcomes of which 1 is favorable. Then p(two girls given that it has at least one girl) = $\frac{1}{3}$.

c) The reduced sample space is {GB, GG}. There are 2 equally likely outcomes of which 1 is favorable. Then p(2 girls given that the oldest child is a girl) = $\frac{1}{2}$.

11. The sample space is {HH, HT, TH, TT}. The reduced sample space is {TH, TT}. There are 2 equally likely outcomes of which 1 is favorable. Then p(second is a head given that the first is a tail) = $\frac{1}{2}$.

13. Let P = the event that the product is 6 and
     T = the event that one die is a 2.
   Now P = {(1, 6), (2, 3), (3, 2), (6, 1)} and
   T = {(1, 2), (2, 2), (3, 2), (4, 2), (5, 2), (6, 2), (2, 1), (2, 3), (2, 4), (2, 5), (2, 6)}.
   (See the figure in Example 4.) Then $P \cap T$ = {(2, 3), (3, 2)}, and $p(T|P) = \frac{n(P \cap T)}{n(P)} = \frac{2}{4} = \frac{1}{2}$.

15. A total of 166,478 people (78,570 males + 87,908 females) were surveyed.

$p(M) = \frac{\text{number of males}}{\text{number of people}} = \frac{78,570}{166,478}$

The number who ever smoked is 89,684 (49,799 males + 39,885 females).

$p(S) = \frac{\text{number who ever smoked}}{\text{number of people}} = \frac{89,684}{166,478}$

The number who never smoked is 76,794 (28,771 males + 48,023 females).

$p(N) = \frac{\text{number who never smoked}}{\text{number of people}} = \frac{76,794}{166,478}$

The number of males who ever smoked is 49,799 so

$p(M \cap S) = \frac{49,799}{166,478}$.

The number of males who never smoked is 28,771 so

$p(M \cap N) = \frac{28,771}{166,478}$.

$$p(M|S) = \frac{p(S \cap M)}{p(S)} = \frac{\frac{49,799}{166,478}}{\frac{89,684}{166,478}} = \frac{49,799}{89,684}$$

$$p(S|M) = \frac{p(M \cap S)}{p(M)} = \frac{\frac{49,799}{166,478}}{\frac{78,570}{166,478}} = \frac{49,799}{78,570}$$

$$p(M|N) = \frac{p(N \cap M)}{p(N)} = \frac{\frac{28,771}{166,478}}{\frac{76,794}{166,478}} = \frac{28,771}{76,794}$$

$$p(N|M) = \frac{p(M \cap N)}{p(M)} = \frac{\frac{28,771}{166,478}}{\frac{78,570}{166,478}} = \frac{28,771}{78,570}$$

17. The number of males 18-24 years of age who ever smoked is 5,723, so $p(S \cap M_1) = \frac{5,723}{166,478}$. Then

$$p(M_1|S) = \frac{p(S \cap M_1)}{p(S)} = \frac{\frac{5,723}{166,478}}{\frac{89,684}{166,478}} = \frac{5,723}{89,684}.$$

The number of females 18-24 years of age who ever smoked is 6,682, so $p(S \cap F_1) = \frac{6,682}{166,478}$. Then

$$p(F_1|S) = \frac{p(S \cap F_1)}{p(S)} = \frac{\frac{6,682}{166,478}}{\frac{89,684}{166,478}} = \frac{6,682}{89,684}.$$

19. There are 74 color-blind people (64 males + 10 females). Thus, p(color-blind) = 74/1000.

21. There are 64 color-blind males, so p(male and color-blind) = 64/1000.

23. 500 females are not color-blind, so p(female and not color-blind) = 500/1000.

25. Let M = the event of being male and
    C = the event of being color-blind.

$$p(M|C) = \frac{p(C \cap M)}{p(C)} = \frac{\frac{64}{1000}}{\frac{74}{1000}} = \frac{64}{74}$$

(See Exercises 19 and 21.)

27. Define M and C as in Exercise 25.

$$p(C|M) = \frac{p(M \cap C)}{p(M)} = \frac{\frac{64}{1000}}{\frac{490}{1000}} = \frac{64}{490}.$$

(See Exercise 21.)

29. Let F = the event of being female and
    C = the event of being color-blind.

$$p(C'|F) = \frac{p(F \cap C')}{p(F)} = \frac{\frac{500}{1000}}{\frac{510}{1000}} = \frac{500}{510}$$

(See Exercise 23.)

31. a)

	1/8 → D	$p(A \cap D) = \frac{1}{3} \cdot \frac{1}{8} = \frac{1}{24}$
A	7/8 → N	$p(A \cap N) = \frac{1}{3} \cdot \frac{7}{8} = \frac{7}{24}$
1/3 B	3/11 → D	$p(B \cap D) = \frac{1}{3} \cdot \frac{3}{11} = \frac{1}{11}$
1/3	8/11 → N	$p(B \cap N) = \frac{1}{3} \cdot \frac{8}{11} = \frac{8}{33}$
1/3 C	2/7 → D	$p(C \cap D) = \frac{1}{3} \cdot \frac{2}{7} = \frac{2}{21}$
	5/7 → N	$p(C \cap N) = \frac{1}{3} \cdot \frac{5}{7} = \frac{5}{21}$

b) $p(D) = p(A \cap D) + p(B \cap D) + p(C \cap D)$

$$= \frac{1}{24} + \frac{1}{11} + \frac{2}{21} = \frac{421}{1848}$$

c) $p(N) = 1 - p(D) = 1 - \frac{421}{1848} = \frac{1427}{1848}$

33. There are 30 equally likely outcomes in B1 of which 5 are S, so $p(S|B1) = \frac{5}{30} = \frac{1}{6}$.

There are 33 equally likely outcomes in B2 of which 7 are S, so $P(S|B2) = \frac{7}{33}$.

There are 28 equally likely outcomes in B3 and in B4 of which 3 are S in each box, so $p(S|B3) = \frac{3}{28}$ and $p(S|B4) = \frac{3}{28}$.

There are 28 equally likely outcomes in B5 of which 6 are S, so $p(S|B5) = \frac{6}{28} = \frac{3}{14}$.

35. The reasoning in this exercise is the same as in Exercise 33.

$p(C|B1) = \frac{4}{30} = \frac{2}{15}$, $p(C|B2) = \frac{4}{33}$, $p(C|B3) = \frac{6}{28} = \frac{3}{14}$,

$p(C|B4) = \frac{3}{28}$, $p(C|B5) = \frac{4}{28} = \frac{1}{7}$

37. The reasoning in this exercise is the same as in Exercise 33.

$p(L|B1) = \frac{5}{30} = \frac{1}{6}$, $p(L|B2) = \frac{5}{33}$, $p(L|B3) = \frac{8}{28} = \frac{2}{7}$,

$p(L|B4) = \frac{8}{28} = \frac{2}{7}$, $p(L|B5) = \frac{5}{28}$

39. There are 33 equally likely outcomes in B2 of which 9 are O, so $p(O|B2) = \frac{9}{33} = \frac{3}{11}$.

There are 28 equally likely outcomes in B4 of which 6 are E, so $p(E|B4) = \frac{6}{28} = \frac{3}{14}$.

There are a total of 37 orange gum drops of which 6 are in B3, so $p(B3|O) = \frac{6}{37}$.

There are a total of 34 lemon gum drops of which 6 are in B4, so $p(B4|E) = \frac{6}{34} = \frac{3}{17}$.

There are 28 equally likely outcomes in B5 of which 5 are E, so $p(E|B5) = \frac{5}{28}$.

41. Construct a probability tree that describes the situation.

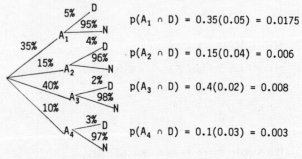

$p(A_1 \cap D) = 0.35(0.05) = 0.0175$

$p(A_2 \cap D) = 0.15(0.04) = 0.006$

$p(A_3 \cap D) = 0.4(0.02) = 0.008$

$p(A_4 \cap D) = 0.1(0.03) = 0.003$

$p(D) = p(A_1 \cap D) + p(A_2 \cap D) + p(A_3 \cap D) + p(A_4 \cap D) = 0.0175 + 0.006 + 0.008 + 0.003 = 0.0345$, or 3.45%

$p(N) = 1 - p(D) = 1 - 0.0345 = 0.9655$, or 96.55%

43. Construct a probability tree describing the situation.

$$p(H \cap R) = \frac{1}{2} \cdot \frac{5}{11} = \frac{5}{22}$$

$$p(T \cap R) = \frac{1}{2} \cdot \frac{4}{11} = \frac{4}{22}$$

$$p(R) = p(H \cap R) + p(T \cap R) = \frac{5}{22} + \frac{4}{22} = \frac{9}{22}$$

$$p(B) = 1 - p(R) = 1 - \frac{9}{22} = \frac{13}{22}$$

45. $p_1 = p(E') = 1 - p(E) = 1 - \frac{3}{8} = \frac{5}{8}$

$p_2 = p(F'|E) = 1 - p(F|E) = 1 - \frac{7}{11} = \frac{4}{11}$

$p_1 \cdot p_3 = \frac{5}{18}$ and $p_1 = \frac{5}{8}$, so $\frac{5}{8} \cdot p_3 = \frac{5}{18}$, or

$p_3 = \frac{5}{18} \cdot \frac{8}{5} = \frac{4}{9}$

$p_4 = p(F'|E') = 1 - p(F|E') = 1 - p_3 = 1 - \frac{4}{9} = \frac{5}{9}$

$p_5 = \frac{3}{8} \cdot \frac{7}{11} = \frac{21}{88}$

$p_6 = \frac{3}{8} \cdot p_2 = \frac{3}{8} \cdot \frac{4}{11} = \frac{3}{22}$

$p_7 = p_1 \cdot p_4 = \frac{5}{8} \cdot \frac{5}{9} = \frac{25}{72}$

47. Consider the following drawing.

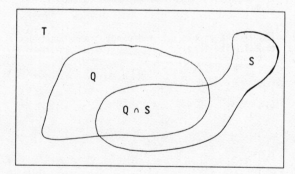

$n(S) = n(S \cup Q) - n(Q \cap S')$, so

$p(S) = \frac{n(S)}{n(T)} = \frac{n(S \cup Q)}{n(T)} - \frac{n(Q \cap S')}{n(T)}$, or

$p(S) = p(S \cup Q) - p(Q \cap S')$.

The exercise tells us that $p(S \cup Q) = \frac{1}{3}$ and $p(Q \cap S') = \frac{1}{9}$, so $p(S) = \frac{1}{3} - \frac{1}{9} = \frac{2}{9}$. The correct answer is choice A.

Exercise Set 6.2

1. E and F are independent, $p(E) = \frac{7}{9}$, $p(F) = \frac{11}{14}$

$p(E \cap F) = p(E)p(F) = \frac{7}{9} \cdot \frac{11}{14} = \frac{11}{18}$

$p(E \cup F) = p(E) + p(F) - p(E)p(F)$

$= \frac{7}{9} + \frac{11}{14} - \frac{11}{18} = \frac{120}{126} = \frac{20}{21}$

$p(E' \cap F) = p(E')p(F) = \frac{2}{9} \cdot \frac{11}{14} = \frac{11}{63}$

$p(E \cap F') = p(E)p(F') = \frac{7}{9} \cdot \frac{3}{14} = \frac{1}{6}$

$p(E' \cap F') = \frac{2}{9} \cdot \frac{3}{14} = \frac{1}{21}$

3. E and F are independent, $p(E) = 48\%$, $p(F) = 33\%$

$p(E \cap F) = p(E)p(F) = 0.48(0.33) = 0.1584$, or 15.84%

$p(E \cup F) = p(E) + p(F) - p(E)p(F)$
$= 0.48 + 0.33 - 0.1584 = 0.6516$, or 65.16%

$p(E' \cap F) = p(E')p(F) = 0.52(0.33) = 0.1716$, or 17.16%

$p(E \cap F') = p(E)p(F') = 0.48(0.67) = 0.3216$, or 32.16%

$p(E' \cap F') = p(E')p(F') = 0.52(0.67) = 0.3484$, or 34.84%

5. E and F are independent, $p(E') = \frac{2}{3}$, $p(F') = \frac{1}{8}$

$p(E \cap F) = p(E)p(F) = \frac{1}{3} \cdot \frac{7}{8} = \frac{7}{24}$

$p(E \cup F) = p(E) + p(F) - p(E)p(F)$

$= \frac{1}{3} + \frac{7}{8} - \frac{7}{24} = \frac{22}{24} = \frac{11}{12}$

$p(E' \cap F) = p(E')p(F) = \frac{2}{3} \cdot \frac{7}{8} = \frac{7}{12}$

$p(E \cap F') = p(E)p(F') = \frac{1}{3} \cdot \frac{1}{8} = \frac{1}{24}$

$p(E' \cap F') = \frac{2}{3} \cdot \frac{1}{8} = \frac{1}{12}$

7. $p(E)p(F) = 0.5(0.6) = 0.3 = p(E \cap F)$, so E and F are independent.

9. $p(E)p(F) = 0.45(0.65) = 0.2925$, or 29.25% but $p(E \cap F) = 29\%$, so E and F are dependent.

11. $p(E)p(F) = \frac{1}{6} \cdot \frac{2}{3} = \frac{1}{9} = p(E \cap F)$, so E and F are independent.

13. a) There are 535 members of Congress.

300 are Democrats, so $p(D) = \frac{300}{535}$.

100 are Senators, so $p(S) = \frac{100}{535}$.

47 Democrats are Senators, so $p(D \cap S) = \frac{47}{535}$.

Of the 100 Senators, 47 are Democrats, so

$p(D|S) = \frac{47}{100}$.

Of the 300 Democrats, 47 are Senators, so

$p(S|D) = \frac{47}{300}$.

b) $p(S)p(D) = \frac{100}{535} \cdot \frac{300}{535} = \frac{30,000}{286,225}$, but $p(S \cap D) =$

$\frac{47}{535} \neq \frac{30,000}{286,225}$.   Thus, S and D are not

independent.

15. $p(H) = \frac{435}{535}$, $p(R) = \frac{235}{535}$, and $p(H \cap R) = \frac{182}{535}$.

Now $p(H)p(R) = \frac{435}{535} \cdot \frac{235}{535} = \frac{102,225}{286,225}$, but $p(H \cap R) =$

$\frac{182}{535} \neq \frac{102,225}{286,225}$.   Thus, H and R are not independent.

17. When the first ball is drawn there are 5 balls in

the box. 3 are red and 2 are white, so $p(R_1) = \frac{3}{5}$

and $p(W_1) = \frac{2}{5}$. There are 4 balls in the box when

the second ball is drawn. Now $p(R_2|R_1) = \frac{2}{4} = \frac{1}{2}$,

$p(W_2|R_1) = \frac{2}{4} = \frac{1}{2}$, $p(R_2|W_1) = \frac{3}{4}$, and $p(W_2|W_1) = \frac{1}{4}$.

We construct a probability tree.

a)

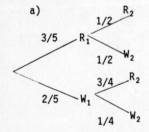

b) We found $p(R_1) = \frac{3}{5}$ in part a).

$p(R_1 \cap R_2) = \frac{3}{5} \cdot \frac{1}{2} = \frac{3}{10}$

$p(R_2) = p(R_1 \cap R_2) + p(W_1 \cap R_2) = \frac{3}{5} \cdot \frac{1}{2} + \frac{2}{5} \cdot \frac{3}{4}$

$= \frac{3}{10} + \frac{3}{10} = \frac{6}{10} = \frac{3}{5}$

c) $p(R_1)p(R_2) = \frac{3}{5} \cdot \frac{3}{5} = \frac{9}{25}$, but $p(R_1 \cap R_2)$

$= \frac{3}{10} \neq \frac{9}{25}$.

Thus R_1 and R_2 are not independent.

17. (continued)

d) p(both balls are red) $= p(R_1 \cap R_2) = \frac{3}{10}$. (We

found this in part b).

p(one is red and the other is white) =

$p(R_1 \cap W_2) \cup p(W_1 \cap R_2) = p(R_1 \cap W_2) +$

$p(W_1 \cap R_2) = \frac{3}{5} \cdot \frac{1}{2} + \frac{2}{5} \cdot \frac{3}{4} = \frac{3}{10} + \frac{3}{10} = \frac{6}{10} = \frac{3}{5}$

p(both balls are white) $= p(W_1 \cap W_2) = \frac{2}{5} \cdot \frac{1}{4}$

$= \frac{1}{10}$

19. Let E = event of having an eye defect and

C = event of having a color variation.

p(eye defect or color variation) $= p(E \cup C) =$
$p(E) + p(C) - p(E)p(C)$ since E and C are
independent.

Then $p(E \cup C) = 0.3 + 0.6 - 0.3(0.6) = 0.72$, or
72%.

21. It seems reasonable to assume that the events are
independent. Let S = event of snow and F = event
of fire in your home. Then $p(S \text{ or } F) = p(S \cup F) =$
$p(S) + p(F) - p(S)p(F) = 0.7 + 0.0006 -$
$0.7(0.0006) = 0.70018$

23. It seems reasonable to assume that the events are
independent. Let S = event of a snowstorm and F =
event of an employee getting the flu. Then
$p(S \text{ or } F) = p(S \cup F) = p(S) + p(F) - p(S)p(F) =$
$0.4 + 0.15 - 0.4(0.15) = 0.49$, or 49%. Thus, 49%
of the employees will miss work because of a
snowstorm or the flu, and 100% - 49%, or 51%, of
the employees should be expected to show up for
work.

25. Assume the events are independent and construct a
probability tree.

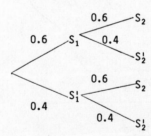

p(neither driver was wearing a seat belt) =
$p(S_1' \cap S_2') = p(S_1')p(S_2') = 0.4(0.4) = 0.16$
p(just one driver was wearing a seat belt) =
$p(S_1 \cap S_2') + p(S_1' \cap S_2) = p(S_1)p(S_2') +$
$p(S_1')p(S_2) = 0.6(0.4) + 0.4(0.6) = 0.48$

27. Assume that the events are independent. Let $C_i =$
event getting through checkpoint i. Then
$p(C_1 \cap C_2 \cap C_3) = p(C_1)p(C_2)p(C_3) =$
$0.02(0.02)(0.02) = 0.000008$, or 0.0008%.

29. Assume the events are independent.

    a) $p(B_1 \cap B_2) = p(B_1)p(B_2) = \frac{18}{38} \cdot \frac{18}{38} = \frac{81}{361}$

    b) $p(B_1 \cap B_2 \cap B_3) = p(B_1)p(B_2)p(B_3) =$
    $\frac{18}{38} \cdot \frac{18}{38} \cdot \frac{18}{38} = \frac{729}{6859}$

    c) $p(B_1 \cap B_2 \cap B_3 \cap B_4 \cap B_5) =$
    $p(B_1)p(B_2)p(B_3)p(B_4)p(B_5) =$
    $\frac{18}{38} \cdot \frac{18}{38} \cdot \frac{18}{38} \cdot \frac{18}{38} \cdot \frac{18}{38} = \left(\frac{18}{38}\right)^5 = \left(\frac{9}{19}\right)^5$
    $= \frac{59,049}{2,476,099}$

    d) The events are independent. Thus, the probability of red on the next spin is always $\frac{18}{38}$, or $\frac{9}{19}$.

31. 157,478 people were surveyed (78,569 males + 78,909 females). 78,569 were male, so $p(M) = \frac{78,569}{157,478}$.

    The number who never smoked is 76,794 (28,770 males + 48,024 females), so $p(N) = \frac{76,794}{157,478}$.

    The number of males who never smoked is 28,770, so $p(M \cap N) = \frac{28,770}{157,478} \approx 0.1827$. Now $p(M)p(N) = \frac{78,569}{157,478} \cdot \frac{76,794}{157,478} \approx 0.2433 \neq 0.1827$, so M and N are not independent.

33. $p(M) = \frac{78,569}{157,478}$ (See Exercise 31.)

    The number who ever smoked is 80,684 (49,799 males + 30,885 females), so $p(S) = \frac{80,684}{157,478}$.

    The number of males who ever smoked is 49,799, so $p(M \cap S) = \frac{49,799}{157,478} \approx 0.3162$. Now $p(M)p(S) = \frac{78,569}{157,478} \cdot \frac{80,684}{157,478} \approx 0.2556 \neq 0.3162$, so M and S are not independent.

35. $p(A|B) = \frac{p(B \cap A)}{p(B)} = \frac{\frac{1}{2}}{\frac{2}{3}} = \frac{1}{2} \cdot \frac{3}{2} = \frac{3}{4}$

    $p(A)p(B) = \frac{3}{5} \cdot \frac{2}{3} = \frac{2}{5} \neq \frac{3}{4}$, so the events are not independent.

37. If the events were independent, 51% of the employees should show up. (See Exercise 23.) Therefore, the employees were taking advantage of the situation. If 50% show up, this is closer to what is expected, but the events still are not independent.

39. Let S = event of food spoiled and
    F = event of a fuel shortage.

    $p(S)p(F) = 0.5(0.8) = 0.4$, or 40%. Since the buyer received only 40% of her order, $p(S \cap F) = 40\%$, and the events are independent.

41.

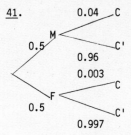

p(a person selected at random is color-blind) = $p(M \cap C) + p(F \cap C) = 0.5(0.04) + 0.5(0.003) = 0.0215$, or 2.15%.

43. Let R1 = event stock comes from the first ranch,
    R2 = event stock comes from second ranch,
    R3 = event stock comes from third ranch,
    D = event stock has hoof-and-mouth disease,
    and N = event stock does not have hoof-and mouth disease.

Consider the stock from the first and third ranches. Construct a tree diagram.

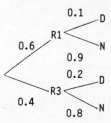

p(stock is infected) = $p(R1 \cap D) + p(R3 \cap D) = 0.6(0.1) + 0.4(0.2) = 0.06 + 0.08 = 0.14$

Thus, 14% of the combined stock from the first and third ranches is infected.

Next consider the stock from all three ranches, a total of 1000 cattle. Construct a tree diagram.

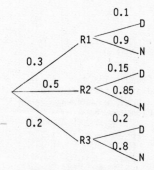

p(stock is infected) = $p(R1 \cap D) + p(R2 \cap D) + p(R3 \cap D) = 0.3(0.1) + 0.5(0.15) + 0.2(0.2) = 0.03 + 0.075 + 0.04 = 0.145$

Thus, 14.5% of the combined stock from all three ranches is infected.

45. Assume the two-tailed coin is in the right hand.

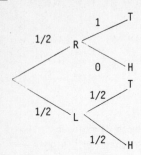

$$p(T) = p(R \cap T) + p(L \cap T) = \frac{1}{2} \cdot 1 + \frac{1}{2} \cdot \frac{1}{2} = \frac{3}{4}$$

47. The probability tree for the first choice of a box and a washer is shown below.

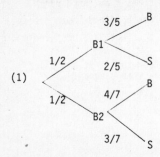

(1)

Case I: B1 is chosen and then a brass washer is chosen and put into B2. We have another probability tree describing the situation for the second choice of a box and washer.

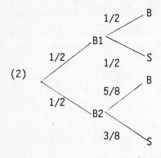

(2)

Let p_1 be a probability found using tree (1) and p_2 be a probability found using tree (2). The probability of choosing a brass washer the second time in this case is $p_1(B1 \cap B)p_2(B1 \cap B) + p_1(B1 \cap B)p_2(B2 \cap B) = \frac{1}{2} \cdot \frac{3}{5} \cdot \frac{1}{2} \cdot \frac{1}{2} + \frac{1}{2} \cdot \frac{3}{5} \cdot \frac{1}{2} \cdot \frac{5}{8} = \frac{3}{40} + \frac{3}{32} = \frac{27}{160}$.

Case II: B1 is chosen and then a steel washer is chosen and put into B2. We have another probability tree describing the second choice.

47. (continued)

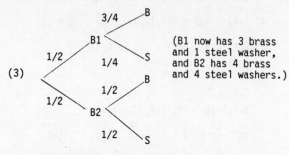

(3)   (B1 now has 3 brass and 1 steel washer, and B2 has 4 brass and 4 steel washers.)

The probability of choosing a brass washer the second time in this case is $p_1(B1 \cap S)p_3(B1 \cap B) + p_1(B1 \cap S)p_3(B2 \cap B) = \frac{1}{2} \cdot \frac{2}{5} \cdot \frac{1}{2} \cdot \frac{3}{4} + \frac{1}{2} \cdot \frac{2}{5} \cdot \frac{1}{2} \cdot \frac{1}{2} = \frac{3}{40} + \frac{2}{40} = \frac{1}{8}$

Case III: B2 is chosen and then a brass washer is chosen and put into B1. We have another probability tree describing the second choice.

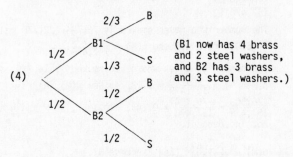

(4)   (B1 now has 4 brass and 2 steel washers, and B2 has 3 brass and 3 steel washers.)

The probability of choosing a brass washer the second time in this case is $p_1(B2 \cap B)p_4(B1 \cap B) + p_1(B2 \cap B)p_4(B2 \cap B) = \frac{1}{2} \cdot \frac{4}{7} \cdot \frac{1}{2} \cdot \frac{2}{3} + \frac{1}{2} \cdot \frac{4}{7} \cdot \frac{1}{2} \cdot \frac{1}{2} = \frac{2}{21} + \frac{1}{14} = \frac{1}{6}$

Case IV: B2 is chosen and then a steel worker is chosen and put into B1. We have another probability tree describing the second choice.

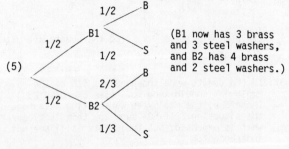

(5)   (B1 now has 3 brass and 3 steel washers, and B2 has 4 brass and 2 steel washers.)

The probability of choosing a brass washer the second time in this case is $p_1(B2 \cap S)p_5(B1 \cap B) + p_1(B2 \cap S)p_5(B2 \cap B) = \frac{1}{2} \cdot \frac{3}{7} \cdot \frac{1}{2} \cdot \frac{1}{2} + \frac{1}{2} \cdot \frac{3}{7} \cdot \frac{1}{2} \cdot \frac{2}{3} = \frac{3}{56} + \frac{1}{14} = \frac{1}{8}$

47. (continued)

The probability the second washer chosen is brass is the sum of the probabilities in the four cases:

$$\frac{27}{160} + \frac{1}{8} + \frac{1}{6} + \frac{1}{8} = \frac{281}{480}$$

To find the probability the same washer is picked both times we consider the same four cases as before.

Case I: Use trees (1) and (2). Remember, in tree (2) only 1 of the 8 washers in B2 is the same washer chosen from B1. $p_1(B1 \cap B)p_2(B2 \cap$ same brass washer is chosen from B2 as from B1) =
$$\frac{1}{2} \cdot \frac{3}{5} \cdot \frac{1}{2} \cdot \frac{1}{8} = \frac{3}{160}$$

Case II: Use trees (1) and (3). Remember, in tree (3) only 1 of the 8 washers in B2 is the same washer chosen from B1.

$p_2(B1 \cap S)p_3(B2 \cap$ same steel washer is chosen from B2 as from B1) = $\frac{1}{2} \cdot \frac{2}{5} \cdot \frac{1}{2} \cdot \frac{1}{8} = \frac{1}{80}$

Case III: Use trees (1) and (4). Remember, in tree (4) only 1 of the 6 washers in B1 is the same washer chosen from B2.

$p_1(B2 \cap B)p_4(B1 \cap$ same brass washer is chosen from B1 as from B2) = $\frac{1}{2} \cdot \frac{4}{7} \cdot \frac{1}{2} \cdot \frac{1}{6} = \frac{1}{42}$

Case IV: Use trees (1) and (5). Remember, in tree (5) only 1 of the 6 washers in B1 is the same washer chosen from B2.

$p_1(B2 \cap S)p_5(B1 \cap$ same steel washer is chosen from B1 as from B2) = $\frac{1}{2} \cdot \frac{3}{7} \cdot \frac{1}{2} \cdot \frac{1}{6} = \frac{1}{56}$

The probability the same washer is picked both times is the sum of the probabilities in the four cases:

$$\frac{3}{160} + \frac{1}{80} + \frac{1}{42} + \frac{1}{56} = \frac{245}{3360} = \frac{7}{96}$$

49. Let K = event of knowing the answer,

  I = event of "isn't sure,"

  D = event of doesn't know,

  R = event of getting the answer right, and

  W = event of getting the answer wrong.

If the student knows the answer, he or she will always get it right. Then $p(R|K) = 1$ and $p(W|K) = 0$. If the student "isn't sure" but has it narrowed down to one of two choices, then $p(R|I) = 0.5$ and $p(W|I) = 0.5$.

If the student doesn't know the answer, he or she picks at random any one of the four possible answers. Then $p(R|D) = 0.25$ and $p(W|D) = 0.75$. Also, if the student knows 80% of the answers and doesn't know 10%, he or she isn't sure of 10% of the answers. Construct a tree.

49. (continued)

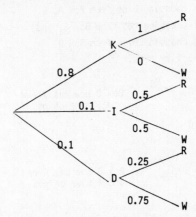

$$p(R) = p(K \cap R) + p(I \cap R) + p(D \cap R)$$
$$= 0.8(0.1) + 0.1(0.5) + 0.1(0.25) = 0.875$$

51. In a), b), and c), $p(M) = \frac{70}{100} = \frac{7}{10}$, $p(F) = \frac{30}{100} = \frac{3}{10}$, $p(S) = \frac{60}{100} = \frac{3}{5}$, and $p(Y) = \frac{40}{100} = \frac{2}{5}$. Then $p(M)p(S) = \frac{7}{10} \cdot \frac{3}{5} = \frac{21}{50}$, $p(F)p(S) = \frac{3}{10} \cdot \frac{3}{5} = \frac{9}{50}$, $p(M)p(Y) = \frac{7}{10} \cdot \frac{2}{5} = \frac{7}{25}$, and $p(F)p(Y) = \frac{3}{10} \cdot \frac{2}{5} = \frac{3}{25}$.

a) $p(M \cap S) = \frac{30}{100} = \frac{3}{10} \neq p(M)p(S)$, so M and S are not independent.

  $p(F \cap S) = \frac{30}{100} = \frac{3}{10} \neq p(F)p(S)$, so F and S are not independent.

  $p(M \cap Y) = \frac{40}{100} = \frac{2}{5} \neq p(M)p(Y)$, so M and Y are not independent.

  $p(F \cap Y) = \frac{0}{100} = 0 \neq p(F)p(Y)$, so F and Y are not independent.

b) $P(M \cap S) = \frac{60}{100} = \frac{3}{5} \neq p(M)p(S)$, so M and S are not independent.

  $p(F \cap S) = \frac{0}{100} = 0 \neq p(F)p(S)$, so F and S are not independent.

  $p(M \cap Y) = \frac{10}{100} = \frac{1}{10} \neq p(M)p(Y)$, so M and Y are not independent.

  $p(F \cap Y) = \frac{40}{100} = \frac{2}{5} \neq p(F)p(Y)$, so F and Y are not independent.

c) $p(M \cap S) = \frac{42}{100} = \frac{21}{50} = p(M)p(S)$, so M and S are independent.

  $p(F \cap S) = \frac{18}{100} = \frac{9}{50} = p(F)p(S)$, so F and S are independent.

  $p(M \cap Y) = \frac{28}{100} = \frac{7}{25} = p(M)p(Y)$, so M and Y are independent.

  $p(F \cap Y) = \frac{12}{100} = \frac{3}{25} = p(F)p(Y)$, so F and Y are independent.

53. Let R_A = event of choosing red from Box A,

   L_A = event of choosing blue from Box A,

   R_B = event of choosing red from Box B, and

   L_B = event of choosing blue from Box B.

Construct a probability tree.

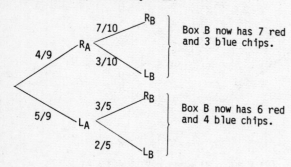

Box B now has 7 red and 3 blue chips.

Box B now has 6 red and 4 blue chips.

$$p(L_A|R_B) = \frac{p(L_A \cap R_B)}{p(L_A \cap R_B) + p(R_A \cap R_B)} =$$

$$\frac{\frac{5}{9} \cdot \frac{3}{5}}{\frac{5}{9} \cdot \frac{3}{5} + \frac{4}{9} \cdot \frac{7}{10}} = \frac{\frac{1}{3}}{\frac{1}{3} + \frac{14}{45}} = \frac{\frac{1}{3}}{\frac{29}{45}} =$$

$$\frac{1}{3} \cdot \frac{45}{29} = \frac{15}{29}.$$

Answer C is the correct choice.

Exercise Set 6.3

1. a) $p(E|F) = 0.1$ (We read this from the tree.)

   b) $p(E'|F) = 0.9$ (We read this from the tree.)

   c) $p(E'|F') = 0.6$ (We read this from the tree.)

   d) $p(E|F') = 0.4$ (We read this from the tree.)

   e) $p(E)$ = the sum of the product probabilities along arcs ending at $E = p(F)p(E|F) + p(F')p(E|F') = 0.2(0.1) + 0.8(0.4) = 0.34.$

   f) Use Bayes' Theorem.

   $$p(F|E) = \frac{p(F \cap E)}{p(E)} = \frac{p(F)p(E|F)}{p(E)} = \frac{0.2(0.1)}{0.34} =$$

   $$\frac{0.02}{0.34} = \frac{2}{34} = \frac{1}{17}$$

   g) $F'|E$ and $F|E$ are complements in the reduced sample space E. Thus

   $$p(F'|E) = 1 - p(F|E) = 1 - \frac{1}{17} = \frac{16}{17}.$$

   h) $p(E') = 1 - p(E) = 1 - 0.34 = 0.66$

   i) $p(F|E') = \frac{p(F \cap E')}{p(E')} = \frac{p(F)p(E'|F)}{p(E')} = \frac{0.2(0.9)}{0.66} =$

   $$\frac{0.18}{0.66} = \frac{18}{66} = \frac{3}{11}$$

   j) $p(F'|E') = 1 - p(F|E') = 1 - \frac{3}{11} = \frac{8}{11}$

3. Exercise 1 gives the following results:

   $p(E) = 0.34$, $p(E') = 0.66$, $p(F|E) = \frac{1}{17}$, $p(F'|E) = \frac{16}{17}$, $p(F|E') = \frac{3}{11}$, $p(F'|E') = \frac{8}{10}$

   Place these values on the probability tree.

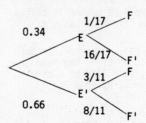

5. Let S be the sample space of 10,000 people, and let

   H = event of having hepatitis,

   H' = event of not having hepatitis,

   P = event of test being positive, and

   P' = event of test being negative.

Construct a probability tree.

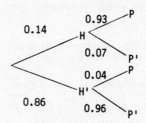

a) Use Bayes' Theorem.

$$p(H|P) = \frac{p(H)p(P|H)}{p(P)} = \frac{p(H)p(P|H)}{p(H)p(P|H) + p(H')p(P|H')}$$

$$= \frac{0.14(0.93)}{0.14(0.93) + 0.86(0.04)} = \frac{0.1302}{0.1646} = \frac{1302}{1646}$$

$$= \frac{651}{823}$$

b) H and H' are complements in the reduced sample space P, so $p(H'|P) = 1 - p(H|P) = 1 - \frac{651}{823} = \frac{172}{823}.$

c) Use Bayes' Theorem.

$$p(H|P') = \frac{p(H)p(P'|H)}{p(H)p(P'|H) + p(H')p(P'|H')}$$

$$= \frac{0.14(0.07)}{0.14(0.07) + 0.86(0.96)} = \frac{0.0098}{0.8354}$$

$$= \frac{98}{8354} = \frac{49}{4177}$$

d) $p(H'|P') = 1 - p(H|P') = 1 - \frac{49}{4177} = \frac{4128}{4177}$

<u>7</u>. Let D = event speaker is defective,

E_1 = event it came from I,

E_2 = event it came from II, and

E_3 = event it came from III.

Construct a probability tree.

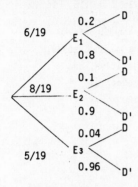

a) p(D) = the sum of all product probabilities

ending at D = $p(E_1)p(D|E_1) + p(E_2)p(D|E_2) +$

$p(E_3)p(D|E_3) = \frac{6}{19}(0.2) + \frac{8}{19}(0.1) + \frac{5}{19}(0.04)$

$= \frac{6}{19} \cdot \frac{2}{10} + \frac{8}{19} \cdot \frac{1}{10} + \frac{5}{19} \cdot \frac{4}{100} = \frac{220}{1900} = \frac{11}{95}$

b) p(D') = 1 - p(D) = $1 - \frac{11}{95} = \frac{84}{95}$

c) $p(E_1|D) = \frac{p(E_1)p(D|E_1)}{p(D)} = \frac{\frac{6}{19} \cdot \frac{2}{10}}{\frac{11}{95}} = \frac{\frac{6}{95}}{\frac{11}{95}} = \frac{6}{11}$

$p(E_2|D) = \frac{p(E_2)p(D|E_2)}{p(D)} = \frac{\frac{8}{19} \cdot \frac{1}{10}}{\frac{11}{95}} = \frac{\frac{4}{95}}{\frac{11}{95}} = \frac{4}{11}$

$p(E_3|D) = \frac{p(E_3)p(D|E_3)}{p(D)} = \frac{\frac{5}{19} \cdot \frac{4}{100}}{\frac{11}{95}} = \frac{\frac{1}{95}}{\frac{11}{95}} = \frac{1}{11}$

<u>9</u>. In Exercise 41, Exercise Set 6.2, we found p(C) = 2.15%.

$p(F|C) = \frac{p(F)p(C|F)}{p(C)} = \frac{0.5(0.003)}{0.0215} = \frac{0.0015}{0.0215} = \frac{15}{215}$

$= \frac{3}{43}$

<u>11</u>. If an animal is selected at random from those shipped by the first and third ranches, use the first probability tree in Exercise 43, Exercise Set 6.2. In that exercise we found p(D) = 14%.

$p(R3|D) = \frac{p(R3)p(D|R3)}{p(D)} = \frac{0.4(0.2)}{0.14} = \frac{0.08}{0.14} = \frac{8}{14} = \frac{4}{7}$

If an animal is selected at random from those shipped by all three ranches, use the second probability tree constructed in Exercise 43, Exercise Set 6.2. In that exercise we found p(D) = 14.5%.

$p(R3|D) = \frac{p(R3)p(D|R3)}{p(D)} = \frac{0.2(0.2)}{0.145} = \frac{0.04}{0.145} = \frac{40}{145}$

$= \frac{8}{29}$

<u>13</u>. Use the probability tree in Exercise 45, Exercise Set 6.2. In that exercise we found p(T) = $\frac{3}{4}$.

$p(R|T) = \frac{p(R)p(T|R)}{p(T)} = \frac{\frac{1}{2} \cdot 1}{\frac{3}{4}} = \frac{1}{2} \cdot \frac{4}{3} = \frac{2}{3}$

<u>15</u>. Let A_D = event of dropping the two-tailed coin in the urn with exactly 3 fair coins,

B_D = event of dropping the two-tailed coin in the urn with 4 fair coins,

A_C = event of choosing the urn with exactly 3 fair coins and the two-tailed coin,

B = event of choosing the urn with exactly 4 coins all of which are fair,

A = event of choosing the urn with exactly 3 coins all of which are fair,

B_C = event of choosing the urn with 4 fair coins and the two-tailed coin,

F = event of choosing a fair coin,

U = event of choosing the two-tailed coin,

H = event of heads, and

T = event of tails.

Construct a probability tree.

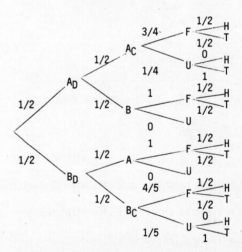

p(coin came from the urn with the two-tailed coin given that tails shows) = $p(A_C|T) \cup p(B_C|T) =$

$p(A_C|T) + p(B_C|T) = \frac{p(A_C)p(T|A_C)}{p(T)} + \frac{p(B_C)p(T|B_C)}{p(T)}$

First find p(T) = the sum of all the product probabilities along arcs ending at T.

$p(T) = \frac{1}{2} \cdot \frac{1}{2} \cdot \frac{3}{4} \cdot \frac{1}{2} + \frac{1}{2} \cdot \frac{1}{2} \cdot \frac{1}{4} \cdot 1 +$

$\frac{1}{2} \cdot \frac{1}{2} \cdot 1 \cdot \frac{1}{2} + \frac{1}{2} \cdot \frac{1}{2} \cdot 1 \cdot \frac{1}{2} +$

$\frac{1}{2} \cdot \frac{1}{2} \cdot \frac{4}{5} \cdot \frac{1}{2} + \frac{1}{2} \cdot \frac{1}{2} \cdot \frac{1}{5} \cdot 1 =$

$\frac{3}{32} + \frac{1}{16} + \frac{1}{8} + \frac{1}{8} + \frac{1}{10} + \frac{1}{20} = \frac{89}{160}$

Now $p(A_C) = \frac{1}{4}$, $p(T|A_C) = \frac{3}{4} \cdot \frac{1}{2} + \frac{1}{4} \cdot 1 = \frac{5}{8}$,

$p(B_C) = \frac{1}{4}$, and $p(T|B_C) = \frac{4}{5} \cdot \frac{1}{2} + \frac{1}{5} \cdot 1 = \frac{3}{5}$.

15. (continued)

Then p(coin came from the urn with the two-tailed

coin given that tails shows) = $\dfrac{\dfrac{1}{4}\cdot\dfrac{5}{8}}{\dfrac{89}{160}} + \dfrac{\dfrac{1}{4}\cdot\dfrac{3}{5}}{\dfrac{89}{160}}$ =

$\dfrac{\dfrac{5}{32}}{\dfrac{89}{160}} + \dfrac{\dfrac{3}{20}}{\dfrac{89}{160}} = \dfrac{5}{32}\cdot\dfrac{160}{89} + \dfrac{3}{20}\cdot\dfrac{160}{89} = \dfrac{25}{89} + \dfrac{24}{89} = \dfrac{49}{89}.$

17. Let AAA = event of a AAA risk,

     AA = event of a AA risk,

      A = event of an A risk,

      D = event of default, and

     D' = event of not defaulting.

Construct a probability tree.

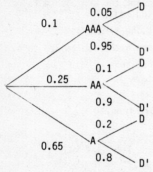

$p(AAA|D) = \dfrac{p(AAA)p(D|AAA)}{p(D)}$

$= \dfrac{0.1(0.05)}{0.1(0.05) + 0.25(0.1) + 0.65(0.2)}$

$= \dfrac{0.005}{0.16} = \dfrac{5}{160} = \dfrac{1}{32}$

19. Let A = event of choosing a site with characteristics A,

    B = event of choosing a site with characteristics B,

    C = event of choosing a site with characteristics C,

    S = event of striking oil, and

    S' = event of not striking oil.

Construct a probability tree.

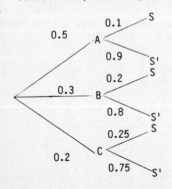

19. (continued)

p(S) = the sum of the product probabilities along all the arcs ending at S = 0.5(0.1) + 0.3(0.2) + 0.2(0.25) = 0.16, or 16%.

$p(A|S) = \dfrac{p(A)p(S|A)}{p(S)} = \dfrac{0.5(0.1)}{0.16} = \dfrac{0.05}{0.16} = \dfrac{5}{16}$

21. Let M = event of having mononucleosis,

    M' = event of not having mononucleosis,

     P = event of testing positive, and

    P' = event of testing negative.

Construct a probability tree.

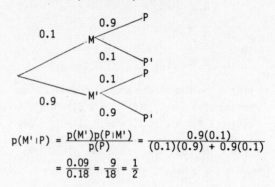

$p(M'|P) = \dfrac{p(M')p(P|M')}{p(P)} = \dfrac{0.9(0.1)}{(0.1)(0.9) + 0.9(0.1)}$

$= \dfrac{0.09}{0.18} = \dfrac{9}{18} = \dfrac{1}{2}$

23. Let W = event a white mouse escapes,

    B = event a black mouse escapes,

    B_i = event the mouse escaped from box i, and

    R_j = event the mouse is returned to box j.

Construct a probability tree.

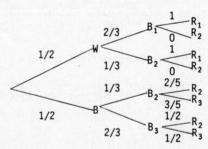

p(the mouse is returned to the box from which it escaped) = $p(W \cap B_1 \cap R_1) + p(W \cap B_2 \cap R_2) + p(B \cap B_2 \cap R_2) + p(B \cap B_3 \cap R_3) =$
$\dfrac{1}{2}\cdot\dfrac{2}{3}\cdot 1 + \dfrac{1}{2}\cdot\dfrac{1}{3}\cdot 0 + \dfrac{1}{2}\cdot\dfrac{1}{3}\cdot\dfrac{2}{5} + \dfrac{1}{2}\cdot\dfrac{2}{3}\cdot\dfrac{1}{2} =$
$\dfrac{1}{3} + 0 + \dfrac{1}{15} + \dfrac{1}{6} = \dfrac{17}{30}$

$p(B_3|R_3) = \dfrac{p(B_3)p(R_3|B_3)}{p(R_3)} = \dfrac{\left(\dfrac{1}{2}\cdot\dfrac{2}{3}\right)\left(\dfrac{1}{2}\right)}{\dfrac{1}{2}\cdot\dfrac{1}{3}\cdot\dfrac{3}{5} + \dfrac{1}{2}\cdot\dfrac{2}{3}\cdot\dfrac{1}{2}}$

$= \dfrac{\dfrac{1}{6}}{\dfrac{1}{10} + \dfrac{1}{6}} = \dfrac{\dfrac{1}{6}}{\dfrac{8}{30}} = \dfrac{5}{8}$

25. p(more than one defective) = p(2 defective) +

$$p(3\text{ defective}) = \frac{\binom{3}{2}\binom{3}{1}}{\binom{6}{3}} + \frac{\binom{3}{3}}{\binom{6}{3}} = \frac{\frac{3\cdot2}{2\cdot1}\cdot\frac{3}{1}}{\frac{6\cdot5\cdot4}{3\cdot2\cdot1}} +$$

$$\frac{\frac{3\cdot2\cdot1}{3\cdot2\cdot1}}{\frac{6\cdot5\cdot4}{3\cdot2\cdot1}} = \frac{9}{20} + \frac{1}{20} = \frac{1}{2}$$

Construct a probability tree for the second part of the exercise. Let

C_i = event i defective clocks are drawn in the first sample,

S = event second sample has exactly one defective clock, and

S' = event second sample does not have exactly one defective clock.

Note that a second sample is taken only when i = 2 or i = 3.

```
1/20          C_0
      9/20  C_1
      9/20  C_2   3/5   S
                        S'
                  2/5
                  9/20  S
      1/20  C_3         S'
                  11/20
```

To find $p(S|C_2)$ we computed $\frac{\binom{3}{1}\binom{3}{1}}{\binom{6}{2}}$, and to find

$p(S|C_3)$ we computed $\frac{\binom{3}{1}\binom{3}{2}}{\binom{6}{3}}$.

Now p(only one clock in the sample displayed is defective) = $p(C_1) + p(C_2 \cap S) + p(C_3 \cap S)$ = $\frac{9}{20} + \frac{9}{20}\cdot\frac{3}{5} + \frac{1}{20}\cdot\frac{9}{20} = \frac{9}{20} + \frac{27}{100} + \frac{9}{400} = \frac{297}{400}$.

p(some clocks have been redrawn given that only one clock in the sample displayed is defective) =

$$\frac{p(C_2 \cap S) + p(C_3 \cap S)}{p(S)} = \frac{\frac{9}{20}\cdot\frac{3}{5} + \frac{1}{20}\cdot\frac{9}{20}}{\frac{297}{400}} =$$

$$\frac{\frac{27}{100} + \frac{9}{400}}{\frac{297}{400}} = \frac{117}{297} = \frac{13}{33}$$

27. a) There are 2 possible outcomes for each of the 4 possible games remaining. (Team A wins or loses). Thus, there are 2^4, or 16, possible outcomes. If Team A is to win the playoffs it must win either all 3 of the next 3 or 3 of the next 4 games. Then p(Team A will win) =

$$\frac{\binom{3}{3} + \binom{4}{3}}{16} = \frac{1+4}{16} = \frac{5}{16}$$

27. (continued)

b) There are 2^3, or 8, possible outcomes for the 3 possible remaining games. If Team A is to win the playoffs, it must win either both of the next 2 or 2 of the next 3 games. The probability this will happen is

$$\frac{\binom{2}{2} + \binom{3}{2}}{8} = \frac{1+3}{8} = \frac{1}{2}.$$

c) There are 2^3, or 8, possible outcomes for the remaining 3 games. If Team A is to win the playoffs it must win 3 of the next 3 games. The probability this will happen is

$$\frac{\binom{3}{3}}{8} = \frac{1}{8}.$$

Exercise Set 6.4

1. $q = 1 - p = 1 - \frac{1}{4} = \frac{3}{4}$

$\binom{10}{3}\left(\frac{1}{4}\right)^3\left(\frac{3}{4}\right)^7 = \frac{10\cdot9\cdot8}{3\cdot2\cdot1}\left(\frac{1}{4}\right)^3\left(\frac{3}{4}\right)^7 = 120\left(\frac{1}{4}\right)^3\left(\frac{3}{4}\right)^7 \approx 0.2503$

3. $q = 100\% - 20\%$, or $1 - 0.2 = 0.8$

$\binom{8}{2}(0.2)^2(0.8)^6 = \frac{8\cdot7}{2\cdot1}(0.2)^2(0.8)^6 = 28(0.2)^2(0.8)^6 \approx 0.2936$

5. $q = 1 - p = 1 - 0.14 = 0.86$

$\binom{11}{6}(0.14)^6(0.86)^5 = \frac{11\cdot10\cdot9\cdot8\cdot7\cdot6}{6\cdot5\cdot4\cdot3\cdot2\cdot1}(0.14)^6(0.86)^5 = 462(0.14)^6(0.86)^5 \approx 0.0016$

7. p = probability of getting a six on any one trial = $\frac{1}{6}$. The probability of 8 sixes in 8 trials is found using Theorem 6 with n = 8, k = 8, $p = \frac{1}{6}$, and $q = 1 - \frac{1}{6}$, or $\frac{5}{6}$:

$\binom{8}{8}\left(\frac{1}{6}\right)^8\left(\frac{5}{6}\right)^0 = 1\cdot\left(\frac{1}{6}\right)^8\cdot1 = \left(\frac{1}{6}\right)^8 = \frac{1}{1,679,616} \approx 0.0000006$

9. Use Theorem 6 with n = 8, k = 5, $p = \frac{1}{6}$, and $q = \frac{5}{6}$:
$\binom{8}{5}\left(\frac{1}{6}\right)^5\left(\frac{5}{6}\right)^3 = 56\left(\frac{1}{6}\right)^5\left(\frac{5}{6}\right)^3 \approx 0.0042$

11. $p_{k\geqslant5} = p_5 + p_6 + p_7 + p_8$

$= p_5 + \binom{8}{6}\left(\frac{1}{6}\right)^6\left(\frac{5}{6}\right)^2 + \binom{8}{7}\left(\frac{1}{6}\right)^7\left(\frac{5}{6}\right) + p_8 \approx 0.0046$

(Note that we already computed P_5 and P_8 in Exercises 9 and 7, respectively.)

13. $p_{k\leqslant2} = p_0 + p_1 + p_2 = \binom{8}{0}\left(\frac{1}{6}\right)^0\left(\frac{5}{6}\right)^8 +$

$\binom{8}{1}\left(\frac{1}{6}\right)\left(\frac{5}{6}\right)^7 + \binom{8}{2}\left(\frac{1}{6}\right)^2\left(\frac{5}{6}\right)^6$

≈ 0.8653 (Answers may vary slightly due to rounding differences.)

15. $p_5 = \binom{5}{5}(0.78)^5 \approx 0.2887$

17. $p_3 = \binom{5}{3}(0.78)^3(0.22)^2$ $(q = 1 - 0.78 = 0.22)$

    ≈ 0.2297

19. $p_k \geqslant 3 = p_3 + p_4 + p_5$

    $= p_3 + \binom{5}{4}(0.78)^4(0.22) + p_5$

    ≈ 0.9256

    (Note that we already computed p_3 and p_5 in Exercises 17 and 15, respectively.)

21. $p_4 = \binom{4}{4}(0.85)^4 \approx 0.5220$

23. $p_k \geqslant 3 = p_3 + p_4$ $(q = 1 - 0.85 = 0.15)$

    $= \binom{4}{3}(0.85)^3(0.15) + p_4$

    ≈ 0.8905

    (Note that we already computed p_4 in Exercise 21.)

25. $p_0 = \binom{10}{0}(0.021)^0(0.979)^{10}$ $(q = 1 - 0.021 = 0.979)$

    ≈ 0.8088

27. $p_3 = \binom{10}{3}(0.021)^3(0.979)^7$

    ≈ 0.0010

29. $p_k \geqslant 1 = 1 - p_k < 1 = 1 - p_0 \approx 1 - 0.8088 \approx 0.1912$
    (We computed p_0 in Exercise 25.)

31. Up to that point Montana had completed 230/346 of his passes, so p = 230/346.

    $p_{22} = \binom{22}{22}\left[\frac{230}{346}\right]^{22} \approx 0.000125$, or 0.0125%

33. a) $p_5 = \binom{5}{5}(0.6)^5 \approx 0.0778$

    b) p(all 5 prefer Bud's) = p(0 prefer Koor's) = $\binom{5}{0}(0.6)^0(0.4)^5 \approx 0.0102$

    c) $p_3 = \binom{5}{3}(0.6)^3(0.4)^2 = 0.3456$

    d) $p_k \geqslant 3 = p_3 + p_4 + p_5 = p_3 + \binom{5}{4}(0.6)^4(0.4) +$

       $p_5 \approx 0.6826$ ✓

    (We computed p_3 and p_5 in parts c) and a), respectively.)

35. a) $p_1 = \binom{6}{1}\left[\frac{1}{4}\right]\left[\frac{3}{4}\right]^5$ $\left[q = 1 - \frac{1}{4} = \frac{3}{4}\right]$

       ≈ 0.3560

    b) $p_k \geqslant 2 = 1 - p_k < 2 = 1 - (p_0 + p_1)$

       $= 1 - \binom{6}{0}\left[\frac{3}{4}\right]^6 - p_1$

       ≈ 0.4661

    (We computed p_1 in part a). Answers may vary slightly due to rounding differences.)

37. a) $p_6 = \binom{12}{6}(0.53)^6(0.47)^6$ $(q = 1 - 0.53 = 0.47)$

       ≈ 0.2208

    b) p(10 prefer Sleazz) = p(2 prefer Slyme) = $p_2 =$
       $\binom{12}{2}(0.53)^2(0.47)^{10} \approx 0.0098$

    c) $p_k \geqslant 3 = 1 - p_k < 3 = 1 - (p_0 + p_1 + p_2) =$

       $1 - \binom{12}{0}(0.47)^{12} - \binom{12}{1}(0.53)(0.47)^{11} - p_2$

       ≈ 0.9886

       (We computed p_2 in part b). Answers may vary slightly due to rounding differences.)

    d) p(at most 7 prefer Sleazz) = p(at least 5 prefer Slyme) = $p_k \geqslant 5 = 1 - p_k < 5 =$

       $1 - (p_0 + p_1 + p_2 + p_3 + p_4) =$

       $1 - p_0 - p_1 - p_2 - p_3 - p_4$

       In part c) we found $1 - p_0 - p_1 - p_2 \approx 0.9886$

       $p_3 = \binom{12}{3}(0.53)^3(0.47)^9 \approx 0.0367$

       $p_4 = \binom{12}{4}(0.53)^4(0.47)^8 \approx 0.0930$

       Thus, $p_k \geqslant 5 \approx 0.9886 - 0.0367 - 0.0930 \approx 0.8589$

39. p = p(rolling a six) = $\frac{1}{6}$

    $q = 1 - p = 1 - \frac{1}{6} = \frac{5}{6}$

    a) $p_5 = \binom{5}{5}\left[\frac{1}{6}\right]^5 \approx 0.00013$

    b) $p_3 = \binom{5}{3}\left[\frac{1}{6}\right]^3\left[\frac{5}{6}\right]^2 \approx 0.0322$

    c) $p_k \geqslant 4 = p_4 + p_5 = \binom{5}{4}\left[\frac{1}{6}\right]^4\left[\frac{5}{6}\right] + p_5 \approx 0.0033$

       (We computed p_5 in part a)).

41. a) There is one path from A to A: A
    b) There is one path from A to B: AB
    c) There is one path from A to C: AC
    d) There is one path from A to D: ABD
    e) There are 2 paths from A to E: ABE, ACE
    f) There is 1 path from A to F: ACF
    g) There is 1 path from A to G: ABDG
    h) There are 3 paths from A to H: ABDH, ABEH, ACEH
    i) There are 3 paths from A to I: ABEI, ACEI, ACFI
    j) There is 1 path from A to J: ACFJ
    k) There is 1 path from A to K: ABDGK
    l) There are 4 paths from A to L: ABDGL, ABDHL, ABEHL, ACEHL
    m) There are 6 paths from A to M: ABDHM, ABEHM, ABEIM, ACEHM, ACEIM, ACFIM
    n) There are 4 paths from A to N: ABEIN, ACEIN, ACFIN, ACFJN
    o) There is 1 path from A to O: ACFJO
    p) There is 1 path from A to P: ABDGKP
    q) There are 5 paths from A to Q: ABDGKQ, ABDGLQ, ABDHLQ, ABEHLQ, ACEHLQ

41. (continued)

r) There are 10 paths from A to R: ABDGLR, ABDHLR, ABDHMR, ABEHLR, ABEHMR, ABEIMR, ACEHLR, ACEHMR, ACEIMR, ACFIMR

s) There are 10 paths from A to S: ABDHMS, ABEHMS, ABEIMS, ABEINS, ACFJNS, ACFIMS, ACFINS, ACEHMS, ACEIMS, ACEINS

t) There are 5 paths from A to T: ABEINT, ACEINT, ACFINT, ACFJNT, ACFJOT

u) There is 1 path from A to U: ACFJOU

43. We have binomial probability. Let p = p(left turn) = $\frac{2}{5}$ and q = p(right turn) = $\frac{3}{5}$. The salesman must make 5 right turns to go from A to P, so p(taking all right turns to get from A to P) = $\binom{5}{0}\left(\frac{2}{5}\right)\left(\frac{3}{5}\right)^5$ = 0.07776.

45. We have binomial probability. Let p = p(left turn) = 2/5 and q = p(right turn) = 3/5.

a) The salesman must make 5 right turns to go from A to P, so p(getting from A to P) = $\binom{5}{0}\left(\frac{3}{5}\right)^5$.

b) Any combination of 1 left turn and 4 right turns will take the salesman from A to Q. Then p(getting from A to Q) = $\binom{5}{1}\left(\frac{2}{5}\right)\left(\frac{3}{5}\right)^4$.

c) Any combination of 2 left turns and 3 right turns will take the salesman from A to R. Then p(getting from A to R) = $\binom{5}{2}\left(\frac{2}{5}\right)^2\left(\frac{3}{5}\right)^3$.

d) Any combination of 3 left turns and 2 right turns will take the salesman from A to S. Then p(getting from A to S) = $\binom{5}{3}\left(\frac{2}{5}\right)^3\left(\frac{3}{5}\right)^2$.

e) Any combination of 4 left turns and 1 right turn will take the salesman from A to T. Then p(getting from A to T) = $\binom{5}{4}\left(\frac{2}{5}\right)^4\left(\frac{3}{5}\right)$.

f) The salesman must take 5 left turns to get from A to U, so p(getting from A to U) = $\binom{5}{5}\left(\frac{2}{5}\right)^5$.

47. $(a + b)^2 = \binom{2}{0}a^2 + \binom{2}{1}a^{2-1}b + \binom{2}{2}b^2$
$= a^2 + 2ab + b^2$

49. $(a + b)^4 = \binom{4}{0}a^4 + \binom{4}{1}a^{4-1}b + \binom{4}{2}a^{4-2}b^2 +$
$\binom{4}{3}a^{4-3}b^3 + \binom{4}{4}b^4 = a^2 + 4a^3b + 6a^2b^2 + 4ab^3 + b^4$

51. $(a + b)^6 = \binom{6}{0}a^6 + \binom{6}{1}a^{6-1}b + \binom{6}{2}a^{6-2}b^2 +$
$\binom{6}{3}a^{6-3}b^3 + \binom{6}{4}a^{6-4}b^4 + \binom{6}{5}a^{6-5}b^5 + \binom{6}{6}b^6 =$
$a^6 + 6a^5b + 15a^4b^2 + 20a^3b^3 + 15a^2b^4 + 6ab^5 + b^6$

53. $(a + b)^9 = \binom{9}{0}a^9 + \binom{9}{1}a^{9-1}b + \binom{9}{2}a^{9-2}b^2 +$
$\binom{9}{3}a^{9-3}b^3 + \binom{9}{4}a^{9-4}b^4 + \binom{9}{5}a^{9-5}b^5 + \binom{9}{6}a^{9-6}b^6 +$
$\binom{9}{7}a^{9-7}b^7 + \binom{9}{8}a^{9-8}b^8 + \binom{9}{9}b^9 = a^9 + 9a^8b +$
$36a^7b^2 + 84a^6b^3 + 126a^5b^4 + 126a^4b^5 + 84a^3b^6 +$
$36a^2b^7 + 9ab^8 + b^9$

55. The coefficients correspond to the rows of Pascal's triangle.

57. a) The ways in which five 5's can be obtained using three rolls and their binomial probabilities are shown in the table below.

Number of 5's on Roll 1	Number of 5's on Roll 2	Number of 5's on Roll 3	Binomial probability
5			$\binom{5}{5}\left(\frac{1}{6}\right)^5 \approx 0.000129$
4	1		$\binom{5}{4}\left(\frac{1}{6}\right)^4\left(\frac{5}{6}\right)\binom{1}{1}\left(\frac{1}{6}\right) \approx 0.000536$
4	0	1	$\binom{5}{4}\left(\frac{1}{6}\right)^4\left(\frac{5}{6}\right)\binom{1}{0}\left(\frac{5}{6}\right)\binom{1}{1}\left(\frac{1}{6}\right) \approx 0.000447$
3	2		$\binom{5}{3}\left(\frac{1}{6}\right)^3\left(\frac{5}{6}\right)^2\binom{2}{2}\left(\frac{1}{6}\right)^2 \approx 0.000893$
3	0	2	$\binom{5}{3}\left(\frac{1}{6}\right)^3\left(\frac{5}{6}\right)^2\binom{2}{0}\left(\frac{5}{6}\right)^2\binom{2}{2}\left(\frac{1}{6}\right)^2$ ≈ 0.000620
3	1	1	$\binom{5}{3}\left(\frac{1}{6}\right)^3\left(\frac{5}{6}\right)^2\binom{2}{1}\left(\frac{1}{6}\right)\left(\frac{5}{6}\right)\binom{1}{1}\left(\frac{1}{6}\right)$ ≈ 0.001488
2	3		$\binom{5}{2}\left(\frac{1}{6}\right)^2\left(\frac{5}{6}\right)^3\binom{3}{3}\left(\frac{1}{6}\right)^3 \approx 0.000744$
2	0	3	$\binom{5}{2}\left(\frac{1}{6}\right)^2\left(\frac{5}{6}\right)^3\binom{3}{0}\left(\frac{5}{6}\right)^3\binom{3}{3}\left(\frac{1}{6}\right)^3$ ≈ 0.000431
2	2	1	$\binom{5}{2}\left(\frac{1}{6}\right)^2\left(\frac{5}{6}\right)^3\binom{3}{2}\left(\frac{1}{6}\right)^2\left(\frac{5}{6}\right)\binom{1}{1}\left(\frac{1}{6}\right)$ ≈ 0.001861
2	1	2	$\binom{5}{2}\left(\frac{1}{6}\right)^2\left(\frac{5}{6}\right)^3\binom{3}{1}\left(\frac{1}{6}\right)\left(\frac{5}{6}\right)^2\binom{2}{2}\left(\frac{1}{6}\right)^2$ ≈ 0.001550
1	4		$\binom{5}{1}\left(\frac{1}{6}\right)\left(\frac{5}{6}\right)^4\binom{4}{4}\left(\frac{1}{6}\right)^4 \approx 0.000310$
1	0	4	$\binom{5}{1}\left(\frac{1}{6}\right)\left(\frac{5}{6}\right)^4\binom{4}{0}\left(\frac{5}{6}\right)^4\binom{4}{4}\left(\frac{1}{6}\right)^4$ ≈ 0.000150
1	3	1	$\binom{5}{1}\left(\frac{1}{6}\right)\left(\frac{5}{6}\right)^4\binom{4}{3}\left(\frac{1}{6}\right)^3\left(\frac{5}{6}\right)\binom{1}{1}\left(\frac{1}{6}\right)$ ≈ 0.001037
1	1	3	$\binom{5}{1}\left(\frac{1}{6}\right)\left(\frac{5}{6}\right)^4\binom{4}{1}\left(\frac{1}{6}\right)\left(\frac{5}{6}\right)^3\binom{3}{3}\left(\frac{1}{6}\right)^3$ ≈ 0.000718
1	2	2	$\binom{5}{1}\left(\frac{1}{6}\right)\left(\frac{5}{6}\right)^4\binom{4}{2}\left(\frac{1}{6}\right)^2\left(\frac{5}{6}\right)^2\binom{2}{2}\left(\frac{1}{6}\right)^2$ ≈ 0.001292
0	5		$\binom{5}{0}\left(\frac{5}{6}\right)^5\binom{5}{5}\left(\frac{1}{6}\right)^5 \approx 0.000052$
0	4	1	$\binom{5}{0}\left(\frac{5}{6}\right)^5\binom{5}{4}\left(\frac{1}{6}\right)^4\left(\frac{5}{6}\right)\binom{1}{1}\left(\frac{1}{6}\right) \approx 0.000215$
0	1	4	$\binom{5}{0}\left(\frac{5}{6}\right)^5\binom{5}{1}\left(\frac{1}{6}\right)\left(\frac{5}{6}\right)^4\binom{4}{4}\left(\frac{1}{6}\right)^4$ ≈ 0.000127
0	3	2	$\binom{5}{0}\left(\frac{5}{6}\right)^5\binom{5}{3}\left(\frac{1}{6}\right)^3\left(\frac{5}{6}\right)^2\binom{2}{2}\left(\frac{1}{6}\right)^2$ ≈ 0.000359
0	2	3	$\binom{5}{0}\left(\frac{5}{6}\right)^5\binom{5}{2}\left(\frac{1}{6}\right)^2\left(\frac{5}{6}\right)^3\binom{3}{3}\left(\frac{1}{6}\right)^3$ ≈ 0.000299
0	0	5	$\binom{5}{0}\left(\frac{5}{6}\right)^5\binom{5}{0}\left(\frac{5}{6}\right)^5\binom{5}{5}\left(\frac{1}{6}\right)^5 \approx 0.000021$

57. (continued)

p(all 5's) = the sum of the probabilities in the table ≈ 0.012969, or 0.013 rounded to the nearest thousandth.

b) The ways in which exactly four 5's can be obtained using three rolls and their binomial probabilities are shown in the table below.

Number of 5's on Roll 1	Number of 5's on Roll 2	Number of 5's on Roll 3	Binomial probability
4	0	0	$\binom{5}{4}\left(\frac{1}{6}\right)^4\left(\frac{5}{6}\right)\binom{5}{0}\left(\frac{1}{6}\right)^0\binom{5}{0}\left(\frac{1}{6}\right)^0\left(\frac{5}{6}\right)$ ≈ 0.002233
3	0	1	$\binom{5}{3}\left(\frac{1}{6}\right)^3\left(\frac{5}{6}\right)^2\binom{5}{0}\left(\frac{2}{6}\right)\binom{5}{1}\left(\frac{1}{6}\right)\left(\frac{5}{6}\right)$ ≈ 0.003101
3	1	0	$\binom{5}{3}\left(\frac{1}{6}\right)^3\left(\frac{5}{6}\right)^2\binom{5}{1}\left(\frac{1}{6}\right)\binom{5}{0}\left(\frac{1}{6}\right)\left(\frac{5}{6}\right)$ ≈ 0.007442
2	2	0	$\binom{5}{2}\left(\frac{1}{6}\right)^2\left(\frac{5}{6}\right)^3\binom{5}{2}\left(\frac{1}{6}\right)^2\binom{5}{0}\left(\frac{1}{6}\right)\left(\frac{5}{6}\right)$ ≈ 0.009303
2	0	2	$\binom{5}{2}\left(\frac{1}{6}\right)^2\left(\frac{5}{6}\right)^3\binom{5}{0}\left(\frac{3}{6}\right)\binom{5}{2}\left(\frac{1}{6}\right)^2\left(\frac{5}{6}\right)$ ≈ 0.006460
2	1	1	$\binom{5}{2}\left(\frac{1}{6}\right)^2\left(\frac{5}{6}\right)^3\binom{5}{1}\left(\frac{1}{6}\right)\binom{5}{1}\left(\frac{2}{6}\right)\left(\frac{1}{6}\right)\left(\frac{5}{6}\right)$ ≈ 0.015505
1	3	0	$\binom{5}{1}\left(\frac{1}{6}\right)\binom{5}{4}\left(\frac{4}{6}\right)\binom{5}{3}\left(\frac{1}{6}\right)^3\binom{5}{1}\left(\frac{1}{6}\right)\left(\frac{5}{6}\right)$ ≈ 0.005168
1	0	3	$\binom{5}{1}\left(\frac{1}{6}\right)\binom{5}{4}\left(\frac{4}{6}\right)\binom{5}{0}\binom{5}{4}\left(\frac{4}{6}\right)\left(\frac{1}{6}\right)^3\left(\frac{5}{6}\right)$ ≈ 0.002991
1	2	1	$\binom{5}{1}\left(\frac{1}{6}\right)\binom{5}{4}\left(\frac{4}{6}\right)\binom{5}{2}\left(\frac{1}{6}\right)^2\binom{5}{2}\left(\frac{2}{6}\right)\left(\frac{1}{6}\right)\left(\frac{5}{6}\right)$ ≈ 0.012920
1	1	2	$\binom{5}{1}\left(\frac{1}{6}\right)\binom{5}{4}\left(\frac{4}{6}\right)\binom{5}{1}\left(\frac{1}{6}\right)\binom{5}{3}\left(\frac{3}{6}\right)\left(\frac{1}{6}\right)^2\left(\frac{5}{6}\right)$ ≈ 0.010767
0	4	0	$\binom{5}{0}\left(\frac{5}{6}\right)^5\binom{5}{4}\left(\frac{1}{6}\right)^4\binom{5}{0}\left(\frac{1}{6}\right)\left(\frac{5}{6}\right)$ ≈ 0.001077
0	0	4	$\binom{5}{0}\left(\frac{5}{6}\right)^5\binom{5}{0}\left(\frac{5}{6}\right)^5\binom{5}{4}\left(\frac{1}{6}\right)^4\left(\frac{5}{6}\right)$ ≈ 0.000519
0	3	1	$\binom{5}{0}\left(\frac{5}{6}\right)^5\binom{5}{3}\left(\frac{1}{6}\right)^3\binom{5}{2}\left(\frac{2}{6}\right)\left(\frac{1}{6}\right)\left(\frac{5}{6}\right)$ ≈ 0.003589
0	1	3	$\binom{5}{0}\left(\frac{5}{6}\right)^5\binom{5}{1}\left(\frac{1}{6}\right)\binom{5}{4}\left(\frac{4}{6}\right)\left(\frac{1}{6}\right)^3\left(\frac{5}{6}\right)$ ≈ 0.002492
0	2	2	$\binom{5}{0}\left(\frac{5}{6}\right)^5\binom{5}{2}\left(\frac{1}{6}\right)^2\binom{5}{3}\left(\frac{3}{6}\right)\left(\frac{1}{6}\right)^2\left(\frac{5}{6}\right)$ ≈ 0.004486

p(exactly four 5's) = the sum of the probabilities in the table ≈ 0.088053, or 0.088 rounded to the nearest thousandth.

c) The ways in which exactly three 5's can be obtained using three rolls and their binomial probabilities are shown in the table below.

Number of 5's on Roll 1	Number of 5's on Roll 2	Number of 5's on Roll 3	Binomial probability
3	0	0	$\binom{5}{3}\left(\frac{1}{6}\right)^3\left(\frac{5}{6}\right)^2\binom{5}{0}\left(\frac{5}{6}\right)^2\binom{5}{0}\left(\frac{2}{6}\right)\left(\frac{5}{6}\right)^2$ ≈ 0.015505
2	1	0	$\binom{5}{2}\left(\frac{1}{6}\right)^2\left(\frac{5}{6}\right)^3\binom{5}{3}\left(\frac{3}{6}\right)\binom{5}{1}\left(\frac{1}{6}\right)\binom{5}{0}\left(\frac{2}{6}\right)\left(\frac{5}{6}\right)^2$ ≈ 0.038761
2	0	1	$\binom{5}{2}\left(\frac{1}{6}\right)^2\left(\frac{5}{6}\right)^3\binom{5}{3}\left(\frac{3}{6}\right)\binom{5}{0}\left(\frac{5}{6}\right)^3\binom{5}{3}\left(\frac{1}{6}\right)\left(\frac{5}{6}\right)^2$ ≈ 0.032301
1	2	0	$\binom{5}{1}\left(\frac{1}{6}\right)\binom{5}{4}\left(\frac{4}{6}\right)\binom{5}{2}\left(\frac{1}{6}\right)^2\binom{5}{2}\left(\frac{2}{6}\right)\left(\frac{5}{6}\right)^2$ ≈ 0.032301
1	0	2	$\binom{5}{1}\left(\frac{1}{6}\right)\binom{5}{4}\left(\frac{4}{6}\right)\binom{5}{4}\left(\frac{4}{6}\right)\binom{5}{2}\left(\frac{1}{6}\right)^2\left(\frac{5}{6}\right)$ ≈ 0.022431
1	1	1	$\binom{5}{1}\left(\frac{1}{6}\right)\binom{5}{4}\left(\frac{4}{6}\right)\binom{5}{1}\left(\frac{1}{6}\right)\binom{5}{3}\left(\frac{3}{6}\right)\left(\frac{1}{6}\right)\left(\frac{5}{6}\right)^2$ ≈ 0.053835
0	3	0	$\binom{5}{0}\left(\frac{5}{6}\right)^5\binom{5}{3}\left(\frac{1}{6}\right)^3\binom{5}{2}\left(\frac{2}{6}\right)\left(\frac{5}{6}\right)^2$ ≈ 0.008973
0	0	3	$\binom{5}{0}\left(\frac{5}{6}\right)^5\binom{5}{0}\left(\frac{5}{6}\right)^5\binom{5}{3}\left(\frac{1}{6}\right)^3\left(\frac{5}{6}\right)^2$ ≈ 0.005192
0	2	1	$\binom{5}{0}\left(\frac{5}{6}\right)^5\binom{5}{2}\left(\frac{1}{6}\right)^2\binom{5}{3}\left(\frac{3}{6}\right)\left(\frac{1}{6}\right)\left(\frac{5}{6}\right)^2$ ≈ 0.022431
0	1	2	$\binom{5}{0}\left(\frac{5}{6}\right)^5\binom{5}{1}\left(\frac{1}{6}\right)\binom{5}{4}\left(\frac{4}{6}\right)\left(\frac{1}{6}\right)^2\left(\frac{5}{6}\right)^2$ ≈ 0.018693

p(exactly three 5's) = the sum of the probabilities in the table ≈ 0.250423, or 0.250 rounded to the nearest thousandth

d) The ways in which exactly two 5's can be obtained using three rolls and their binomial probabilities are shown in the table below.

Number of 5's on Roll 1	Number of 5's on Roll 2	Number of 5's on Roll 3	Binomial probability
2	0	0	$\binom{5}{2}\left(\frac{1}{6}\right)^2\left(\frac{5}{6}\right)^3\binom{5}{3}\left(\frac{5}{6}\right)^3\binom{5}{0}\left(\frac{5}{6}\right)^3$ ≈ 0.053835
1	1	0	$\binom{5}{1}\left(\frac{1}{6}\right)\binom{5}{4}\left(\frac{4}{6}\right)\binom{5}{1}\left(\frac{1}{6}\right)\binom{5}{3}\left(\frac{3}{6}\right)\binom{5}{0}\left(\frac{5}{6}\right)^3$ ≈ 0.089725
1	0	1	$\binom{5}{1}\left(\frac{1}{6}\right)\binom{5}{4}\left(\frac{4}{6}\right)\binom{5}{0}\binom{5}{4}\left(\frac{4}{6}\right)\binom{5}{1}\left(\frac{1}{6}\right)\left(\frac{5}{6}\right)^3$ ≈ 0.074771
0	2	0	$\binom{5}{0}\left(\frac{5}{6}\right)^5\binom{5}{2}\left(\frac{1}{6}\right)^2\binom{5}{3}\left(\frac{3}{6}\right)\binom{5}{0}\left(\frac{5}{6}\right)^3$ ≈ 0.037386
0	0	2	$\binom{5}{0}\left(\frac{5}{6}\right)^5\binom{5}{0}\left(\frac{5}{6}\right)^5\binom{5}{2}\left(\frac{1}{6}\right)^2\left(\frac{5}{6}\right)^3$ ≈ 0.025962
0	1	1	$\binom{5}{0}\left(\frac{5}{6}\right)^5\binom{5}{1}\left(\frac{1}{6}\right)\binom{5}{4}\left(\frac{4}{6}\right)\binom{5}{1}\left(\frac{1}{6}\right)\left(\frac{5}{6}\right)^3$ ≈ 0.062309

p(exactly two 5's) = the sum of the probabilities in the table ≈ 0.343988, or 0.344 rounded to the nearest thousandth

e) The ways in which exactly one 5 can be obtained using three rolls and their binomial probabilities are shown in the table below.

57. (continued)

Number of 5's on Roll 1	Number of 5's on Roll 2	Number of 5's on Roll 3	Binomial probability
1	0	0	$\binom{5}{1}\left(\frac{1}{6}\right)\left(\frac{5}{6}\right)^4\binom{4}{0}\left(\frac{5}{6}\right)^4\binom{4}{0}\left(\frac{5}{6}\right)^4$ ≈ 0.093464
0	1	0	$\binom{5}{0}\left(\frac{5}{6}\right)^5\binom{5}{1}\left(\frac{1}{6}\right)\left(\frac{5}{6}\right)^4\binom{4}{0}\left(\frac{5}{6}\right)^4$ ≈ 0.0778866
0	0	1	$\binom{5}{0}\left(\frac{5}{6}\right)^5\binom{5}{0}\left(\frac{5}{6}\right)^5\binom{5}{1}\left(\frac{1}{6}\right)\left(\frac{5}{6}\right)^4$ ≈ 0.064905

p(exactly one 5) = the sum of the probabilities in the table ≈ 0.2362556, or 0.236 rounded to the nearest thousandth

f) p(no 5's) = p(no 5's on Roll 1)·p(no 5's on Roll 2)·p(no 5's on Roll 3) =

$\binom{5}{0}\left(\frac{5}{6}\right)^5\binom{5}{0}\left(\frac{5}{6}\right)^5\binom{5}{0}\left(\frac{5}{6}\right)^5 \approx 0.064905$,

or 0.065 rounded to the nearest thousandth

59. Let p_k = the probability of drawing the ace of spades exactly k times. Then p_4 =

$\binom{5}{4}\left(\frac{1}{52}\right)^4\left(\frac{51}{52}\right) = \frac{255}{52^5}$, and $p_5 = \binom{5}{5}\left(\frac{1}{52}\right)^5 = \frac{1}{52^5}$.

Now $p_{k \geqslant 4} = p_4 + p_5 = \frac{256}{52^5}$.

Let F = the event the ace of spades is drawn exactly 4 times and

E = the event the ace of spades is drawn 4 or 5 times.

$p(F|E) = \frac{p(F)p(E|F)}{p(E)}$

Note that $p(F) = p_4$ and $p(E) = p_{k \geqslant 4}$. Also, $p(E|F) = 1$, since it is certain that the ace of spades will be drawn 4 or 5 times (event E) given that the ace of spades is drawn 4 times (event F). Then

$p(F|E) = \frac{\frac{255}{52^5} \cdot 1}{\frac{256}{52^5}} = \frac{255}{256}$.

The correct answer is choice E.

Exercise Set 6.5

1. a) S = {HHH, HHT, HTH, THH, TTT, TTH, THT, HTT}

b) The first outcome in S has 3 heads, the next three outcomes have 2 heads, the fifth outcome has 0 heads, and the last three have 1 head. Therefore, the range set of outputs is {0, 1, 2, 3}.

c) The probability of each outcome is 1/8. There is 1 outcome with 0 heads, so p_0 = 1/8. There are 3 outcomes with 1 head, so p_1 = 3(1/8) = 3/8. There are 3 outcomes with 2 heads and 1 outcome with 3 heads, so p_2 = 3(1/8) = 3/8 and p_3 = 1/8. We can list the probability distribution in a table.

x_i	0	1	2	3
p_i	1/8	3/8	3/8	1/8

---

1. (continued)

Construct a graph.

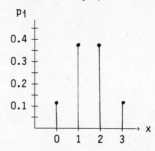

3. Using the array of ordered pairs in Section 6.1, we see that there is 1 pair whose sum is 2, there are 2 pairs whose sum is 3, 3 pairs whose sum is 4, and so on. The probability that a particular pair will occur is 1/36. Use this information to list the probability distribution.

x_i	p_i
2	1/36
3	2/36
4	3/36
5	4/36
6	5/36
7	6/36
8	5/36
9	4/36
10	3/36
11	2/36
12	1/36

(This information could be displayed horizontally as in the text. Space limitations precluded our doing this here, however.)

Construct a graph.

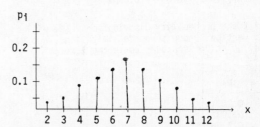

5. Use the array of ordered pairs in Section 6.1. There are 6 pairs with |a - b| = 0, 10 pairs with |a - b| = 1, 8 pairs with |a - b| = 2, and so on. The probability that a particular pair will occur is 1/36. Use this information to list the probability distribution.

d_i	0	1	2	3	4	5
p_i	6/36	10/36	8/36	6/36	4/36	2/36

5. (continued)

Construct a graph.

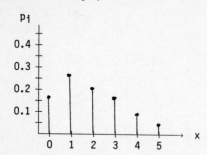

7. a) Let p_i = probability of getting exactly i defective speakers. This is binomial probability with $p = 0.2$ and $q = 1 - p = 0.8$. Then

$p_i = \binom{6}{i}(0.2)^i(0.8)^{6-i}$. We have

$p_0 = \binom{6}{0}(0.2)^0(0.8)^6 \approx 0.2621$

$p_1 = \binom{6}{1}(0.2)^1(0.8)^5 \approx 0.3932$

$p_2 = \binom{6}{2}(0.2)^2(0.8)^4 \approx 0.2458$

$p_3 = \binom{6}{3}(0.2)^3(0.8)^3 \approx 0.0819$

$p_4 = \binom{6}{4}(0.2)^4(0.8)^2 \approx 0.0154$

$p_5 = \binom{6}{5}(0.2)^5(0.8) \approx 0.0015$

$p_6 = \binom{6}{6}(0.2)^6(0.8)^0 \approx 0.0001$

b) List the probability distribution. (We will do it vertically due to space limitations, but you may do it horizontally as in the text.)

x_i	p_i
0	0.2621
1	0.3932
2	0.2458
3	0.0819
4	0.0154
5	0.0015
6	0.0001

Construct a graph.

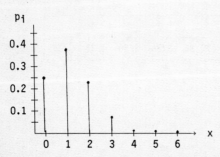

9. Let X be a random variable such that X(s) = number of defective transistors. Let p_i = probability of getting exactly i defective transistors.

$p_0 = \dfrac{\binom{15}{3}}{\binom{20}{3}} = \dfrac{455}{1140} = \dfrac{91}{228}$

$p_1 = \dfrac{\binom{15}{2}\binom{5}{1}}{\binom{20}{3}} = \dfrac{105 \cdot 5}{1140} = \dfrac{105}{228}$

$p_2 = \dfrac{\binom{15}{1}\binom{5}{2}}{\binom{20}{3}} = \dfrac{15 \cdot 10}{1140} = \dfrac{30}{228}$

$p_3 = \dfrac{\binom{5}{3}}{\binom{20}{3}} = \dfrac{10}{1140} = \dfrac{2}{228}$

List the probability distribution.

x_i	0	1	2	3
p_i	91/228	105/228	30/228	2/228

Construct a graph.

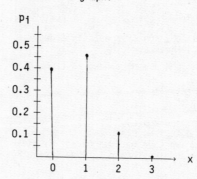

11. $s = 20$, $m = 10$ (women), $s - m = 10$ (men), $n = 4$, and r can assume the values 0, 1, 2, 3, 4.

$p(1) = \dfrac{\binom{10}{0}\binom{10}{4}}{\binom{20}{4}} = \dfrac{1 \cdot 210}{4845} = \dfrac{14}{323}$

$p(2) = \dfrac{\binom{10}{1}\binom{10}{3}}{\binom{20}{4}} = \dfrac{10 \cdot 120}{4845} = \dfrac{80}{323}$

$p(3) = \dfrac{\binom{10}{2}\binom{10}{2}}{\binom{20}{4}} = \dfrac{45 \cdot 45}{4845} = \dfrac{135}{323}$

$p(4) = \dfrac{\binom{10}{3}\binom{10}{1}}{\binom{20}{4}} = \dfrac{120 \cdot 10}{4845} = \dfrac{80}{323}$

$p(5) = \dfrac{\binom{10}{4}\binom{10}{0}}{\binom{20}{4}} = \dfrac{210 \cdot 1}{4845} = \dfrac{14}{323}$

List the probability distribution.

11. (continued)

x_i	0	1	2	3	4
p_i	14/323	80/323	135/323	80/323	14/323

13. $s = 20$, $m = 6$, $s - m = 14$, $n = 5$, and r can assume the values 0, 1, 2, 3, 4, 5.

$$p(0) = \frac{\binom{6}{0}\binom{14}{5}}{\binom{20}{5}} = \frac{1 \cdot 2002}{15,504} = \frac{1001}{7752}$$

$$p(1) = \frac{\binom{6}{1}\binom{14}{4}}{\binom{20}{5}} = \frac{6 \cdot 1001}{15,504} = \frac{3003}{7752}$$

$$p(2) = \frac{\binom{6}{2}\binom{14}{3}}{\binom{20}{5}} = \frac{15 \cdot 364}{15,504} = \frac{2730}{7752}$$

$$p(3) = \frac{\binom{6}{3}\binom{14}{2}}{\binom{20}{5}} = \frac{20 \cdot 91}{15,504} = \frac{910}{7752}$$

$$p(4) = \frac{\binom{6}{4}\binom{14}{1}}{\binom{20}{5}} = \frac{15 \cdot 14}{15,504} = \frac{105}{7752}$$

$$p(5) = \frac{\binom{6}{5}\binom{14}{0}}{\binom{20}{5}} = \frac{6 \cdot 1}{15,504} = \frac{3}{7752}$$

List the probability distribution.

x_i	0	1	2	3	4	5
p_i	$\frac{1001}{7752}$	$\frac{3003}{7752}$	$\frac{2730}{7752}$	$\frac{910}{7752}$	$\frac{105}{7752}$	$\frac{3}{7752}$

15. Let C_1 be the cage with 3 white mice and 2 black ones, and let C_2 be the cage with 2 white mice and 3 black ones. First find p(0 white mice): Since C_1 contains only 2 black mice, any combination of 3 mice from C_1 must contain at least one white mouse. Then p(0 white mice given that C_1 is chosen) = 0. Therefore, p(0 white mice) =

p(0 white mice from C_2) = $\frac{1}{2} \cdot \frac{\binom{3}{3}}{\binom{5}{3}} = \frac{1}{20}.$

p(exactly 1 white mouse) = p(exactly 1 white mouse from C_1) + p(exactly 1 white mouse from C_2) =

$$\frac{1}{2} \cdot \frac{\binom{3}{1}\binom{2}{2}}{\binom{5}{3}} + \frac{1}{2} \cdot \frac{\binom{2}{1}\binom{3}{2}}{\binom{5}{3}} = \frac{9}{20}$$

p(exactly 2 white mice) = p(exactly 2 white mice from C_1) + p(exactly 2 white mice from C_2) =

$$\frac{1}{2} \cdot \frac{\binom{3}{2}\binom{2}{1}}{\binom{5}{3}} + \frac{1}{2} \cdot \frac{\binom{2}{2}\binom{3}{1}}{\binom{5}{3}} = \frac{9}{20}$$

15. (continued)

Since C_2 contains only 2 white mice, p(3 white mice) = p(3 white mice from C_1) =

$$\frac{1}{2} \cdot \frac{\binom{3}{3}}{\binom{5}{3}} = \frac{1}{20}.$$

List the probability distribution.

x_i	0	1	2	3
p_i	1/20	9/20	9/20	1/20

17. The total number of ways in which n numbers can be selected from a group of 80 numbers is $\binom{80}{n}$. The number of ways in which exactly k numbers can be chosen from a group of 20 (and n − k numbers can be chosen from the remaining 60) is $\binom{20}{k}\binom{60}{n-k}$.

Then $p(k) = \dfrac{\binom{20}{k}\binom{60}{n-k}}{\binom{80}{n}}.$

19. We found an expression for p(k) in Exercise 17. Here n = 6.

$$p(0) = \frac{\binom{20}{0}\binom{60}{6}}{\binom{80}{6}} = \frac{1 \cdot 50,063,860}{300,500,200} \approx 0.1666,$$

$$p(1) = \frac{\binom{20}{1}\binom{60}{5}}{\binom{80}{6}} = \frac{20 \cdot 5,461,512}{300,500,200} \approx 0.3635,$$

$$p(2) = \frac{\binom{20}{2}\binom{60}{4}}{\binom{80}{6}} = \frac{190 \cdot 487,635}{300,500,200} \approx 0.3083,$$

$$p(3) = \frac{\binom{20}{3}\binom{60}{3}}{\binom{80}{6}} = \frac{1140 \cdot 34,220}{300,500,200} \approx 0.1298,$$

$$p(4) = \frac{\binom{20}{4}\binom{60}{2}}{\binom{80}{6}} = \frac{4845 \cdot 1770}{300,500,200} \approx 0.0285,$$

$$p(5) = \frac{\binom{20}{5}\binom{60}{1}}{\binom{80}{6}} = \frac{15,504 \cdot 60}{300,500,200} \approx 0.0031,$$

$$p(6) = \frac{\binom{20}{6}\binom{60}{0}}{\binom{80}{6}} = \frac{38,760 \cdot 1}{300,500,200} \approx 0.0001,$$

21. We found an expression for p(k) in Exercise 17. Here n = 12.

$$p(0) = \frac{\binom{20}{0}\binom{60}{12}}{\binom{80}{12}} \approx 0.0232, \quad p(1) = \frac{\binom{20}{1}\binom{60}{11}}{\binom{80}{12}} \approx 0.1138,$$

$$p(2) = \frac{\binom{20}{2}\binom{60}{10}}{\binom{80}{12}} \approx 0.2378, \quad p(3) = \frac{\binom{20}{3}\binom{60}{9}}{\binom{80}{12}} \approx 0.2797,$$

$$p(4) = \frac{\binom{20}{4}\binom{60}{8}}{\binom{80}{12}} \approx 0.2058, \quad p(5) = \frac{\binom{20}{5}\binom{60}{7}}{\binom{80}{12}} \approx 0.0994,$$

$$p(6) = \frac{\binom{20}{6}\binom{60}{6}}{\binom{80}{12}} \approx 0.0322, \quad p(7) = \frac{\binom{20}{7}\binom{60}{5}}{\binom{80}{12}} \approx 0.0070,$$

$$p(8) = \frac{\binom{20}{8}\binom{60}{4}}{\binom{80}{12}} \approx 0.0010, \quad p(9) = \frac{\binom{20}{9}\binom{60}{3}}{\binom{80}{12}} \approx 0.0001,$$

$$p(10) = \frac{\binom{20}{10}\binom{60}{2}}{\binom{80}{12}} \approx 0.0000+,$$

$$p(11) = \frac{\binom{20}{11}\binom{60}{1}}{\binom{80}{12}} \approx 0.0000+,$$

$$p(12) = \frac{\binom{20}{12}\binom{60}{0}}{\binom{80}{12}} \approx 0.0000+$$

25. a) $p(7) = \dfrac{\binom{11}{7}}{\binom{80}{7}} = \dfrac{330}{3,176,716,400} \approx \dfrac{1}{9,626,413}$

b) Since about one ticket in every 9,626,413 is a winner, you would have to buy 9,626,413 tickets to ensure a win.

23. p(1 toss) = 0 since a gambler must win 2 tosses in a row to win the game.

p(2 tosses) = p[(HH ∩ TT) ∪ (HH ∩ HH) ∪ (TT ∩ HH) ∪ (TT ∩ TT) ∪ (HT ∩ HT) ∪ (HT ∩ TH) ∪ (TH ∩ HT) ∪ (TH ∩ TH)] = $8 \cdot \left[\frac{1}{2} \cdot \frac{1}{2} \cdot \frac{1}{2} \cdot \frac{1}{2}\right] = \frac{1}{2}$

There are 16 ways in which 3 tosses will be required, each with probability $\frac{1}{2^6}$, so
p(3 tosses) = $16 \cdot \frac{1}{2^6} = \frac{1}{4}$.

There are 32 ways in which 4 tosses will be required, each with probability $\frac{1}{2^8}$, so
p(4 tosses) = $32 \cdot \frac{1}{2^8} = \frac{1}{8}$.

In general, there are 2^{n+1} ways in which n tosses will be required, each with probability $\frac{1}{2^{2n}}$, so
p(n tosses) = $2^{n+1} \cdot \frac{1}{2^{2n}} = \frac{1}{2^{n-1}}$.

List the probability distribution.

x_i	1	2	3	4	5	$\cdots$
p_i	0	1/2	1/4	1/8	1/16	$\cdots$

Exercise Set 7.1

1. a) Draw and label a vertical axis. The smallest value is $161,000, and the largest is $380,000. We will mark off multiples of $50,000 from $0 through $400,000.

   b) Draw a horizontal axis, and label it with the names of the six television shows.

   c) Draw vertical bars to show the cost of a 30-second commercial on each show.

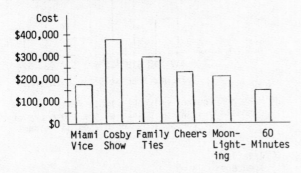

3. a) Draw and label a vertical axis. The smallest value is $9,719, and the largest is $17,155. We will mark off multiples of $2500 from $0 to $20,000.

   b) Draw a horizontal axis, and label it with the names of the states and district.

   c) Draw vertical bars to show the average income per person in each state.

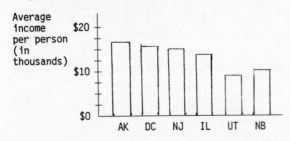

5. a) Draw and label a vertical axis. The smallest value is $3268, and the largest is $3536. We start with $3200 and mark off increments of $100, ending with $3600. (We could go higher, but it is not necessary.)

   b) Draw a horizontal axis, and label it with the repayment times.

   c) Start at each repayment time, move up to its corresponding amount, and mark a point.

   d) Connect adjacent points with a line.

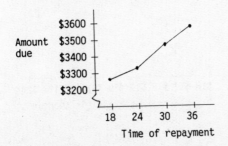

7. a) Draw and label a vertical axis. The smallest weight for either men or women is 122 lb and the largest is 152 lb. We start with 120 lb and mark off increments of 10 lb, ending with 160 lb.

   b) Draw a horizontal axis, and label it with the heights.

   c) Start at each height, move up to the corresponding weight for men, and mark a point.

   d) Connect adjacent points with a solid line.

   e) Start at each height again, move up to the corresponding weight for women, and mark a point.

   f) Connect adjacent points with a dashed line.

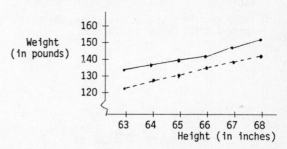

9. a) First multiply each percent by 360° to determine the angle to which it corresponds.

   $38.7\% \times 360° = 0.387 \times 360° \approx 139°$

   $37.1\% \times 360° = 0.371 \times 360° \approx 134°$

   $16.1\% \times 360° = 0.161 \times 360° \approx 58°$

   $8.1\% \times 360° = 0.081 \times 360° \approx 29°$

   b) Draw a circle and a radius.

   c) Draw the angles found in a).

   d) Label each part of the circle with the corresponding item and percent.

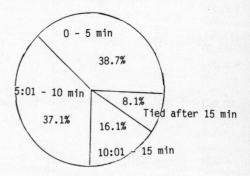

11. a) Complete the table.

   $\frac{34}{100} = 34\%$, $34\% \times 360° = 0.34 \times 360° \approx 122°$

   $\frac{26}{100} = 26\%$, $26\% \times 360° = 0.26 \times 360° \approx 94°$

   $\frac{38}{100} = 38\%$, $38\% \times 360° = 0.38 \times 360° \approx 137°$

   $\frac{2}{100} = 2\%$, $2\% \times 360° = 0.02 \times 360° \approx 7°$

11. (continued)

Deodorant	Number Preferring	Percent	Angle
A	34	34%	122°
B	26	26%	94°
C	38	38%	137°
D	2	2%	7°
Total	100	100%	360°

b) Draw a circle and a radius.

c) Draw the angles with the measures given in the table.

d) Label each part of the circle with the corresponding letter and percent.

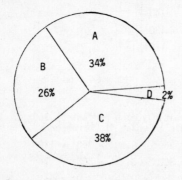

13. a), b) Tally the data and complete a frequency distribution table.

Interval	Tally	Frequency	Relative Frequency	Cumulative Frequency	Cumulative Relative Frequency
$16,001-17,000	III	3	≈ 0.09	3 0.09	.09
$17,001-18,000	IIII	4	≈ 0.11	7 0.20	.2
$18,001-19,000	IIII	4	≈ 0.11	11 0.31	.31
$19,001-20,000	THL IIII	9	≈ 0.26	20 0.57	.57
$20,001-21,000	THL II	7	0.20	27 0.77	.77
$21,001-22,000	II	2	≈ 0.06	29 0.83	.83
$22,001-23,000	IIII	4	≈ 0.11	33 0.94	.94
$23,001-24,000	II	2	≈ 0.06	35 1.00	1.00
Sum = 35			1.00		

Note that we divided each frequency by the total number of starting salaries, 35, to get the relative frequency. We added successively down the relative frequency column to obtain the cumulative relative frequencies.

13. (continued)

c) Make a histogram. It is a compressed bar graph.

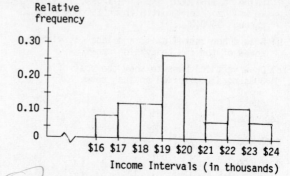

d) Make a frequency polygon. It is formed by connecting adjacent points at the midpoints of the tops of the bars on the histogram in part c).

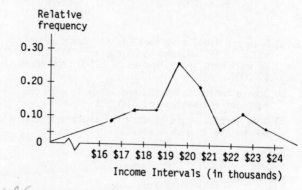

e) Make a cumulative frequency distribution line graph.

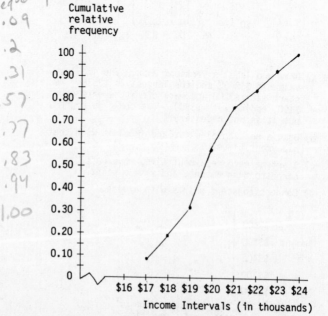

15. a), b) Tally the data and complete a frequency
distribution table.

Interval	Tally	Fre-quency	Relative Frequency	Cumulative Frequency
352-352.9	I	1	0.02	0.02
353-353.9	卌 II	7	0.14	0.16
354-354.9	卌 III	8	0.16	0.32
355-355.9	卌 卌 卌 卌 卌	25	0.50	0.82
356-356.9	卌 IIII	9	0.18	1.00
	Sum = 50		1.00	

Note that we divided each frequency by the
total number of cars, 50, to get the relative
frequency. We added successively down the
relative frequency column to obtain the
cumulative relative frequency.

c) Make a histogram. This is a compressed bar
graph.

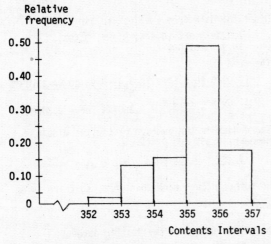

d) Make a frequency polygon by connecting adjacent
points at the midpoints of the tops of the bars
on the histogram in part c).

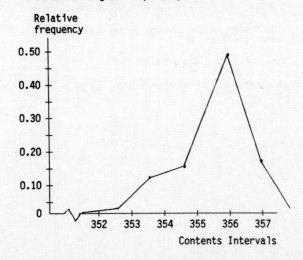

15. (continued)

e) Make a cumulative frequency distribution line
graph.

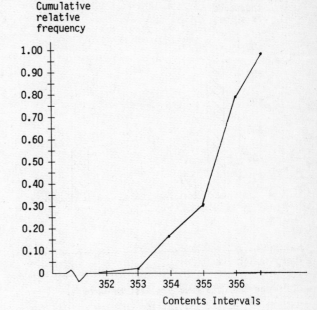

17. Divide the population of each state by its area to
find the population densities.

Alaska: 479,000 people/570,833 sq mi ≈
0.8391 people/sq mi

California: 25,174,000 people/156,299 sq mi ≈
161.0631 people/sq mi

New Jersey: 7,468,000 people/7,468 sq mi =
1000 people/sq mi

New York: 17,667,000 people/47,377 sq mi ≈
372.9025 people/sq mi

Rhode Island: 955,000 people/1,055 sq mi ≈
905.2133 people/sq mi

Texas: 15,724,000 people/262,017 sq mi ≈
60.0114 people/sq mi

19. a) Draw and label a vertical axis. The smallest
area is 1,055 sq mi and the largest is 570,833
sq mi. We will mark off multiples of 100,000 sq
mi from 0 sq mi through 600,000 sq mi.

b) Draw a horizontal axis and label it with the
names of the states.

19. (continued)

c) Draw vertical bars to show the area of each state.

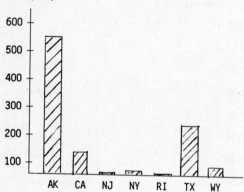

Area (in
thousands
of sq mi)

Exercise Set 7.2

1. Order the data from smallest to largest.

    7, 8, 12, 15, 15, 15

    The mean is

    $$\bar{x} = \frac{7 + 8 + 12 + 15 + 15 + 15}{6} = \frac{72}{6} = 12.$$

    The median is halfway between the two middle numbers, 12 and 15.

    $$\text{Median} = \frac{12 + 15}{2} = \frac{27}{2} = 13.5$$

    The number 15 occurs most often, so 15 is the mode.

3. The data is already ordered from smallest to largest.

    $5, $10, $15, $20, $25, $30, $35

    The mean is

    $$\bar{x} = \frac{\$5 + \$10 + \$15 + \$20 + \$25 + \$30 + \$35}{7} = \frac{\$140}{7} = \$20.$$

    The median is the middle number, $20. No number occurs more than once, so there is no mode.

5. The data is already ordered from smallest to largest.

    1.2m, 4.3m, 5.7m, 7.4m, 7.4m

    The mean is

    $$\bar{x} = \frac{1.2m + 4.3m + 5.7m + 7.4m + 7.4m}{5} = \frac{26m}{5} = 5.2m.$$

    The median is the middle number, 5.7m. The number 7.4m occurs most often, so the mode is 7.4m.

7. The data is already ordered from smallest to largest.

    250 lb, 255 lb, 260 lb, 260 lb, 270 lb, 270 lb, 290 lb

    The mean is

    $$\bar{x} = \frac{(250 + 255 + 260 + 260 + 270 + 270 + 290)\ lb}{7} = \frac{1855\ lb}{7} = 265\ lb.$$

    The median is the middle number, 260 lb. The numbers 260 lb and 270 lb both occur most often. The modes are 260 lb and 270 lb. (The data is bimodal.)

9. Order the data from smallest to largest.

    44.6, 44.7, 45.5, 46.8, 47.5

    The mean is

    $$\bar{x} = \frac{44.6 + 44.7 + 45.5 + 46.8 + 47.5}{5} = \frac{229.1}{5} = 45.82$$

    The median is the middle number, 45.5. No number occurs more than once, so there is no mode.

11. Order the data from smallest to largest.

    7.02%, 8.36%, 8.64%, 9.38%, 9.49%, 9.95%, 10.94%, 11.09%

    The mean is

    $$\bar{x} = \frac{(7.02+8.36+8.64+9.38+9.49+9.95+10.94+11.09)\%}{8} = \frac{74.87\%}{8} = 9.35875\% \ (or \approx 9.3588\%).$$

    The median is the average of the two middle numbers, 9.38% and 9.49%.

    $$\frac{9.38\% + 9.49\%}{2} = \frac{18.87\%}{2} = 9.435\%$$

    No number occurs more than once, so there is no mode.

13. Order the data from smallest to largest.

    69, 72, 74, 75

    The average (or mean) is

    $$\bar{x} = \frac{69 + 72 + 74 + 75}{4} = \frac{290}{4} = 72.5$$

    The median is the average of the two middle numbers, 72 and 74.

    $$\frac{72 + 74}{2} = \frac{146}{2} = 73$$

    No number occurs more than once, so there is no mode.

15. For each course, multiply the grade point value by the number of hours in the course and add.

$$3.00(4) = 12$$
$$3.00(5) = 15$$
$$3.00(3) = 9$$
$$2.00(4) = \underline{8}$$
$$44 \quad \text{(Total grade points)}$$

The total number of hours taken is $4 + 5 + 3 + 4$, or 16. Divide 44 by 16 to find the GPA.

$$\frac{44}{16} = 2.75$$

The student's GPA is 2.75.

17. To find the average number of strawberry gum drops, find the total number of strawberry gum drops and divide by the number of boxes.

$$\bar{x} = \frac{5 + 7 + 3 + 3 + 6}{5} = \frac{24}{5} = 4.8$$

19. To find the average number of orange gum drops, find the total number of orange gum drops and divide by the number of boxes.

$$\bar{x} = \frac{6 + 9 + 6 + 8 + 8}{5} = \frac{37}{5} = 7.4$$

21. To find the average number of lemon gum drops, find the total number of lemon gum drops and divide by the number of boxes.

$$\bar{x} = \frac{10 + 8 + 5 + 6 + 5}{5} = \frac{34}{5} = 6.8$$

23. To find the mean of the ages, find the total of the ages and divide by the number of presidents, 40.

$$\bar{x} = \frac{2193}{40} = 54.825$$

25. Order the data from smallest to largest.

42, 43, 46, 47, 48, 49, 49, 50, 50, 51, 51,
51, 51, 52, 52, 54, 54, 54, 54, 55, 55, 55,
55, 56, 56, 56, 57, 57, 57, 57, 58, 60, 61,
61, 61, 62, 64, 65, 68, 69

There are four numbers that occur most often (4 times each). The modes are 51, 54, 55, and 57.

27. To find the mean, add the salaries of the state governors and divide by the number of salaries, 13.

$$\bar{x} = \frac{\$896,851}{13} \approx \$68,988.54$$

29. To find the mean, add the salaries of the basketball coaches and divide by the number of salaries, 13.

$$\bar{x} = \frac{\$876,707}{13} = \$67,439$$

31. In 11 situations (all except N.C. and N.Y.) the university president makes more than the governor.

33. In 2 situations (N.C. and N.Y.) the governor makes more than the university president.

35. In 4 situations (Ariz., Ill., N.C., and N.Y.) the governor makes more than the football coach.

37. The average salaries are:

University president	$102,485.54
State governor	$ 68,988.54
Football coach	$ 80,642.38
Basketball coach	$ 67,439.00
Athletic director	$ 75,997.08

The average is highest for university presidents.

39. To find the average of the winning scores, add all the winning scores and divide by the number of games, 49.

$$\bar{x} = \frac{3416}{49} \approx 69.71$$

Exercise Set 7.3

1. $E(x) = \sum p_i x_i$
$$= 0.93(\$0) + 0.04(\$100) + 0.02(\$700) +$$
$$0.01(\$5000) = \$68$$

3. $E(x) = \sum p_i x_i$
$$= 0.47(-2) + 0.03(-1) + 0.04(1) + 0.46(2) =$$
$$-0.01$$

5. In Exercise 1, we found $E(x) = \$68$. In Exercise 2, $E(x) = 0.78(\$0) + 0.11(\$100) + 0.07(\$700) + 0.04(\$5000) = \$260$. You would choose the distribution in Exercise 2 to maximize expected value.

7. a) In Exercise Set 6.5, Exercise 1, we found the probability distribution.

x_i	0	1	2	3
p_i	1/8	3/8	3/8	1/8

$$E(x) = \frac{1}{8}(0) + \frac{3}{8}(1) + \frac{3}{8}(2) + \frac{1}{8}(3) = \frac{12}{8} = \frac{3}{2}, \text{ or}$$
$$1.5$$

b) The probability distribution for the random variable Y is given below. (See Exercise 6.5, Exercise 1.)

y_i	0	1	2	3
p_i	1/8	3/8	3/8	1/8

$$E(Y) = \frac{1}{8}(0) + \frac{3}{8}(1) + \frac{3}{8}(2) + \frac{1}{8}(3) = \frac{12}{8} = \frac{3}{2}, \text{ or}$$
$$1.5$$

Note that $E(Y) = 3 - E(x)$.

9. Using the array of ordered pairs in Section 6.1, we see that there is 1 pair with a - b = -5, 2 pairs with a - b = -4, 3 pairs with a - b = -3, 4 pairs with a - b = -2, 5 pairs with a - b = -1, 6 pairs with a - b = 0, 5 pairs with a - b = 1, 4 pairs with a - b = 2, 3 pairs with a - b = 3, 2 pairs with a - b = 4, and 1 pair with a - b = 5. The probability that a particular pair will occur is 1/36. We get the following probability distribution.

x_i	-5	-4	-3	-2	-1	0	1	2	3	4	5
p_i	$\frac{1}{36}$	$\frac{2}{36}$	$\frac{3}{36}$	$\frac{4}{36}$	$\frac{5}{36}$	$\frac{6}{36}$	$\frac{5}{36}$	$\frac{4}{36}$	$\frac{3}{36}$	$\frac{2}{36}$	$\frac{1}{36}$

$E(X) = \frac{1}{36}(-5) + \frac{2}{36}(-4) + \frac{3}{36}(-3) + \frac{4}{36}(-2) + \frac{5}{36}(-1) +$

$\frac{6}{36}(0) + \frac{5}{36}(1) + \frac{4}{36}(2) + \frac{3}{36}(3) + \frac{2}{36}(4) +$

$\frac{1}{36}(5) = 0$

11. In Exercise Set 6.5, Exercise 5, we found the following probability distribution.

d_i	0	1	2	3	4	5
x_i	6/36	10/36	8/36	6/36	4/36	2/36

$E(D) = \frac{6}{36}(0) + \frac{10}{36}(1) + \frac{8}{36}(2) + \frac{6}{36}(3) + \frac{4}{36}(4) +$

$\frac{2}{36}(5) = \frac{70}{36} \approx 1.94$

13. a) In Exercise Set 6.5, Exercise 7, we found the following probability distribution.

x_i	p_i
0	0.2621
1	0.3932
2	0.2458
3	0.0819
4	0.0154
5	0.0015
6	0.0001

$E(X) = 0.2621(0) + 0.3932(1) + 0.2458(2) +$
$0.0819(3) + 0.0154(4) + 0.0015(5) +$
$0.0001(6) = 1.2002 \approx 1.2$

b) $y_i = x_{6-i}$ (that is, 6 defective speakers means 0 nondefective and so on), so $p(y_i) = p(x_{6-i})$. We have the following probability distribution for Y.

13. (continued)

y_i	p_i
0	0.0001
1	0.0015
2	0.0154
3	0.0819
4	0.2458
5	0.3932
6	0.2621

$E(Y) = 0.0001(0) + 0.0015(1) + 0.0154(2) +$
$0.0819(3) + 0.2458(4) + 0.3932(5) +$
$0.2621(6) = 4.7998 \approx 4.8$

Note that $E(Y) = 6 - E(X)$.

15. a) We have binomial probability with p = 0.357 and q = 0.643.

$p_0 = \binom{5}{0}(0.357)^0(0.643)^5 \approx 0.110$

$p_1 = \binom{5}{1}(0.357)(0.643)^4 \approx 0.305$

$p_2 = \binom{5}{2}(0.357)^2(0.643)^3 \approx 0.339$

$p_3 = \binom{5}{3}(0.357)^3(0.643)^2 \approx 0.188$

$p_4 = \binom{5}{4}(0.357)^4(0.643) \approx 0.052$

$p_5 = \binom{5}{5}(0.357)^5(0.643)^0 = 0.006$

We have the following probability distribution.

h_i	0	1	2	3	4	5
p_i	0.110	0.305	0.339	0.188	0.052	0.006

$E(H) = 0.110(0) + 0.305(1) + 0.339(2) +$
$0.188(3) + 0.052(4) + 0.006(5) = 1.785$

b) $f_i = h_{5-i}$ (that is, if Boggs gets 5 hits in five at bats, the number of times a hit is not achieved is 0 and so on), so $p(f_i) = p(h_{5-i})$. We have the following probability distribution.

f_i	0	1	2	3	4	5
p_i	0.006	0.052	0.188	0.339	0.305	0.110

$E(F) = 0.006(0) + 0.052(1) + 0.188(2) +$
$0.339(3) + 0.305(4) + 0.110(5) = 3.215$

Note that $E(F) = 5 - E(H)$.

17. $p(0) = \binom{5}{0}(0.2)^0(0.8)^5 = 0.32768$

$p(1) = \binom{5}{1}(0.2)(0.8)^4 = 0.4096$

$p(2) = \binom{5}{2}(0.2)^2(0.8)^3 = 0.2048$

$p(3) = \binom{5}{3}(0.2)^3(0.8)^2 = 0.0512$

17. (continued)

$p(4) = \begin{bmatrix} 5 \\ 4 \end{bmatrix}(0.2)^4(0.8) = 0.0064$

$p(5) = \begin{bmatrix} 5 \\ 5 \end{bmatrix}(0.2)^5(0.8)^0 = 0.00032$

$E(X) = 0.32768(0) + 0.4096(1) + 0.2048(2) + 0.0512(3) + 0.0064(4) + 0.00032(5) = 1$

19. 40% = 0.4 and 60% = 0.6

$p(0) = \begin{bmatrix} 9 \\ 0 \end{bmatrix}(0.4)^0(0.6)^9 \approx 0.010$

$p(1) = \begin{bmatrix} 9 \\ 1 \end{bmatrix}(0.4)(0.6)^8 \approx 0.060$

$p(2) = \begin{bmatrix} 9 \\ 2 \end{bmatrix}(0.4)^2(0.6)^7 \approx 0.161$

$p(3) = \begin{bmatrix} 9 \\ 3 \end{bmatrix}(0.4)^3(0.6)^6 \approx 0.251$

$p(4) = \begin{bmatrix} 9 \\ 4 \end{bmatrix}(0.4)^4(0.6)^5 \approx 0.251$

$p(5) = \begin{bmatrix} 9 \\ 5 \end{bmatrix}(0.4)^5(0.6)^4 \approx 0.167$

$p(6) = \begin{bmatrix} 9 \\ 6 \end{bmatrix}(0.4)^6(0.6)^3 \approx 0.074$

$p(7) = \begin{bmatrix} 9 \\ 7 \end{bmatrix}(0.4)^7(0.6)^2 \approx 0.021$

$p(8) = \begin{bmatrix} 9 \\ 8 \end{bmatrix}(0.4)^8(0.6) \approx 0.004$

$p(9) = \begin{bmatrix} 9 \\ 9 \end{bmatrix}(0.4)^9(0.6)^0 \approx 0.000+$

$E(X) = 0.010(0) + 0.060(1) + 0.161(2) + 0.251(3) + 0.251(4) + 0.167(5) + 0.074(6) + 0.021(7) + 0.004(8) + (0.000+)(9) = 3.597$, or 3.6 rounded to the nearest tenth.

21. This is binomial probability with p = 0.78 and q = 0.22.

$p(0) = \begin{bmatrix} 5 \\ 0 \end{bmatrix}(0.78)^0(0.22)^5 \approx 0.001$

$p(1) = \begin{bmatrix} 5 \\ 1 \end{bmatrix}(0.78)(0.22)^4 \approx 0.009$

$p(2) = \begin{bmatrix} 5 \\ 2 \end{bmatrix}(0.78)^2(0.22)^3 \approx 0.065$

$p(3) = \begin{bmatrix} 5 \\ 3 \end{bmatrix}(0.78)^3(0.22)^2 \approx 0.230$

$p(4) = \begin{bmatrix} 5 \\ 4 \end{bmatrix}(0.78)^4(0.22) \approx 0.407$

$p(5) = \begin{bmatrix} 5 \\ 5 \end{bmatrix}(0.78)^5(0.22)^0 = 0.289$

$E(X) = 0.001(0) + 0.009(1) + 0.065(2) + 0.230(3) + 0.407(4) + 0.289(5) = 3.902$, or 3.9 rounded to the nearest tenth.

23. Let T = the amount of winnings per ticket. We have the following probability distribution.

t_i	$0	$400	$7000	$15,000
p_i	$\frac{49,993}{50,000}$	$\frac{5}{50,000}$	$\frac{1}{50,000}$	$\frac{1}{50,000}$

$E(T) = \frac{49,993}{50,000}(\$0) + \frac{5}{50,000}(\$400) + \frac{1}{50,000}(\$7000) + \frac{1}{50,000}(\$15,000) = \$\frac{24,000}{50,000} = \$0.48.$

The expected value of the game is

$E(G) = E(T - 1) = E(T) - \$1 = \$0.48 - \$1 = -\$0.52.$

25. The woman has a probability of 0.999 of living in which case the devil will receive $135. She has a probability of 0.001 of dying in which case the devil will lose $50,000. We have the following probability distribution.

x_i	$135	-$50,000
p_i	0.999	0.001

The expected value to the devil is

$E(X) = 0.999(\$135) + 0.001(-\$50,000) = \$84.865.$

27. Let $b = the amount your friend must bet so the game is fair. Then the probability that your friend will lose $b is 5/8, and the probability that your friend will win $1(the amount of your bet) is 3/8. We have the following distribution.

x_i	-$b	$1
p_i	5/8	3/8

In order for the game to be fair, E(X) must be 0.

$E(X) = \frac{5}{8}(-\$b) + \frac{3}{8}(\$1)$, so

$-\$\frac{5}{8}b + \$\frac{3}{8} = 0$

$-\$\frac{5}{8}b = -\$\frac{3}{8}$

$b = \$\frac{3}{5}$, or \$0.60

29. Let X = the event of drawing a defective transistor. This event has probability $p = \frac{5}{20}$, or $\frac{1}{4}$. The probability of drawing a nondefective transistor $q = 1 - p = 1 - \frac{1}{4}$, or $\frac{3}{4}$. The probability of drawing x defective transistors is given by $p(x) = \begin{bmatrix} 3 \\ x \end{bmatrix}\left(\frac{1}{4}\right)^x\left(\frac{3}{4}\right)^{3-x}$.

29. (continued)

The probability distribution is

x_i	p_i
0	$\binom{3}{0}\left(\frac{1}{4}\right)^0\left(\frac{3}{4}\right)^3 = \frac{27}{64}$
1	$\binom{3}{1}\left(\frac{1}{4}\right)\left(\frac{3}{4}\right)^2 = \frac{27}{64}$
2	$\binom{3}{2}\left(\frac{1}{4}\right)^2\left(\frac{3}{4}\right) = \frac{9}{64}$
3	$\binom{3}{3}\left(\frac{1}{4}\right)^3\left(\frac{3}{4}\right)^0 = \frac{1}{64}$

$$E(x) = \frac{27}{64}(0) + \frac{27}{64}(1) + \frac{9}{64}(2) + \frac{1}{64}(3) = \frac{48}{64} = \frac{3}{4}$$

31. We have a hypergeometric probability distribution with s = 20, m = 10, and n = 4. Then

$$E(R) = n \cdot \frac{m}{s} = 4 \cdot \frac{10}{20} = 2.$$

33. We have a hypergeometric probability distribution with s = 20, m = 6, and n = 5. Then

$$E(R) = n \cdot \frac{m}{s} = 5 \cdot \frac{6}{20} = \frac{3}{2}, \text{ or } 1.5.$$

35. Let X = the winnings. There are two outcomes, X = \$2 if the ball falls into a slot of the player's color and X = -\$1 if it does not. The probability distribution is

x_i	\$2	-\$1
p_i	$\frac{18}{38}$	$\frac{20}{38}$

$$E(X) = \frac{18}{38}(\$2) + \frac{20}{38}(-\$1) = -\$\frac{2}{38} \approx -\$0.053$$

37. In Exercise Set 6.5, Exercise 15, we found the probability distribution.

x_i	0	1	2	3
p_i	1/20	9/20	9/20	1/20

$$E(X) = \frac{1}{20}(0) + \frac{9}{20}(1) + \frac{9}{20}(2) + \frac{1}{20}(3) = \frac{30}{20}, \text{ or } 1.5$$

39. In Exercise Set 6.5, Exercise 17, we found that the probability of exactly k numbers turning up if n numbers are originally selected (k ⩽ n) is

$$p(k) = \frac{\binom{20}{k}\binom{60}{n-k}}{\binom{80}{n}}.$$ Using this expression with n = 14 we can find the following probabilities.

39. (continued)

k	p(k)
6	$\dfrac{\binom{20}{6}\binom{60}{8}}{\binom{80}{14}} \approx 0.0658$
7	$\dfrac{\binom{20}{7}\binom{60}{7}}{\binom{80}{14}} \approx 0.0199$
8	$\dfrac{\binom{20}{8}\binom{60}{6}}{\binom{80}{14}} \approx 0.0042$
9	$\dfrac{\binom{20}{9}\binom{60}{5}}{\binom{80}{14}} \approx 0.0006$
10	$\dfrac{\binom{20}{10}\binom{60}{4}}{\binom{80}{14}} \approx 0.0001$
11	$\dfrac{\binom{20}{11}\binom{60}{3}}{\binom{80}{14}} \approx 0.0000+$
12	$\dfrac{\binom{20}{12}\binom{60}{2}}{\binom{80}{14}} \approx 0.0000+$
13	$\dfrac{\binom{20}{13}\binom{60}{1}}{\binom{80}{14}} \approx 0.0000+$
14	$\dfrac{\binom{20}{14}\binom{60}{0}}{\binom{80}{14}} \approx 0.0000+$

We have the following probability distribution.

x_i	p_i
\$1	0.0658
\$10	0.0199
\$40	0.0042
\$300	0.0006
\$1000	0.0001
\$3200	0.0000+
\$16,000	0.0000+
\$25,000	0.0000+
\$40,000	0.0000+

$$E(X) = 0.0658(\$1) + 0.0199(\$10) + 0.0042(\$40) +$$
0.0006(\$300) + 0.0001(\$1000) +
0.0000+(\$3200) + 0.0000+(\$16,000) +
0.0000+(\$25,000) + 0.0000+(\$40,000) ≈ \$0.71
(Answers may vary slightly due to rounding differences.)

41. The expected value of the game in Exercise 39 is approximately \$0.71 while in Exercise 40 it is approximately \$0.70. Therefore, it is best to play the game in Exercise 39.

43. Let X = winnings per visit. The probability distribution is

x_i	$1000	$100	$20	$5	$2	$1
p_i	$\frac{1}{147,620}$	$\frac{1}{13,779}$	$\frac{1}{6928}$	$\frac{1}{2601}$	$\frac{1}{833}$	$\frac{1}{105}$

$E(X) = \frac{1}{147,620}(\$1000) + \frac{1}{13,779}(\$100) + \frac{1}{6928}(\$20) +$

$\frac{1}{2601}(\$5) + \frac{1}{833}(\$2) + \frac{1}{105}(\$1) \approx \$0.031$

45. If the client holds the investment for three days before selling, we have the following probability distribution.

x_i	$5000	$8000	$12,000	$30,000
p_i	0.4	0.2	0.3	0.1

The expected value is

$E(X) = 0.4(\$5000) + 0.2(\$8000) + 0.3(\$12,000) + 0.1(\$30,000) = \$10,200.$

Since the expected value of holding the investment exceeds the current selling price, the most reasonable statement is (a).

Exercise Set 7.4

1. $\mu = E(X) = 0.1(0) + 0.1(1) + 0.2(2) + 0.3(3) + 0.2(4) + 0.1(5) = 2.7$

We find the variance and standard deviation as follows:

x_i	p_i	$x_i - \mu$	$(x_i - \mu)^2$	$p_i(x_i - \mu)^2$
0	0.1	-2.7	7.29	0.729
1	0.1	-1.7	2.89	0.289
2	0.2	-0.7	0.49	0.098
3	0.3	0.3	0.09	0.027
4	0.2	1.3	1.69	0.338
5	0.1	2.3	5.29	0.529
				Sum = 2.01

$\sigma^2 = $ Variance $= 2.01$

$\sigma = $ Standard deviation $= \sqrt{2.01} \approx 1.418$

3. In Exercise Set 7.3, Exercise 1, we found $\mu = E(X) = \$68$.

We find the variance and standard deviation as follows.

x_i	p_i	$x_i - \mu$	$(x_i - \mu)^2$	$p_i(x_i - \mu)^2$
$0	0.93	-$68	$4624	$4300.32
$100	0.04	$32	$1024	$40.96
$700	0.02	$632	$399,424	$7988.48
$5000	0.01	$4932	$24,324,624	$243,246.24
				Sum = $255,576

3. (continued)

$\sigma^2 = $ Variance $= \$255,576$

$\sigma = $ Standard deviation $= \sqrt{\$255,576} \approx \$505.545$

5. In Exercise Set 7.3, Exercise 3, we found $\mu = E(X) = -0.01$.

We find the variance and standard deviation as follows.

x_i	p_i	$x_i - \mu$	$(x_i - \mu)^2$	$p_i(x_i - \mu)^2$
-2	0.47	-1.99	3.9601	1.861247
-1	0.03	-0.99	0.9801	0.029403
1	0.04	1.01	1.0201	0.040804
2	0.46	2.01	4.0401	1.858446
				Sum = 3.7899

$\sigma^2 = $ Variance $= 3.7899$

$\sigma = $ Standard deviation $= \sqrt{3.7899} \approx 1.9468$

7. a) In Exercise Set 6.5, Exercise 1, we found the probability distribution and in Exercise Set 7.3, Exercise 7, we found $\mu = E(X) = 1.5$. We find the variance and standard deviation as follows.

x_i	p_i	$x_i - \mu$	$(x_i - \mu)^2$	$p_i(x_i - \mu)^2$
0	$\frac{1}{8}$	-1.5	2.25	0.28125
1	$\frac{3}{8}$	-0.5	0.25	0.09375
2	$\frac{3}{8}$	0.5	0.25	0.09375
3	$\frac{1}{8}$	1.5	2.25	0.28125
				Sum = 0.75

$\sigma^2 = 0.75$

$\sigma = \sqrt{0.75} \approx 0.866$

b) In Exercise Set 6.5, Exercise 1, we found the probability distribution and in Exercise Set 7.3, Exercise 7, we found $\mu = E(Y) = 1.5$. Both are the same as in (a), so $\sigma^2 = 0.75$ and $\sigma \approx 0.866$.

9. Using the array of ordered pairs in Section 6.1, we see that there is 1 pair with $a + b = 2$, 2 pairs with $a + b = 3$, 3 pairs with $a + b = 4$, 4 pairs with $a + b = 5$, 5 pairs with $a + b = 6$, 6 pairs with $a + b = 7$, 5 pairs with $a + b = 8$, 4 pairs with $a + b = 9$, 3 pairs with $a + b = 10$, 2 pairs with $a + b = 11$, and 1 pair with $a + b = 12$. The probability that a particular pair will occur is 1/36. We get the following probability distribution.

x_i	2	3	4	5	6	7	8	9	10	11	12
p_i	$\frac{1}{36}$	$\frac{2}{36}$	$\frac{3}{36}$	$\frac{4}{36}$	$\frac{5}{36}$	$\frac{6}{36}$	$\frac{5}{36}$	$\frac{4}{36}$	$\frac{3}{36}$	$\frac{2}{36}$	$\frac{1}{36}$

9. (continued)

$$\mu = E(X) = \frac{1}{36}(2) + \frac{2}{36}(3) + \frac{3}{36}(4) + \frac{4}{36}(5) + \frac{5}{36}(6) +$$

$$\frac{6}{36}(7) + \frac{5}{36}(8) + \frac{4}{36}(9) + \frac{3}{36}(10) +$$

$$\frac{2}{36}(11) + \frac{1}{36}(12) = 7$$

We find the variance and standard deviation as follows.

x_i	p_i	$x_i - \mu$	$(x_i - \mu)^2$	$p_i(x_i - \mu)^2$
2	$\frac{1}{36}$	-5	25	$\frac{25}{36}$
3	$\frac{2}{36}$	-4	16	$\frac{32}{36}$
4	$\frac{3}{36}$	-3	9	$\frac{27}{36}$
5	$\frac{4}{36}$	-2	4	$\frac{16}{36}$
6	$\frac{5}{36}$	-1	1	$\frac{5}{36}$
7	$\frac{6}{36}$	0	0	0
8	$\frac{5}{36}$	1	1	$\frac{5}{36}$
9	$\frac{4}{36}$	2	4	$\frac{16}{36}$
10	$\frac{3}{36}$	3	9	$\frac{27}{36}$
11	$\frac{2}{36}$	4	16	$\frac{32}{36}$
12	$\frac{1}{36}$	5	25	$\frac{25}{36}$

$$\text{Sum} = \frac{210}{36} \approx 5.8333$$

$\sigma^2 = 5.8333$

$\sigma = \sqrt{5.8333} \approx 2.42$

11. In Exercise Set 6.5, Exercise 5, we found the following probability distribution and in Exercise Set 7.3, Exercise 11, we found $\mu = E(D) = 1.94$.

We find the variance and standard deviation as follows.

d_i	p_i	$d_i - \mu$	$(d_i - \mu)^2$	$p_i(d_i - \mu)^2$
0	$\frac{6}{36}$	-1.94	3.7636	0.6273
1	$\frac{10}{36}$	-0.94	0.8836	0.2454
2	$\frac{8}{36}$	-0.06	0.0036	0.0008
3	$\frac{6}{36}$	-1.06	1.1236	0.1873
4	$\frac{4}{36}$	-2.06	4.2436	0.4715
5	$\frac{2}{36}$	-3.06	9.3636	0.5202

$$\text{Sum} = 2.0525$$

$\sigma^2 \approx 2.0525$

$\sigma \approx \sqrt{2.0525} \approx 1.43$

13. a) In Exercise Set 6.5, Exercise 7, we found the probability distribution, and in Exercise Set 7.3, Exercise 13 a), we found $\mu = E(X) \approx 1.2$. We find the variance and standard deviation as follows.

x_i	p_i	$x_i - \mu$	$(x_i - \mu)^2$	$p_i(x_i - \mu)^2$
0	0.2621	-1.2	1.44	0.3774
1	0.3932	-0.2	0.04	0.0157
2	0.2458	0.8	0.64	0.1573
3	0.0819	1.8	3.24	0.2654
4	0.0154	2.8	7.84	0.1207
5	0.0015	3.8	14.44	0.0217
6	0.0001	4.8	23.04	0.0023

$$\text{Sum} = 0.9605$$

$\sigma^2 \approx 0.9605$

$\sigma \approx \sqrt{0.9605} \approx 0.98$

b) In Exercise Set 7.3, Exercise 13 b), we found the probability distribution and we found $\mu = E(Y) \approx 4.8$. We find the variance and standard deviation as follows.

y_i	p_i	$y_i - \mu$	$(y_i - \mu)^2$	$p_i(y_i - \mu)^2$
0	0.0001	-4.8	23.04	0.0023
1	0.0015	-3.8	14.44	0.0217
2	0.0154	-2.8	7.84	0.1207
3	0.0819	-1.8	3.24	0.2654
4	0.2458	-0.8	0.64	0.1573
5	0.3932	0.2	0.04	0.0157
6	0.2621	1.2	1.44	0.3774

$$\text{Sum} = 0.9605$$

$\sigma^2 \approx 0.9605$

$\sigma \approx \sqrt{0.9605} \approx 0.98$

Note that this is the same result as in part a) since $E(Y) = 6 - E(X)$ and $p(y_i) = p(x_{6-i})$. (See Exercise Set 7.3, Exercise 13 b).)

15. a) In Exercise Set 7.3, Exercise 15, we found the probability distribution, and we found $\mu = E(H) = 1.785$. We find the variance and standard deviation as follows.

h_i	p_i	$h_i - \mu$	$(h_i - \mu)^2$	$p_i(h_i - \mu)^2$
0	0.110	-1.785	3.186225	0.3505
1	0.305	-0.785	0.616225	0.1879
2	0.339	0.215	0.046225	0.0157
3	0.188	1.215	1.476225	0.2775
4	0.052	2.215	4.906225	0.2551
5	0.006	3.215	10.336225	0.0620

$$\text{Sum} = 1.1487$$

15. (continued)

$\sigma^2 \approx 1.1487$

$\sigma \approx \sqrt{1.1487} \approx 1.07$

b) Since $E(F) = 5 - E(H)$ and $p(f_i) = p(h_{5-i})$, the variance and the standard deviation are the same as in part a). (See Exercise Set 7.3, Exercise 15 b).)

17. In Exercise Set 7.3, Exercise 17, we found the probability distribution, and we found $\mu = E(X) = 1$. We find the variance and standard deviation as follows.

x_i	p_i	$x_i - \mu$	$(x_i - \mu)^2$	$p_i(x_i - \mu)^2$
0	0.32768	-1	1	0.32768
1	0.4096	0	0	0
2	0.2048	1	1	0.2048
3	0.0512	2	4	0.2048
4	0.0064	3	9	0.0576
5	0.00032	4	16	0.00512
			Sum =	0.8

$\sigma^2 = 0.8$

$\sigma = \sqrt{0.8} \approx 0.89$

19. In Exercise Set 7.3, Exercise 19, we found the probability distribution, and we found $\mu = E(X) \approx 3.6$. We find the variance and standard deviation as follows.

x_i	p_i	$x_i - \mu$	$(x_i - \mu)^2$	$p_i(x_i - \mu)^2$
0	0.010	-3.6	12.96	0.1296
1	0.060	-2.6	6.76	0.4056
2	0.161	-1.6	2.56	0.4122
3	0.251	-0.6	0.36	0.0904
4	0.251	0.4	0.16	0.0402
5	0.167	1.4	1.96	0.3273
6	0.074	2.4	5.76	0.4262
7	0.021	3.4	11.56	0.2428
8	0.004	4.4	19.36	0.0774
9	0.000+	5.4	29.16	0.0076
			Sum =	2.1593

$\sigma^2 \approx 2.16$

$\sigma \approx \sqrt{2.16} \approx 1.47$

21. In Exercise Set 7.3, Exercise 21, we found the probability distribution, and we found $\mu \approx 3.9$. We find the variance and standard deviation as follows.

x_i	p_i	$x_i - \mu$	$(x_i - \mu)^2$	$p_i(x_i - \mu)^2$
0	0.001	-3.9	15.21	0.0152
1	0.009	-2.9	8.41	0.0757
2	0.065	-1.9	3.61	0.2347
3	0.230	-0.9	0.81	0.1863
4	0.407	0.1	0.01	0.0041
5	0.289	1.1	1.21	0.3497
			Sum =	0.8657

$\sigma^2 \approx 0.866$

$\sigma \approx \sqrt{0.866} \approx 0.931$

(Answers may vary slightly due to rounding differences.)

23. The average score is given by $\bar{x} = \frac{1}{N} \Sigma f_i x_i$ where each data point x_i occurs with frequency f_i and $N = \Sigma f_i$.

$N = 1 + 29 + 176 + 30 + 2 = 238$

$\bar{x} = \frac{1}{238}(1 \cdot 3 + 29 \cdot 4 + 176 \cdot 5 + 30 \cdot 6 + 2 \cdot 7)$

$= \frac{1}{238}(1193) \approx 5.01$

We find the variance and standard deviation as follows.

x_i	f_i	$x_i - \bar{x}$	$(x_i - \bar{x})^2$	$(x_i - \bar{x}) f_i$
3	1	-2.01	4.0401	4.0401
4	29	-1.01	1.0201	29.5829
5	176	-0.01	0.0001	0.0176
6	30	0.99	0.9801	29.403
7	2	1.99	3.9601	7.9202
			Sum =	70.9638

$s^2 = \frac{70.9638}{238 - 1} = \frac{70.9638}{237} \approx 0.29943$

$s \approx \sqrt{0.29943} \approx 0.5472$

25. In Exercise Set 7.3, Exercise 29, we found the probability distribution, and we found $\mu = E(X) = \frac{3}{4}$. We find the variance and standard deviation as follows.

25. (continued)

x_i	p_i	$x_i - \mu$	$(x_i - \mu)^2$	$p_i(x_i - \mu)^2$
0	$\frac{27}{64}$	$-\frac{3}{4}$	$\frac{9}{16}$	$\frac{243}{1024}$
1	$\frac{27}{64}$	$\frac{1}{4}$	$\frac{1}{16}$	$\frac{27}{1024}$
2	$\frac{9}{64}$	$\frac{5}{4}$	$\frac{25}{16}$	$\frac{225}{1024}$
3	$\frac{1}{64}$	$\frac{9}{4}$	$\frac{81}{16}$	$\frac{81}{1024}$

$$\text{Sum} = \frac{576}{1024} = \frac{9}{16}$$

$\sigma^2 = \frac{9}{16}$

$\sigma = \sqrt{\frac{9}{16}} = \frac{3}{4}$

27. This is a hypergeometric probability distribution with s = 20, m = 10, and n = 4. Then $p = \frac{m}{s} = \frac{10}{20} = \frac{1}{2}$ and $q = 1 - p = 1 - \frac{1}{2} = \frac{1}{2}$.

$\sigma^2 = npq \frac{s - n}{s - 1} = 4 \cdot \frac{1}{2} \cdot \frac{1}{2} \cdot \frac{20 - 4}{20 - 1} = \frac{16}{19}$

$\sigma = \sqrt{\frac{16}{19}} \approx 0.91766$

29. This is a hypergeometric probability distribution with s = 20, m = 6, and n = 5. Then $p = \frac{m}{s} = \frac{6}{20} = \frac{3}{10}$ and $q = 1 - p = 1 - \frac{3}{10} = \frac{7}{10}$.

$\sigma^2 = npq \frac{s - n}{s - 1} = 5 \cdot \frac{3}{10} \cdot \frac{7}{10} \cdot \frac{20 - 5}{20 - 1} = \frac{63}{76}$

$\sigma = \sqrt{\frac{63}{76}} \approx 0.91047$

31. In Exercise Set 7.2, Exercise 1, we found $\bar{x} = 12$. We find the variance and standard deviation as follows.

Value x_i	Deviation $x_i - \bar{x}$	Deviation Squared $(x_i - \bar{x})^2$
8	-4	16
7	-5	25
15	3	9
15	3	9
15	3	9
12	0	0

Sum of the Squares of the Deviations = 68

$s^2 = \frac{68}{6 - 1} = \frac{68}{5} = 13.6$

$s = \sqrt{13.6} \approx 3.69$

33. In Exercise Set 7.2, Exercise 3, we found $\bar{x} = 20$. We find the variance and standard deviation as follows.

Value x_i	Deviation $x_i - \bar{x}$	Deviation Squared $(x_i - \bar{x})^2$
5	-15	225
10	-10	100
15	-5	25
20	0	0
25	5	25
30	10	100
35	15	225

Sum of the Squares of the Deviations = 700

$s^2 = \frac{700}{7 - 1} = \frac{700}{6} \approx 116.6667$

$s = \sqrt{116.6667} \approx 10.801$

35. In Exercise Set 7.2, Exercise 5, we found $\bar{x} = 5.2$. We find the variance and standard deviation as follows.

Value x_i	Deviation $x_i - \bar{x}$	Deviation Squared $(x_i - \bar{x})^2$
1.2	-4.0	16.0
4.3	-0.9	0.81
5.7	0.5	0.25
7.4	2.2	4.84
7.4	2.2	4.84

Sum of the Squares of the Deviations = 26.74

$s^2 = \frac{26.74}{5 - 1} = \frac{26.74}{4} = 6.685$

$s = \sqrt{6.685} \approx 2.586$

37. In Exercise Set 7.2, Exercise 7, we found $\bar{x} = 265$. We find the variance and standard deviation as follows.

Value x_i	Deviation $x_i - \bar{x}$	Deviation Squared $(x_i - \bar{x})^2$
250	-15	225
255	-10	100
260	-5	25
260	-5	25
270	5	25
270	5	25
290	25	625

Sum of the Squares of the Deviations = 1050

$s^2 = \frac{1050}{7 - 1} = \frac{1050}{6} = 175$

$s = \sqrt{175} \approx 13.2$

39. In Exercise Set 7.2, Exercise 9, we found $\bar{x}$ = 45.82. We find the variance and standard deviation as follows.

Value x_i	Deviation $x_i - \bar{x}$	Deviation Squared $(x_i - \bar{x})^2$
44.6	-1.22	1.4884
45.5	-0.32	0.1024
44.7	-1.12	1.2544
46.8	0.98	0.9604
47.5	1.68	2.8224

Sum of the Squares of the
Deviations = 6.628

$s^2 = \dfrac{6.628}{5-1} = \dfrac{6.628}{4} = 1.657$

$s = \sqrt{1.657} \approx 1.287$

41. In Exercise Set 7.2, Exercise 11, we found $\bar{x}$ = 9.35875. We find the variance and standard deviation as follows.

Value x_i	Deviation $x_i - \bar{x}$	Deviation Squared $(x_i - \bar{x})^2$
11.09	1.73125	2.99723
8.64	-0.71875	0.51660
9.38	0.02125	0.00045
9.95	0.59125	0.34958
10.94	1.58125	2.50035
9.49	0.13125	0.01723
8.36	-0.99875	0.99750
7.02	-2.33875	5.46975

Sum of the Squares of the
Deviations = 12.84869

$s^2 = \dfrac{12.84869}{7} \approx 1.8355$

$s = \sqrt{1.8355} \approx 1.3548$

43. In Exercise Set 7.2, Exercise 13, we found $\bar{x}$ = 72.5. We find the variance and standard deviation as follows.

Value x_i	Deviation $x_i - \bar{x}$	Deviation Squared $(x_i - \bar{x})^2$
74	1.5	2.25
72	-0.5	0.25
75	2.5	6.25
69	-3.5	12.25

Sum of the Squares of the
Deviations = 21

$s^2 = \dfrac{21}{4} = 5.25$ (We divided by n, or 4, rather than n - 1 since we are dealing with population data.)

$s = \sqrt{5.25} \approx 2.291$

45. $\bar{x} = \dfrac{4 + 4 + 6 + 3 + 4}{5} = \dfrac{21}{5} = 4.2$

We find the standard deviation as follows.

Value x_i	Deviation $x_i - \bar{x}$	Deviation Squared $(x_i - \bar{x})^2$
4	-0.2	0.04
4	-0.2	0.04
6	1.8	3.24
3	-1.2	1.44
4	-0.2	0.04

Sum of the Squares of the
Deviations = 4.8

$s = \sqrt{\dfrac{4.8}{5}} = \sqrt{0.96} \approx 0.98$

47. $\bar{x} = \dfrac{5 + 5 + 8 + 8 + 5}{5} = \dfrac{31}{5} = 6.2$

We find the standard deviation as follows.

Value x_i	Deviation $x_i - \bar{x}$	Deviation Squared $(x_i - \bar{x})^2$
5	-1.2	1.44
5	-1.2	1.44
8	1.8	3.24
8	1.8	3.24
5	-1.2	1.44

Sum of the Squares of the
Deviations = 10.8

$s = \sqrt{\dfrac{10.8}{5}} = \sqrt{2.16} \approx 1.47$

49. $\bar{x} = \dfrac{30 + 33 + 28 + 28 + 28}{5} = \dfrac{147}{5} = 29.4$

We find the standard deviation as follows.

Value x_i	Deviation $x_i - \bar{x}$	Deviation Squared $(x_i - \bar{x})^2$
30	0.6	0.36
33	3.6	12.96
28	-1.4	1.96
28	-1.4	1.96
28	-1.4	1.96

Sum of the Squares of the
Deviations = 19.2

$s = \sqrt{\dfrac{19.2}{5}} = \sqrt{3.84} \approx 1.96$

51. $\bar{x} \approx 102,486$

We find the variance and standard deviation as follows.

Value x_1	Deviation $x_1 - \bar{x}$	Deviation Squared $(x_1 - \bar{x})^2$
110,000	7514	56,460,196
101,000	-1486	2,208,196
83,555	-18,931	358,382,761
120,000	17,514	306,740,196
100,000	-2486	6,180,196
110,000	7514	56,460,196
115,000	12,514	156,600,196
112,500	10,014	100,280,196
99,800	-2686	7,214,596
97,725	-4761	22,667,121
92,000	-10,486	109,956,196
95,680	-6806	46,321,636
95,052	-7434	55,264,356

Sum of the Squares of the Deviations = 1,284,736,038

$$s^2 = \frac{1,284,736,038}{13 - 1} = 107,061,336.50$$

$$s = \sqrt{107,061,336.50} \approx 10,347.04$$

53. $\bar{x} \approx 80,642$

We find the variance and standard deviation as follows.

Value x_1	Deviation $x_1 - \bar{x}$	Deviation Squared $(x_1 - \bar{x})^2$
90,000	9358	87,572,164
85,000	4358	18,992,164
71,531	-9111	83,010,321
80,000	-642	412,164
65,000	-15,642	244,672,164
50,000	-30,642	938,932,164
90,100	9458	89,453,764
94,000	13,358	178,436,164
100,000	19,358	374,732,164
85,000	4358	18,992,164
85,000	4358	18,992,164
49,376	-31,266	977,562,756
103,344	22,702	515,380,804

Sum of the Squares of the Deviations = 3,547,141,121

$$s^2 = \frac{3,547,141,121}{13 - 1} \approx 295,595,093.42$$

$$s = \sqrt{295,595,093.42} \approx 17,192.88$$

55. $\bar{x} \approx 75,997$

We find the variance and standard deviation as follows.

Value x_1	Deviation $x_1 - \bar{x}$	Deviation Squared $(x_1 - \bar{x})^2$
90,000	14,003	196,084,009
76,000	3	9
71,531	-4466	19,945,156
85,000	9003	81,054,009
67,200	-8797	77,387,209
86,000	10,003	100,060,009
83,000	7003	49,042,009
66,100	-9897	97,950,609
76,500	503	253,009
85,000	9003	81,054,009
74,000	-1997	3,988,009
50,447	-25,550	652,802,500
77,184	1187	1,408,969

Sum of the Squares of the Deviations = 1,361,029,515

$$s^2 = \frac{1,361,029,515}{13 - 1} = 113,419,126.25$$

$$s = \sqrt{113,419,126.25} \approx 10,649.84$$

__57.__ $\bar{x} = \dfrac{2939}{49} \approx 60$

We find the standard deviation as follows.

Value x_1	Deviation $x_1 - \bar{x}$	Deviation Squared $(x_1 - \bar{x})^2$
73	13	169
69	9	81
64	4	16
75	15	225
52	-8	64
62	2	4
50	-10	100
54	-6	36
64	4	16
88	28	784
59	-1	1
58	-2	4
85	25	625
64	4	16
66	6	36
76	16	256
62	2	4
69	9	81
72	12	144
55	-5	25
64	4	16
65	5	25
80	20	400
83	23	529
58	-2	4
59	-1	1
65	5	25
55	-5	25
70	10	100
72	12	144
53	-7	49
71	11	121
63	3	9
76	16	256
68	8	64
63	3	9
58	-2	4
68	8	64
36	-24	576
42	-18	324
47	-13	169
40	-20	400
45	-15	225
40	-20	400
34	-26	676
38	-22	484
34	-26	676
42	-18	324
33	-27	729

Sum of the Squares of the Deviations = 9515

$s = \sqrt{\dfrac{9515}{49 - 1}} = \sqrt{\dfrac{9515}{48}} \approx 14.079$

Exercise Set 7.5

__1.__ To find $p(0 \leqslant Z \leqslant 2.69)$, go down the left column of Table 2 to 2.6. Then move to the right to the column headed 0.09, and read the number there.

$p(0 \leqslant Z \leqslant 2.69) = 0.4964$

__3.__ $p(-1.11 \leqslant Z \leqslant 0)$

$= p(0 \leqslant Z \leqslant 1.11) = 0.3665$

__5.__ $p(-1.89 \leqslant Z \leqslant 0.45)$

$= p(-1.89 \leqslant Z \leqslant 0) + p(0 \leqslant Z \leqslant 0.45)$

$= p(0 \leqslant Z \leqslant 1.89) + p(0 \leqslant Z \leqslant 0.45)$

$= 0.4706 + 0.1736$

$= 0.6442$

__7.__ $p(1.76 \leqslant Z \leqslant 1.86)$

$= p(0 \leqslant Z \leqslant 1.86) - p(0 \leqslant Z \leqslant 1.76)$

$= 0.4686 - 0.4608$

$= 0.0078$

__9.__ $p(-1.45 \leqslant Z \leqslant -0.69)$

$= p(0.69 \leqslant Z \leqslant 1.45)$

$= p(0 \leqslant Z \leqslant 1.45) - p(0 \leqslant Z \leqslant 0.69)$

$= 0.4265 - 0.2549$

$= 0.1716$

__11.__ $p(Z \geqslant 3.01)$

$= p(Z \geqslant 0) - p(0 \leqslant Z \leqslant 3.01)$

$= 0.5000 - 0.4987$

$= 0.0013$

__13.__ a) $p(-1 \leqslant Z \leqslant 1)$

$= p(-1 \leqslant Z \leqslant 0) + p(0 \leqslant Z \leqslant 1)$

$= p(0 \leqslant Z \leqslant 1) + p(0 \leqslant Z \leqslant 1)$

$= 0.3413 + 0.3413$

$= 0.6826$

b) The entire area is 1, so 0.6826, or 68.26%, of the area is from -1 to 1.

__15.__ a) $p(-3 \leqslant Z \leqslant 3)$

$= p(-3 \leqslant Z \leqslant 0) + p(0 \leqslant Z \leqslant 3)$

$= p(0 \leqslant Z \leqslant 3) + p(0 \leqslant Z \leqslant 3)$

$= 0.4987 + 0.4987$

$= 0.9974$

b) The entire area is 1, so 0.9974, or 99.74%, of the area is from -3 to 3.

__17.__ First standardize 24 and 30.

$Z = \dfrac{X - \mu}{\sigma} = \dfrac{24 - 22}{5} = \dfrac{2}{5} = 0.4$

$Z = \dfrac{X - \mu}{\sigma} = \dfrac{30 - 22}{5} = \dfrac{8}{5} = 1.6$

$p(24 \leqslant X \leqslant 30)$

$= p(0.4 \leqslant Z \leqslant 1.6)$

$= p(0 \leqslant Z \leqslant 1.6) - p(0 \leqslant Z \leqslant 0.4)$

$= 0.4452 - 0.1554$

$= 0.2898$

<u>19.</u> First standardize 19 and 25.

$$Z = \frac{X - \mu}{\sigma} = \frac{19 - 22}{5} = -\frac{3}{5} = -0.6$$

$$Z = \frac{X - \mu}{\sigma} = \frac{25 - 22}{5} = \frac{3}{5} = 0.6$$

$p(19 \leqslant X \leqslant 25)$

$= p(-0.6 \leqslant Z \leqslant 0.6)$

$= p(-0.6 \leqslant Z \leqslant 0) + p(0 \leqslant Z \leqslant 0.6)$

$= p(0 \leqslant Z \leqslant 0.6) + p(0 \leqslant Z \leqslant 0.6)$

$= 0.2257 + 0.2257$

$= 0.4514$

<u>21.</u> a) First standardize 190 and 213.

$$Z = \frac{S - \mu}{\sigma} = \frac{190 - 194}{23} = \frac{-4}{23} \approx -0.17$$

$$Z = \frac{S - \mu}{\sigma} = \frac{213 - 194}{23} = \frac{19}{23} \approx 0.83$$

$p(190 \leqslant S \leqslant 213)$

$= p(-0.17 \leqslant Z \leqslant 0.83)$

$= p(-0.17 \leqslant Z \leqslant 0) + p(0 \leqslant Z \leqslant 0.83)$

$= p(0 \leqslant Z \leqslant 0.17) + p(0 \leqslant Z \leqslant 0.83)$

$= 0.0675 + 0.2967$

$= 0.3642$

b) First standardize 160 and 175.

$$Z = \frac{S - \mu}{\sigma} = \frac{160 - 194}{23} = \frac{-34}{23} \approx -1.48$$

$$Z = \frac{S - \mu}{\sigma} = \frac{175 - 194}{23} = -\frac{19}{23} \approx -0.83$$

$p(160 \leqslant S \leqslant 175)$

$= p(-1.48 \leqslant Z \leqslant -0.83)$

$= p(0.83 \leqslant Z \leqslant 1.48)$

$= p(0 \leqslant Z \leqslant 1.48) - p(0 \leqslant Z \leqslant 0.83)$

$= 0.4306 - 0.2967$

$= 0.1339$

c) First standardize 200.

$$Z = \frac{S - \mu}{\sigma} = \frac{200 - 194}{23} = \frac{6}{23} \approx 0.26$$

$p(S > 200)$

$= p(Z \geqslant 0.26)$

$= p(Z \geqslant 0) - p(0 \leqslant Z \leqslant 0.26)$

$= 0.5000 - 0.1026$

$= 0.3974$

d) First standardize 148.

$$Z = \frac{S - \mu}{\sigma} = \frac{148 - 194}{23} = -\frac{46}{23} = -2$$

$p(S < 148)$

$= p(Z \leqslant -2)$

$= p(Z \geqslant 2)$

$= p(Z \geqslant 0) - p(0 \leqslant Z \leqslant 2)$

$= 0.5000 - 0.4772$

$= 0.0228$

<u>23.</u> We need to find $p(N \geqslant 300)$.

First standardize 300.

$$Z = \frac{N - \mu}{\sigma} = \frac{300 - 250}{20} = \frac{50}{20} = 2.5$$

$p(N \geqslant 300)$

$= p(Z \geqslant 2.5)$

$= p(Z \geqslant 0) - p(0 \leqslant Z \leqslant 2.5)$

$= 0.5000 - 0.4938$

$= 0.0062$

The company will have to hire extra help or pay overtime on 0.62% of the days.

<u>25.</u> a) First standardize 200.

$$Z = \frac{W - \mu}{\sigma} = \frac{200 - 150}{25} = \frac{50}{25} = 2$$

$p(W > 200)$

$= p(Z \geqslant 2)$

$= p(Z \geqslant 0) - p(0 \leqslant Z \leqslant 2)$

$= 0.5000 - 0.4772$

$= 0.0228$

b) $p(W \geqslant 200) = p(W > 200)$ since $p(W = 200) = 0$. (See Theorem 2.)

Thus, $p(W \geqslant 200) = 0.0228$ as in part a).

c) First standardize 160 and 180.

$$Z = \frac{W - \mu}{\sigma} = \frac{160 - 150}{25} = \frac{10}{25} = 0.4$$

$$Z = \frac{W - \mu}{\sigma} = \frac{180 - 150}{25} = \frac{30}{25} = 1.2$$

$p(160 \leqslant W \leqslant 180)$

$= p(0.4 \leqslant Z \leqslant 1.2)$

$= p(0 \leqslant Z \leqslant 1.2) - p(0 \leqslant Z \leqslant 0.4)$

$= 0.3849 - 0.1554$

$= 0.2295$

<u>27.</u> a) We need to find $p(A \geqslant 300)$.

First standardize 300.

$$Z = \frac{A - \mu}{\sigma} = \frac{300 - 490}{170} = -\frac{190}{170} \approx -1.12$$

$p(A \geqslant 300)$

$= p(Z \geqslant -1.12)$

$= p(-1.12 \leqslant Z \leqslant 0) + p(Z \geqslant 0)$

$= p(0 \leqslant Z \leqslant 1.12) + p(Z \geqslant 0)$

$= 0.3686 + 0.5000$

$= 0.8686$

The owner should expect to cover expenses 86.86% of the time.

27. (continued)

  b) We need to find p(A > 720).
     First standardize 720.

     $Z = \dfrac{A - \mu}{\sigma} = \dfrac{720 - 490}{170} = \dfrac{230}{170} \approx 1.35$

     p(A > 720)

     = p(Z ⩾ 1.35)

     = p(Z ⩾ 0) - p(0 ⩽ Z ⩽ 1.35)

     = 0.5000 - 0.4115

     = 0.0885

     They should expect more to attend than they can seat 8.85% of the time.

29. a) To find the mean speed, find the total of all the speeds and divide by the number of speeds, 8.

     $\bar{x} = \dfrac{677.827 \text{ mph}}{8} \approx 84.728 \text{ mph}$

  b)
Value x_i	Deviation $x_i - \bar{x}$	Deviation Squared $(x_i - \bar{x})^2$
91.101	6.373	40.615
90.973	6.245	39.000
89.257	4.529	20.512
85.879	1.151	1.325
82.438	-2.29	5.244
81.962	2.766	7.651
78.113	-6.615	43.758
78.104	-6.624	43.877

     Sum of the Squares of the Deviations = 201.982

     $s = \sqrt{\dfrac{201.982}{8 - 1}} = \sqrt{\dfrac{201.982}{7}} \approx 5.372$

  c) $Z = \dfrac{84 - 84.728}{5.372} = -\dfrac{0.728}{5.372} \approx -0.136$

     A Z-score of approximately -0.136 corresponds to 84 mph.

     $Z = \dfrac{85 - 84.728}{5.372} = \dfrac{0.272}{5.372} \approx 0.051$

     A Z-score of approximately 0.051 corresponds to 85 mph.

  d) $0.5 = \dfrac{S - 84.728}{5.372}$

     $2.686 = S - 84.728$

     $87.414 = S$

     87.414 mph corresponds to Z = 0.5.

     $-0.5 = \dfrac{S - 84.728}{5.372}$

     $-2.686 = S - 84.728$

     $82.042 = S$

     82.042 mph corresponds to Z = -0.5.

29. (continued)

  e) 0.5 standard deviation of the mean = 0.5(5.372) = 2.686

     $\bar{x} - 2.686 = 84.728 - 2.686 = 82.042$

     $\bar{x} + 2.686 = 84.728 + 2.686 = 87.414$

     There are two lap speeds between 82.042 mph and 87.414 mph. (They are 85.879 mph and 82.438 mph.) Thus, $\frac{2}{8}$ or 25% of the cars had lap speeds within 0.5 standard deviation of the mean.

  f) Assuming the speeds are normally distributed and considering Z-scores,

     p(-0.5 ⩽ Z ⩽ 0.5)

     = p(-0.5 ⩽ Z ⩽ 0) + p(0 ⩽ Z ⩽ 0.5)

     = p(0 ⩽ Z ⩽ 0.5) + p(0 ⩽ Z ⩽ 0.5)

     = 0.1915 + 0.1915

     = 0.3830

     You would expect 38.3% of the cars to have lap speeds within 0.5 standard deviation of the mean.

31. To use the normal distribution to approximate the probabilities, first find the mean and standard deviation of the binomial distribution.

     $\mu = np = 1000(0.2) = 200$

     $\sigma = \sqrt{npq} = \sqrt{1000(0.2)(0.8)} = \sqrt{160} \approx 12.65$

  a) We first standardize 200.

     $Z = \dfrac{200 - 200}{12.65} = 0$

     p(D ⩾ 200) = p(Z ⩾ 0) = 0.5

  b) Standardize 190 and 210.

     $Z = \dfrac{190 - 200}{12.65} = -\dfrac{10}{12.65} \approx -0.79$

     $Z = \dfrac{210 - 200}{12.65} = \dfrac{10}{12.65} \approx 0.79$

     p(190 ⩽ D ⩽ 210)

     = p(-0.79 ⩽ Z ⩽ 0.79)

     = p(-0.79 ⩽ Z ⩽ 0) + p(0 ⩽ Z ⩽ 0.79)

     = p(0 ⩽ Z ⩽ 0.79) + p(0 ⩽ Z ⩽ 0.79)

     = 0.2852 + 0.2852

     = 0.5704

  c) The probability is about 68% that the number of defective speakers is 1 standard deviation above and below the mean. One standard deviation below the mean is about 200 - 12.65, or 187.35. One standard deviation above the mean is 200 + 12.65, or 212.65.

33. To use the normal distribution to approximate the probabilities, first find the mean and standard deviation of the binomial distribution.

     $\mu = np = 35(0.357) = 12.495$

     $\sigma = \sqrt{npq} = \sqrt{35(0.357)(0.643)} =$

     $\sqrt{8.034285} \approx 2.834$

33. (continued)

   a) We first standardize 10.

   $$Z = \frac{10 - 12.495}{2.834} = -\frac{2.495}{2.834} \approx -0.88$$

   $p(H \geqslant 10)$

   $= p(Z \geqslant -0.88)$

   $= p(-0.88 \leqslant Z \leqslant 0) + p(Z \geqslant 0)$

   $= p(0 \leqslant Z \leqslant 0.88) + p(Z \geqslant 0)$

   $= 0.3106 + 0.5000$

   $= 0.8106$

   b) Standardize 12.

   $$Z = \frac{12 - 12.495}{2.834} = -\frac{0.495}{2.834} \approx -0.17$$

   $p(H \leqslant 12)$

   $= p(Z \leqslant -0.17)$

   $= p(Z \geqslant 0.17)$

   $= p(Z \geqslant 0) - p(0 \leqslant Z \leqslant 0.17)$

   $= 0.5000 - 0.0675$

   $= 0.4325$

35. $\mu = np = 500(0.2) = 100$

   $\sigma = \sqrt{npq} = \sqrt{500(0.2)(0.8)} = \sqrt{80} \approx 8.94$

   Standardize 75 and 127.

   $$Z = \frac{75 - 100}{8.94} = -\frac{25}{8.94} \approx -2.80$$

   $$Z = \frac{127 - 100}{8.94} = \frac{27}{8.94} \approx 3.02$$

   $p(75 \leqslant X \leqslant 127)$

   $= p(-2.80 \leqslant Z \leqslant 3.02)$

   $= p(-2.80 \leqslant Z \leqslant 0) + p(0 \leqslant Z \leqslant 3.02)$

   $= p(0 \leqslant Z \leqslant 2.80) + p(0 \leqslant Z \leqslant 3.02)$

   $= 0.4974 + 0.4987$

   $= 0.9961$

37. $\mu = np = 909(0.4) = 363.6$

   $\sigma = \sqrt{npq} = \sqrt{909(0.4)(0.6)} = \sqrt{218.16} \approx 14.77$

   Standardize 400.

   $$Z = \frac{400 - 363.6}{14.77} = \frac{36.4}{14.77} \approx 2.46$$

   $p(X \leqslant 400)$

   $= p(Z \leqslant 2.46)$

   $= p(Z \leqslant 0) + p(0 \leqslant Z \leqslant 2.46)$

   $= p(Z \geqslant 0) + p(0 \leqslant Z \leqslant 2.46)$

   $= 0.5000 + 0.4931$

   $= 0.9931$

39. $\mu = np = 20(0.78) = 15.6$

   $\sigma = \sqrt{npq} = \sqrt{20(0.78)(0.22)} = \sqrt{3.432} \approx 1.85$

   Standardize 15.

   $$Z = \frac{15 - 15.6}{1.85} = -\frac{0.6}{1.85} \approx -0.32$$

   $p(J \geqslant 15)$

   $= p(Z \geqslant -0.32)$

   $= p(-0.32 \leqslant Z \leqslant 0) + p(Z \geqslant 0)$

   $= p(0 \leqslant Z \leqslant 0.32) + p(Z \geqslant 0)$

   $= 0.1255 + 0.5000$

   $= 0.6255$

41. Let S = test score. Standardize S.

   $$Z = \frac{S - \mu}{\sigma} = \frac{S - 68}{19}$$

   We are looking for a value of t for which
   $p(Z \geqslant t) = 10\% = 0.1000$. Now if $t \geqslant 0$, then
   $p(Z \geqslant 0) - p(0 \leqslant Z \leqslant t) = 0.5000 - p(0 \leqslant Z \leqslant t) = 0.1000$. Then $p(0 \leqslant Z \leqslant t) = 0.4000$. Use Table 2
   to estimate t. Look in the body of the table for
   the number closest to 0.4000. That number is
   0.3997, and it corresponds to t = 1.28 on the
   outside of the table. Then t = 1.28. Set Z equal
   to 1.28 and solve for S.

   $$1.28 = \frac{S - 68}{19}$$

   $24.32 = S - 68$

   $92.32 = S$

   It takes a score of 92.32 or higher to make an A.

43. Standardize 60.

   $$Z = \frac{60 - \mu}{\sigma} = \frac{60 - \mu}{0.1}$$

   We are looking for a value of t for which
   $p(Z < t) = \frac{1}{20} = 0.05$. Now if $t \geqslant 0$, then
   $p(Z < t) = p(Z \leqslant 0) + p(0 \leqslant Z \leqslant t) =$
   $p(Z \geqslant 0) + p(0 \leqslant Z \leqslant t) = 0.5000 + p(0 \leqslant Z \leqslant t)$.
   Then $p(0 \leqslant Z \leqslant t) = 0.05 - 0.5000 = -0.4500$.
   This is not possible. Therefore, $t \leqslant 0$, and
   $p(Z < t) = p(Z \leqslant 0) - p(t \leqslant Z \leqslant 0) = p(Z \geqslant 0) -$
   $p(0 \leqslant Z \leqslant t) = 0.5000 - p(0 \leqslant Z \leqslant t)$. Then
   $p(0 \leqslant Z \leqslant t) = 0.5000 - 0.05 = 0.4500$. Looking
   in the body of the table for the number closest
   to 0.4500, we find that 0.4495 and 0.4505 are
   equally close. We will choose 0.4495 and find
   the value of t that corresponds to it on the
   outside of the table. We see that t = 1.64, but
   in our case $t \leqslant 0$ so we use t = -1.64. Set Z equal
   to -1.64 and solve for μ.

   $$-1.64 = \frac{60 - \mu}{0.1}$$

   $$-0.164 = 60 - \mu$$

   $$\mu = 60.164$$

   The setting of μ should be 60.164 oz.

<u>45.</u> To use the normal distribution we first find the mean and the standard deviation of the binomial distribution.

$\mu = np = 100(0.8) = 80$

$\sigma = \sqrt{npq} = \sqrt{100(0.8)(0.2)} = \sqrt{16} = 4$

Let U = number of reservations actually used. We need to find p(U > 90). Standardize U.

$Z = \dfrac{90 - 80}{4} = \dfrac{10}{4} = 2.5$

p(U > 90)

$= p(Z \geqslant 2.5)$

$= p(Z \geqslant 0) - p(0 \leqslant Z \leqslant 2.5)$

$= 0.5000 - 0.4938$

$= 0.0062$

The probability that a flight will be overlooked if the airline accepts 100 reservations is 0.62%, or nearly 0.

<u>47.</u> $np = 411(0.73) = 300.03$

$\sqrt{npq} = \sqrt{411(0.73)(0.27)} = \sqrt{81.0081} \approx 9.00$

Use the continuity correction formula:

$$p(a \leqslant Z \leqslant b) = p\left[\frac{a - \frac{1}{2} - np}{\sqrt{npq}} \leqslant Z \leqslant \frac{b + \frac{1}{2} - np}{\sqrt{npq}} \right]$$

p(291 ≤ Z ≤ 309)

$$= p\left[\frac{291 - \frac{1}{2} - 300.03}{9.00} \leqslant Z \leqslant \frac{309 + \frac{1}{2} - 300.03}{9.00} \right]$$

$= p(-1.06 \leqslant Z \leqslant 1.05)$

$= p(-1.06 \leqslant Z \leqslant 0) + p(0 \leqslant Z \leqslant 1.05)$

$= p(0 \leqslant Z \leqslant 1.06) + p(0 \leqslant Z \leqslant 1.05)$

$= 0.3554 + 0.3531$

$= 0.7085$

<u>49.</u> In Exercise 37 we found $np = 363.6$ and $\sqrt{npq} \approx 14.77$. Use the continuity correction formula:

$$p(Z \leqslant 400) = p\left[Z \leqslant \frac{400 + \frac{1}{2} - 363.6}{14.77} \right]$$

$= p(Z \leqslant 2.50)$

$= p(Z \leqslant 0) + p(0 \leqslant Z \leqslant 2.50)$

$= p(Z \geqslant 0) + p(0 \leqslant Z \leqslant 2.50)$

$= 0.5000 + 0.4938$

$= 0.9938$

Exercise Set 8.1

1. $a_n = \dfrac{n}{n+1}$

   $a_1 = \dfrac{1}{1+1} = \dfrac{1}{2}$

   $a_2 = \dfrac{2}{2+1} = \dfrac{2}{3}$

   $a_3 = \dfrac{3}{3+1} = \dfrac{3}{4}$

   $a_4 = \dfrac{4}{4+1} = \dfrac{4}{5}$

   $a_{15} = \dfrac{15}{15+1} = \dfrac{15}{16}$

3. $a_n = \dfrac{n^2-1}{n^3+1}$

   $a_1 = \dfrac{1^2-1}{1^3+1} = \dfrac{1-1}{1+1} = 0$

   $a_2 = \dfrac{2^2-1}{2^3+1} = \dfrac{4-1}{8+1} = \dfrac{3}{9} = \dfrac{1}{3}$

   $a_3 = \dfrac{3^2-1}{3^3+1} = \dfrac{9-1}{27+1} = \dfrac{8}{28} = \dfrac{2}{7}$

   $a_4 = \dfrac{4^2-1}{4^3+1} = \dfrac{16-1}{64+1} = \dfrac{15}{65} = \dfrac{3}{13}$

   $a_{15} = \dfrac{15^2-1}{15^3+1} = \dfrac{225-1}{3375+1} = \dfrac{224}{3376} = \dfrac{14}{211}$

5. 2, 7, 12, 17, $\cdots$

   The first term is 2.

   Subtract any term from the one that follows it to find the common difference, d:

   $7 - 2 = 5$

   The common difference is 5.

7. \$1.06, \$1.12, \$1.18, \$1.24, $\cdots$

   The first term is \$1.06.

   Subtract any term from the one that follows it to find the common difference, d:

   $\$1.12 - \$1.06 = \$0.06$

   The common difference is \$0.06.

9. 5, $4\frac{1}{3}$, $3\frac{2}{3}$, 3, $2\frac{1}{3}$, $\cdots$

   The first term is 5.

   The common difference is $4\frac{1}{3} - 5$, or $-\frac{2}{3}$.

11. 3, 7, 11, $\cdots$

    $a_1 = 3$

    $d = 7 - 3$, or 4

    $a_{12} = a_1 + (12-1)d$

    $\quad = 3 + 11 \cdot 4$

    $\quad = 47$

13. \$1200, \$964.32, \$728.64, $\cdots$

    $a_1 = \$1200$

    $d = \$964.32 - \$1200$, or $-\$235.68$

    $a_{13} = a_1 + (13-1)d$

    $\quad = \$1200 + 12(-\$235.68)$

    $\quad = -\$1628.16$

15. The sum is $1 + 2 + 3 + \cdots + 299 + 300$.

    This is the sum of the first 300 terms of the arithmetic sequence 1, 2, 3, $\cdots$, 299, 300 for which $a_1 = 1$, $a_n = 300$, and $n = 300$. Substituting in the formula

    $$S_n = \dfrac{n}{2}(a_1 + a_n)$$

    we get

    $$S_{300} = \dfrac{300}{2}(1 + 300) = 150(301) = 45{,}150.$$

17. 6, 9, 12, 15, $\cdots$

    $a_1 = 6$; $d = 9 - 6$, or 3; $n = 20$

    Substituting in the formula

    $$S_n = \dfrac{n}{2}\big[2a_1 + (n-1)d\big]$$

    we get

    $$S_{20} = \dfrac{20}{2}\big[2 \cdot 6 + (20-1)3\big] = 10(12 + 19 \cdot 3)$$

    $$= 10 \cdot 69 = 690$$

19. $1 + 2 + 3 + \cdots + n$

    $a_1 = 1$, $a_n = n$

    Substituting in the formula

    $$S_n = \dfrac{n}{2}(a_1 + a_n)$$

    we get

    $$S_n = \dfrac{n}{2}(1 + n), \text{ or } \dfrac{n(n+1)}{2}.$$

21. The sum is $\$0.01 + \$0.02 + \$0.03 + \cdots + \$0.31$. This is the sum of the first 31 terms of the arithmetic sequence \$0.01, \$0.02, \$0.03, $\cdots$, \$0.31 for which $a_1 = \$0.01$, $a_n = \$0.31$, and $n = 31$.

    Substituting in the formula

    $$S_n = \dfrac{n}{2}(a_1 + a_n)$$

    we get

    $$S_{31} = \dfrac{31}{2}(\$0.01 + \$0.31) = \dfrac{31}{2}(\$0.32) = \$4.96.$$

23. $512.50, $1025.00, $1537.50, $\cdots$

a_1 = $512.50; d = $1025.00 - $512.50, or $512.50; n = 8

Substituting in the formula

$$S_n = \frac{n}{2}\left[2a_1 + (n-1)d\right]$$

we get

$$S_8 = \frac{8}{2}[2(\$512.50) + (8-1)(\$512.50)]$$
$$= 4[\$1025.00 + 7(\$512.50)]$$
$$= 4(\$4612.50)$$
$$= \$18,450.$$

25. a) C = $5200, N = 8, S = $1100

Substituting in the formula

$$a_t = C - t\left[\frac{C-S}{N}\right]$$

we get

$$a_t = \$5200 - t\left[\frac{\$5200 - \$1100}{8}\right].$$

b) $a_0 = \$5200 - 0\left[\frac{\$5200 - \$1100}{8}\right]$

$= \$5200$

$a_1 = \$5200 - 1 \cdot \left[\frac{\$5200 - \$1100}{8}\right]$

$= \$4687.50$

$a_2 = \$5200 - 2 \cdot \left[\frac{\$5200 - \$1100}{8}\right]$

$= \$4175$

$a_3 = \$5200 - 3 \cdot \left[\frac{\$5200 - \$1100}{8}\right]$

$= \$3662.50$

$a_4 = \$5200 - 4 \cdot \left[\frac{\$5200 - \$1100}{8}\right]$

$= \$3150$

$a_7 = \$5200 - 7 \cdot \left[\frac{\$5200 - \$1100}{8}\right]$

$= \$1612.50$

$a_8 = \$5200 - 8 \cdot \left[\frac{\$5200 - \$1100}{8}\right]$

$= \$1100$

Exercise Set 8.2

1. 7, 14, 28, 56, $\cdots$

Find the common ratio, r, by dividing any term by the preceding term:

$$r = \frac{14}{7}, \text{ or } 2$$

3. 12, -4, $\frac{4}{3}$, $-\frac{4}{9}$, $\cdots$

Find the common ratio, r, by dividing any term by the preceding term:

$$r = \frac{-4}{12}, \text{ or } -\frac{1}{3}$$

5. $5600, $5320, $5054, $4801.30, $\cdots$

$$r = \frac{\$5320}{\$5600} = 0.95$$

7. 1, 3, 9, $\cdots$

a_1 = 1, n = 8, r = $\frac{3}{1}$, or 3

Substituting in the formula

$$a_n = a_1 r^{n-1}$$

we get

$$a_8 = 1 \cdot 3^{8-1} = 1 \cdot 3^7 = 2187.$$

9. 25, 5, 1, $\frac{1}{5}$, $\frac{1}{25}$, $\cdots$

a_1 = 25, n = 9, r = $\frac{5}{25}$, or $\frac{1}{5}$

Substituting in the formula

$$a_n = a_1 r^{n-1}$$

we get

$$a_9 = 25\left[\frac{1}{5}\right]^{9-1} = 25\left[\frac{1}{5}\right]^8 = 5^2 \cdot \frac{1}{5^8} = \frac{1}{5^6}$$
$$= \frac{1}{15,625}.$$

11. $1000, $1080, $1166.40, $\cdots$

a_1 = $1000, n = 12, r = $\frac{\$1080}{\$1000}$ = 1.08

Substituting in the formula

$$a_n = a_1 r^{n-1}$$

we get

$$a_{12} = \$1000(1.08)^{12-1} = \$1000(1.08)^{11}$$
$$\approx \$2331.64.$$

13. 8, 16, 32, $\cdots$

a_1 = 8, n = 7, r = $\frac{16}{8}$, or 2

Substituting in the formula

$$S_n = \frac{a_1(r^n - 1)}{r - 1}$$

we get

$$S_7 = \frac{8(2^7 - 1)}{2 - 1} = \frac{8(128 - 1)}{1} = 8 \cdot 127 = 1016.$$

15. $1000, $1000(1.08), $1000(1.08)^2, $\cdots$

a_1 = $1000, n = 5, d = $\frac{\$1000(1.08)}{\$1000}$ = 1.08

Substituting in the formula

$$S_n = \frac{a_1(r^n - 1)}{r - 1}$$

we get

$$S_5 = \frac{\$1000(1.08^5 - 1)}{1.08 - 1}$$
$$\approx \frac{\$1000(1.469328077 - 1)}{0.08}$$
$$\approx \$5866.60.$$

17. The amount earned is the series

$\$0.01 + \$0.01(2) + \$0.01(2^2) + \$0.01(2^3) + \cdots + \$0.01(2^{27})$, where $a_1 = \$0.01$, $n = 28$, and $r = 2$. Substituting in the formula

$$S_n = \frac{a_1(r^n - 1)}{r - 1}$$

we get

$$S_{28} = \frac{\$0.01(2^{28} - 1)}{2 - 1}$$
$$= \$0.01(268{,}435{,}456 - 1)$$
$$= \$2{,}684{,}354.55.$$

19. $4 + 20 + 100 + 500 + \cdots$

$|r| = \left|\frac{20}{4}\right| = |5| = 5$, and since $|r| < 1$ the series does <u>not</u> have a sum.

21. $10 + 2 + \frac{2}{5} + \frac{2}{25} + \frac{2}{125} + \cdots$

$|r| = \left|\frac{2}{10}\right| = \left|\frac{1}{5}\right| = \frac{1}{5}$, and since $|r| < 1$ the series does have a sum. The sum is given by $S_\infty = $

$\frac{a_1}{1 - r} = \frac{10}{1 - \frac{1}{5}} = \frac{10}{\frac{4}{5}} = \frac{25}{2}$, or $12\frac{1}{2}$.

23. $162 + 108 + 72 + 48 + \cdots$

$|r| = \left|\frac{108}{162}\right| = \left|\frac{2}{3}\right| = \frac{2}{3}$, and since $|r| < 1$ the series does have a sum. The sum is given by

$S_\infty = \frac{a_1}{1 - r} = \frac{162}{1 - \frac{2}{3}} = \frac{162}{\frac{1}{3}} = 486.$

25. $\$1000(1.08)^{-1} + \$1000(1.08)^{-2} + \$1000(1.08)^{-3} + \cdots$

$|r| = \left|\frac{\$1000(1.08)^{-2}}{\$1000(1.08)^{-1}}\right| = \left|\frac{1}{1.08}\right| = \frac{1}{1.08}$, and since $|r| < 1$ the series does have a sum. The sum is given by $S_\infty = \frac{a_1}{1 - r} = \frac{\$1000(1.08)^{-1}}{1 - \frac{1}{1.08}} = \frac{\$1000(1.08)^{-1}}{\frac{0.08}{1.08}}$

$= \frac{\$1000(1.08)^{-1}}{0.08(1.08)^{-1}} = \$12{,}500.$

27. The total effect on the economy can be modeled as the sum of the infinite geometric series

$\$8{,}000{,}000{,}000 + \$8{,}000{,}000{,}000(0.75) + \$8{,}000{,}000{,}000(0.75)^2 + \cdots$

$|r| = \left|\frac{\$8{,}000{,}000{,}000(0.75)}{\$8{,}000{,}000{,}000}\right| = |0.75| = 0.75,$

and since $|r| < 1$ the series does have a sum. The sum is given by $S_\infty = \frac{a_1}{1 - r} = \frac{\$8{,}000{,}000{,}000}{1 - 0.75}$

$= \frac{\$8{,}000{,}000{,}000}{0.25} = \$32{,}000{,}000{,}000.$

29. The total number of people who will buy the product can be modeled as the sum of the infinite geometric series

$5{,}000{,}000(0.4) + 5{,}000{,}000(0.4)^2 + 5{,}000{,}000(0.4)^3 + \cdots$

$|r| = \left|\frac{5{,}000{,}000(0.4)^2}{5{,}000{,}000(0.4)}\right| = |0.4| = 0.4$, and since $|r| < 1$ the series does have a sum. The sum is given by $S_\infty = \frac{a_1}{1 - r} = \frac{5{,}000{,}000(0.4)}{1 - 0.4}$

$= \frac{5{,}000{,}000(0.4)}{0.6} \approx 3{,}333{,}333.$

To find what percentage of the population this is we divide this number by the total population:

$\frac{3{,}333{,}333}{5{,}000{,}000} = 66\frac{2}{3}\%$

31. The initial effect would be $50{,}000 \times \$450$, or $\$22{,}500{,}000$. The economic multiplier effect can be modeled as the sum of the infinite geometric series

$\$22{,}500{,}000 + \$22{,}500{,}000(0.75) + \$22{,}500{,}000(0.75)^2 + \cdots$

Now $|r| = 0.75$ so the series does have a sum which is given by $S_\infty = \frac{a_1}{1 - r} = \frac{\$22{,}500{,}000}{1 - 0.75} = \frac{\$22{,}500{,}000}{0.25}$

$= \$90{,}000{,}000.$

Exercise Set 8.3

1. The amount A to which principal P will grow at simple interest rate i for t years is given by the formula

$$A = P(1 + it)$$

Here $P = \$2000$, $i = 9\%$, or 0.09, and $t = \frac{4}{12}$, or $\frac{1}{3}$ year. Substituting in the formula we have

$A = \$2000\left[1 + 0.09 \times \frac{1}{3}\right] = \$2000(1 + 0.03)$

$= \$2000(1.03) = \$2060.$

3. We will use the formula $A = P(1 + it)$ with $P = \$2000$, $i = 14\%$, or 0.14, and $t = 2$ years. Substituting we have

$A = \$2000(1 + 0.14 \times 2) = \$2000(1 + 0.28)$

$= \$2000(1.28) = \$2560.$

5. $P = \$2000$, $i = 7\%$, or 0.07, and $t = 2$ years.

a) $A = P(1 + it) = \$2000(1 + 0.07 \times 2)$
$= \$2000(1 + 0.14)$
$= \$2000(1.14) = \$2280$

b) $A = P(1 + i)^t = \$2000(1 + 0.07)^2 = \$2000(1.07)^2$
$= \$2000(1.1449) = \$2289.80$

c) $A = P\left(1 + \frac{i}{n}\right)^{nt} = \$2000\left(1 + \frac{0.07}{2}\right)^{2 \times 2}$
$= \$2000(1 + 0.035)^4$
$= \$2000(1.035)^4$
$= \$2000(1.147523001) \approx \$2295.05$

5. (continued)

d) $A = P\left[1 + \dfrac{i}{n}\right]^{nt} = \$2000\left[1 + \dfrac{0.07}{4}\right]^{4\times 2}$

$= \$2000(1 + 0.0175)^8$

$= \$2000(1.0175)^8$

$\approx \$2000(1.148881783) \approx \$2297.76$

e) $A = P\left[1 + \dfrac{i}{n}\right]^{nt} = \$2000\left[1 + \dfrac{0.07}{365}\right]^{365\times 2}$

$\approx \$2000(1 + 0.00019178)^{730}$

$= \$2000(1.00019178)^{730}$

$\approx \$2000(1.150258357) \approx \$2300.52$

7. The present value P of an amount A at interest rate i, compounded annually, for t years is given by

$P = A(1 + i)^{-t}.$

Here A = \$1000, i = 8%, or 0.08, and t = 3 years. Substituting in the formula we get

$P = \$1000(1 + 0.08)^{-3} = \$1000(1.08)^{-3}$

$\approx \$1000(0.793832241) \approx \$793.83.$

9. The present value P of an amount A at interest rate i, compounded n times per year, for t years is given by

$P = A\left[1 + \dfrac{i}{n}\right]^{-nt}.$

Here A = \$1000, i = 8%, or 0.08, n = 4, and t = 3 years. Substituting in the formula we get

$P = \$1000\left[1 + \dfrac{0.08}{4}\right]^{-4\times 3} = \$1000(1 + 0.02)^{-12}$

$= \$1000(1.02)^{-12} \approx \$1000(0.788493175)$

$\approx \$788.49.$

11. We will use the formula

$P = A\left[1 + \dfrac{i}{n}\right]^{-nt}.$

Here A = \$10,000, i = 6%, or 0.06, n = 2, and t = 18 years. Substituting we have

$P = \$10,000\left[1 + \dfrac{0.06}{2}\right]^{-2\times 18}$

$= \$10,000(1 + 0.03)^{-36}$

$= \$10,000(1.03)^{-36} \approx \$10,000(0.345032425)$

$\approx \$3450.32$

13. Use the formula $A = P(1 + i)^t$ with P = \$1000, i = 7%, or 0.07, and t = 2 years.

$A = \$1000(1 + 0.07)^2 = \$1000(1.07)^2$

$= \$1000(1.1449) = \$1144.90$

15. The cost is $\$1(1 + 0.07)^{25} = \$1(1.07)^{25}$
$\approx \$1(5.42743264) \approx \$5.43.$

17. Use the formula $A = (1 + i)^t$ with A = \$2890, P = \$2560, and t = 2 years.

$\$2890 = \$2560(1 + i)^2$

$\dfrac{\$2890}{\$2560} = (1 + i)^2$

$\dfrac{289}{256} = (1 + i)^2$

$\dfrac{17}{16} = 1 + i$ (Taking the square root of both sides)

$\dfrac{17}{16} - 1 = i$

$\dfrac{1}{16} = i$

The interest rate is $\dfrac{1}{16} = 0.0625$, or 6.25%.

19. For the first 4 months of the year the account has principal P of \$1000 and earns interest

$I = P\cdot i\cdot t = \$1000\cdot i \cdot \dfrac{4}{12} = \dfrac{\$1000}{3}i.$

For the next 3 months the principal is \$1000 + \$200, or \$1200, and the interest earned is

$I = P\cdot i\cdot t = \$1200\cdot i\cdot \dfrac{3}{12} = \$300i.$

For the remaining 5 months the principal is \$1200 - \$300, or \$900, and the interest earned is

$I = P\cdot i\cdot t = \$900\cdot i \cdot \dfrac{5}{12} = \$375i.$

At the end of the year the value of the fund, \$1000, is the sum of the interest earned during the year and the principal at the end of the year.

$\dfrac{\$1000}{3}i + \$300i + \$375i + \$900 = \$1000$

$\dfrac{3025}{3}i = \$100$

$i = \$100 \cdot \dfrac{3}{3025}$

$i \approx 0.099$, or 9.9%

The correct answer is A.

Exercise Set 8.4

1. Use the formula $V = \dfrac{P[(1 + i)^N - 1]}{i}$ with

P = \$1000, i = 7%, or 0.07, and N = 4.

$V = \dfrac{\$1000[(1 + 0.07)^4 - 1]}{0.07} = \dfrac{\$1000(1.07^4 - 1)}{0.07}$

$\approx \dfrac{\$1000(1.31079601 - 1)}{0.07} \approx \$4439.94$

3. Use the formula $V = \dfrac{P[(1 + i)^N - 1]}{i}$ with

P = \$1000, i = 7%, or 0.07, and N = 10.

$V = \dfrac{\$1000[(1 + 0.07)^{10} - 1]}{0.07} = \dfrac{\$1000(1.07^{10} - 1)}{0.07}$

$\approx \dfrac{\$1000(1.967151357 - 1)}{0.07} \approx \$13,816.45$

5. Use the formula $V = \dfrac{P\left[\left(1 + \frac{i}{n}\right)^{nN} - 1\right]}{\frac{i}{n}}$ with

P = \$2000, i = 8%, or 0.08, n = 4, and N = 5.

$$V = \frac{\$2000\left[\left(1 + \frac{0.08}{4}\right)^{4\times 5} - 1\right]}{\frac{0.08}{4}}$$

$$= \frac{\$2000[(1 + 0.02)^{20} - 1]}{0.02} = \frac{\$2000(1.02^{20} - 1)}{0.02}$$

$$\approx \frac{\$2000(1.485947396 - 1)}{0.02} \approx \$48{,}594.74$$

7. Use the formula $V = \dfrac{P\left[\left(1 + \frac{i}{n}\right)^{nN} - 1\right]}{\frac{i}{n}}$ with

P = \$10, i = 6%, or 0.06, n = 12, and N = 8.

$$V = \frac{\$10\left[\left(1 + \frac{0.06}{12}\right)^{12\times 8} - 1\right]}{\frac{0.06}{12}}$$

$$= \frac{\$10[(1 + 0.005)^{96} - 1]}{0.005} = \frac{\$10(1.005^{96} - 1)}{0.005}$$

$$\approx \frac{\$10(1.614142780 - 1)}{0.005} \approx \$1228.29$$

9. P = \$1000, i = 7.5%, or 0.075, and N = 30.

$$V = \frac{P[(1 + i)^N - 1]}{i} = \frac{\$1000[(1 + 0.075)^{30} - 1]}{0.075}$$

$$= \frac{\$1000(1.075^{30} - 1)}{0.075} \approx \frac{\$1000(8.754955189 - 1)}{0.075}$$

$$\approx \$103{,}399.40$$

11. P = \$50, i = 8%, or 0.08, n = 12, and N = 5.

$$V = \frac{P\left[\left(1 + \frac{i}{n}\right)^{nN} - 1\right]}{\frac{i}{n}} = \frac{\$50\left[\left(1 + \frac{0.08}{12}\right)^{12\cdot 5} - 1\right]}{\frac{0.08}{12}}$$

$$= \frac{\$50[(1 + 0.006666667)^{60} - 1]}{0.006666667}$$

$$= \frac{\$50(1.006666667^{60} - 1)}{0.006666667}$$

$$\approx \frac{\$50(1.489845709 - 1)}{0.006666667} \approx \$3673.84$$

13. We can use the formula $V = \dfrac{P[(1 + i)^N - 1]}{i}$.

We know that V = \$10,000, i = 6.5%, or 0.065, and N = 8.

$$\$10{,}000 = \frac{P[(1 + 0.065)^8 - 1]}{0.065}$$

$$\$10{,}000(0.065) = P(1.065^8 - 1)$$

$$\$650 = P(1.654995671 - 1)$$

$$= P(0.654995671)$$

$$P = \frac{\$650}{0.654995671} \approx \$992.37$$

15. We can use the formula $V = \dfrac{P\left[\left(1 + \frac{i}{n}\right)^{nN} - 1\right]}{\frac{i}{n}}$.

We know that V = \$7000, i = 6%, or 0.06, n = 12, and N = 5.

$$\$7000 = \frac{P\left[\left(1 + \frac{0.06}{12}\right)^{12\times 5} - 1\right]}{\frac{0.06}{12}}$$

$$\$7000 = \frac{P[(1 + 0.005)^{60} - 1]}{0.005}$$

$$\$7000(0.005) = P(1.005^{60} - 1)$$

$$\$35 = P(1.348850153 - 1) = P(0.348850153)$$

$$P = \frac{\$35}{0.348850153} \approx \$100.33$$

17. $V = \dfrac{P[(1 + i)^N - 1]}{i}$

$Vi = P[(1 + i)^N - 1]$

$P = \dfrac{Vi}{[(1 + i)^N - 1]}$

Exercise Set 8.5

1. We will use the formula $S = \dfrac{P[1 - (1 + i)^{-N}]}{i}$ with

P = \$1000, i = 7%, or 0.07, and N = 4.

$$S = \frac{\$1000[1 - (1 + 0.07)^{-4}]}{0.07} = \frac{\$1000(1 - 1.07^{-4})}{0.07}$$

$$\approx \frac{\$1000(1 - 0.762895212)}{0.07} \approx \$3387.21$$

3. We will use the formula $S = \dfrac{P[1 - (1 + i)^{-N}]}{i}$ with

P = \$1000, i = 7%, or 0.07, and N = 10.

$$S = \frac{\$1000[1 - (1 + 0.07)^{-10}]}{0.07} = \frac{\$1000(1 - 1.07^{-10})}{0.07}$$

$$\approx \frac{\$1000(1 - 0.508349292)}{0.07} \approx \$7023.58$$

5. We will use the formula $S = \dfrac{P\left[1 - \left(1 + \frac{i}{n}\right)^{-nN}\right]}{\frac{i}{n}}$

with P = \$2000, i = 8%, or 0.08, n = 4, and N = 5.

$$S = \frac{\$2000\left[1 - \left(1 + \frac{0.08}{4}\right)^{-4\times 5}\right]}{\frac{0.08}{4}}$$

$$= \frac{\$2000(1 - 1.02^{-20})}{0.02}$$

$$\approx \frac{\$2000(1 - 0.672971333)}{0.02} \approx \$32{,}702.87$$

7. We will use the formula $S = \dfrac{P\left[1 - \left[1 + \frac{1}{n}\right]^{-nN}\right]}{\frac{1}{n}}$

with $P = \$10$, $i = 6\%$, or 0.06, $n = 12$, and $N = 8$.

$S = \dfrac{\$10\left[1 - \left[1 + \frac{0.06}{12}\right]^{-12 \times 8}\right]}{\frac{0.06}{12}} = \dfrac{\$10[1 - 1.005^{-96}]}{0.005}$

$\approx \dfrac{\$10[1 - 0.619523908]}{0.005} \approx \$760.95$

9. a) We can use the formula $S = \dfrac{P\left[1 - \left[1 + \frac{1}{n}\right]^{-nN}\right]}{\frac{1}{n}}$

with $S = \$4200$, $i = 12\%$, or 0.12, $n = 12$, and $N = 3$.

$\$4200 = \dfrac{P\left[1 - \left[1 + \frac{0.12}{12}\right]^{-12 \times 3}\right]}{\frac{0.12}{12}}$

$\$4200\left(\dfrac{0.12}{12}\right) = P[1 - (1 + 0.01)^{-36}]$

$\$42 = P[1 - 0.698924949]$

$P = \dfrac{\$42}{0.30107505} = \$139.50$

b) The total number of payments is 36. To find the total amount to be paid back, multiply the monthly payment by the total number of payments.
$(\$139.50)(36) = \$5022$

11. a) We can use the formula $S = \dfrac{P\left[1 - \left[1 + \frac{1}{n}\right]^{-nN}\right]}{\frac{1}{n}}$

with $S = \$60,000$, $i = 13.5\%$, or 0.135, $n = 12$, and $N = 30$.

$\$60,000 = \dfrac{P\left[1 - \left[1 + \frac{0.135}{12}\right]^{-12 \times 30}\right]}{\frac{0.135}{12}}$

$\$60,000\left(\dfrac{0.135}{12}\right) = P\left[1 - \left[1 + \frac{0.135}{12}\right]^{-360}\right]$

$\$675 = P[1 - 0.017820813]$

$P = \dfrac{\$675}{0.982179186} = \$687.25$

b) The total number of payments is 360. To find the total amount to be paid back, multiply the monthly payment by the total number of payments.
$(\$687.25)(360) = \$247,410$

11. (continued)

c) Compute as in a), but use $N = 35$.

$\$60,000 = \dfrac{P\left[1 - \left[1 + \frac{0.135}{12}\right]\right]^{-12 \times 35}}{\frac{0.135}{12}}$

$\$60,000\left(\dfrac{0.135}{12}\right) = P\left[1 - \left[1 + \frac{0.135}{12}\right]^{-420}\right]$

$\$675 = P[1 - 0.009107841]$

$P = \dfrac{\$675}{0.990892158} = \$681.20$

There are $12 \cdot 35$, or 420, payments. The total amount to be paid back is

$(\$681.20)(420) = \$286,104.$

13. $S = \dfrac{P[1 - (1 + i)^{-N}]}{i}$

$Si = P[1 - (1 + i)^{-N}]$

$P = \dfrac{Si}{[1 - (1 + i)^{-N}]}$

15. The present value of the annuity is the sum $\$1000(1.08)^{-1} + \$1000(1.08)^{-2} + \$1000(1.08)^{-3} + \cdots$. This is an infinite geometric series with $a_1 = \$1000(1.08)^{-1}$ and $r = 1.08^{-1}$. We can find its sum using the formula

$S_\infty = \dfrac{a_1}{1 - r}$

$S_\infty = \dfrac{\$1000(1.08)^{-1}}{1 - 1.08^{-1}}$

Multiplying by 1 using $\dfrac{1.08}{1.08}$ will simplify the calculation.

$S_\infty = \dfrac{1.08}{1.08} \cdot \dfrac{\$1000(1.08)^{-1}}{1 - 1.08^{-1}}$

$= \dfrac{\$1000}{1.08 - 1} = \dfrac{\$1000}{0.08}$

$= \$12,500$

Another method is to think in terms of simple interest. You want to have $\$1000$ interest each year, so solve $I = Pit$ for P when $I = \$1000$, $i = 0.08$, and $t = 1$.

17. Let a = the amount invested each year. The amount of the annuity can be found using the formula

$V = \dfrac{P[(1 + i)^N - 1]}{i}$ where $P = a$, $i = 8\%$, or 0.08, and $N = 18$.

$V = \dfrac{a[(1 + 0.8)^{18} - 1]}{0.08} = \dfrac{a(1.08^{18} - 1)}{0.08}.$

This value is the same as the present value of an annuity with interest rate 8%, compounded annually, such that $\$15,000$ can be withdrawn for each of the next 4 years. The present value can be expressed as $S = \dfrac{P[1 - (1 + i)^{-N}]}{i}$ where $P = \$15,000$, $i = 0.08$, and $N = 4$.

17. (continued)

$$S = \frac{\$15,000[1 - (1 + 0.08)^{-4}]}{0.08} = \frac{\$15,000(1 - 1.08^{-4})}{0.08}$$

Now $V = S$, so we have

$$\frac{a(1.08^{18} - 1)}{0.08} = \frac{\$15,000(1 - 1.08^{-4})}{0.08}$$

$$a(1.08^{18} - 1) = \$15,000(1 - 1.08^{-4})$$

$$a = \frac{\$15,000(1 - 1.08^{-4})}{1.08^{18} - 1}$$

$$\approx \frac{\$15,000(1 - 0.735029852)}{3.996019499 - 1}$$

$$\approx \$1326.61$$

$1326.61 should be invested each year for 18 years.

19. In Exercise 18 we find that the amount of the loan is $34,584.07. At the end of the first six months the amount of interest owed is $34,584.07 × 0.08 × $\frac{1}{2}$ = $1383.36. A payment of $2000 is made. This month it consists of $1383.36 interest and $2000 - $1383.36, or $616.64, principal. For the next six months the new principal is $34,584.07 - $616.64, or $33,967.43. The amount of interest owed is $33,967.43 × 0.08 × $\frac{1}{2}$ = $1358.70. A payment of $2000 is made. It consists of $1358.70 interest and $2000 - $1358.70, or $641.30 principal. The following table shows the amount of interest and principal each of the first 11 payments consists of.

Payment	Current Principal	Interest	Reduction in Principal
1	$34,584.07	$1383.36	$616.64
2	33,967.43	1358.70	641.30
3	33,326.13	1333.05	666.95
4	32,659.18	1306.37	693.63
5	31,965.55	1278.62	721.38
6	31,244.17	1249.77	750.23
7	30,493.94	1219.76	780.24
8	29,713.70	1188.55	811.45
9	28,902.25	1156.09	843.91
10	28,058.34	1122.33	877.67
11	27,180.67	1087.23	912.77

After the eleventh payment the new principal is $27,180.67 - $912.77 = $26,267.90. This is the remaining liability on the purchase. (Answers may vary slightly due to rounding differences.)

21. The present value of each withdrawal is 1500\left[1 + \frac{0.1}{2}\right]^{-2t}$, or $1500(1.05)^{-2t}$.

The first withdrawal is made at the end of $5\frac{1}{2}$ years, or at $t = 5\frac{1}{2}$. Successive withdrawals are made at $\frac{1}{2}$ year intervals until a total of 16 withdrawals have been made. The final withdrawal is made at $t = 13$. The present value of the annuity is the sum $1500(1.05)^{-11} + $1500(1.05)^{-12} + $1500(1.05)^{-13} + \cdots + $1500(1.05)^{-26}$. This is a geometric series with $a_1 = \$1500(1.05)^{-11}$, $n = 16$, and $r = 1.05^{-1}$. We can find the sum using the formula $S_n = \frac{a_1(r^n - 1)}{r - 1}$.

We have

$$S_{16} = \frac{\$1500(1.05)^{-11}[(1.05^{-1})^{16} - 1]}{1.05^{-1} - 1}$$

$$\approx \frac{-\$475.2464551}{-0.047619047}$$

$$\approx \$9980.18$$

23. The present value of each payment is $250(1 + 0.1362)^{-t}$, or $250(1.1362)^{-t}$. The first payment is made at the end of 4 years, or at $t = 4$. Successive payments are made at 4 year intervals until the end of 40 years, or at $t = 40$. The present value of the annuity is the sum $250(1.1362)^{-4} + $250(1.1362)^{-8} + \cdots + $250(1.1362)^{-40}$. This is a geometric series with $a_1 = \$250(1.1362)^{-4}$, $n = 10$, and $r = 1.1362^{-4}$. We can find the sum using the formula $S = \frac{a_1(r^n - 1)}{r - 1}$.

We have

$$S_{10} = \frac{\$250(1.1362)^{-4}[(1.1362^{-4})^{10} - 1]}{1.1362^{-4} - 1}$$

$$= \frac{1.1362^4}{1.1362^4} \cdot \frac{\$250(1.1362)^{-4}(1.1362^{-40} - 1)}{1.1362^{-4} - 1}$$

$$= \frac{\$250(1.1362^{-40} - 1)}{1 - 1.1362^4}$$

$$\approx \frac{\$250(0.006050741 - 1)}{1 - 1.666553039}$$

$$\approx \$373$$

The correct answer is E.

Exercise Set 8.6

1. We will use the formula $E = \left(1 + \frac{i}{n}\right)^n - 1$ with $i = 8\%$, or 0.08 and $n = 2$.

$$E = \left(1 + \frac{0.08}{2}\right)^2 - 1 = 1.04^2 - 1$$

$$= 1.0816 - 1 = 0.0816 = 8.16\%$$

3. We will use the formula $E = \left(1 + \frac{i}{n}\right)^n - 1$ with $i = 9\%$, or 0.09 and $n = 4$.

$$E = \left(1 + \frac{0.09}{4}\right)^4 - 1 = 1.0225^4 - 1$$

$$\approx 1.093083 - 1 = 0.093083 \approx 9.308\%$$

5. We will use the formula $E = \left(1 + \frac{i}{n}\right)^n - 1$ with $i = 8\%$, or 0.08, and $n = 6$.

$$E = \left(1 + \frac{0.08}{6}\right)^6 - 1 \approx 1.013333333^6 - 1$$

$$\approx 1.0827145 - 1 = 0.0827145 \approx 8.271\%$$

7. We will use the formula $E = \left(1 + \frac{i}{n}\right)^n - 1$ with $i = 8\%$, or 0.08 and $n = 365$.

$$E = \left(1 + \frac{0.08}{365}\right)^{365} - 1 \approx 1.000219178^{365} - 1$$

$$\approx 1.083278 - 1 = 0.083278 \approx 8.328\%$$

9. We will use the formula $E = \left(1 + \frac{i}{n}\right)^n - 1$ with $i = 8\%$, or 0.08 and $n = 8760$.

$$E = \left(1 + \frac{0.08}{8760}\right)^{8760} - 1 \approx 1.000009132^{8760} - 1$$

$$\approx 1.083287 - 1 = 0.083287 \approx 8.329\%$$

11. Using the simple interest formula, $I = Pit$, we find that the loan will earn interest of $1000 \times 0.08 \times 1$, or \$80. This amount is added-on to the \$1000, so the amount to be paid back is \$1080. Each of the 12 payments will be \$1080 ÷ 12, or \$90. Now the APR is defined to be the interest rate i such that $S = \dfrac{P\left[1 - \left(1 + \frac{i}{n}\right)^{-nN}\right]}{\frac{i}{n}}$ where, in this case, $S = \$1000$, $P = \$90$, $n = 12$, and $N = 1$. We get the following equation:

$$\$1000 = \frac{\$90\left[1 - \left(1 + \frac{i}{12}\right)^{-12}\right]}{\frac{i}{12}}$$

This equation is not easy to solve, so we make a guess for the value of i, compute the right side of the equation and see how close we get to \$1000, and then refine our guess accordingly. Let's try $i = 14\%$.

$i = 14\%$ implies $S = \dfrac{\$90\left[1 - \left(1 + \frac{0.14}{12}\right)^{-12}\right]}{\frac{0.14}{12}}$

$$\approx \$1002.37$$

Next let's try $i = 15\%$.

$i = 15\%$ implies $S = \dfrac{\$90\left[1 - \left(1 + \frac{0.15}{12}\right)^{-12}\right]}{\frac{0.15}{12}}$

$$\approx \$997.14$$

We see that $i = 14\%$ is too small and $i = 15\%$ is too large. We try $i = 14.5\%$.

11. (continued)

$i = 14.5\%$ implies $S = \dfrac{\$90\left[1 - \left(1 + \frac{0.145}{12}\right)^{-12}\right]}{\frac{0.145}{12}}$

$$\approx \$999.75$$

This is very close to \$1000, so we will use 14.5% as a good estimate of the APR.

13. Using the simple interest formula, $I = Pit$, we find that the loan will earn interest of $\$2000 \times 0.1 \times 1$, or \$200. This amount is added-on to the \$2000, so the amount to be paid back is \$2200. Each of the 12 payments will be \$2200 ÷ 12, or \$183.33. The APR is the interest rate i such that $S = \dfrac{P\left[1 - \left(1 + \frac{i}{n}\right)^{-nN}\right]}{\frac{i}{n}}$ where $S = \$2000$, $P = \$183.33$, $n = 12$, and $N = 1$. We get the following equation:

$$\$2000 = \frac{\$183.33\left[1 - \left(1 + \frac{i}{12}\right)^{-12}\right]}{\frac{i}{12}}$$

We make a guess for the value of i, compute the right side of the equation and see how close we get to \$2000, and then refine our guess accordingly. Let's try $i = 17\%$.

$i = 17\%$ implies $S = \dfrac{\$183.33\left[1 - \left(1 + \frac{0.17}{12}\right)^{-12}\right]}{\frac{0.17}{12}}$

$$\approx \$2010.09$$

Next let's try $i = 18\%$.

$i = 18\%$ implies $S = \dfrac{\$183.33\left[1 - \left(1 + \frac{0.18}{12}\right)^{-12}\right]}{\frac{0.18}{12}}$

$$\approx \$1999.67$$

This is very close to \$2000, so we will use 18% as a good estimate of the APR.

15. A buyer puts \$500 down and then pays \$100.63 per month for 36 months. He pays a total of $\$500 + \$100.63(36) = \$4122.68$. A buyer paying cash would pay \$3530. Therefore, the interest, or finance charge, is $\$4122.68 - \$3530$, or \$592.68. The principal is equal to the cash price less the down payment, $\$3530 - \$500$, or \$3030. To estimate the APR we use the formula

$APR = \dfrac{72I}{3P(n + 1) + I(n - 1)}$ with $I = \$592.68$, $P = \$3030$, and $n = 36$.

$$APR = \frac{72(\$592.68)}{3(3030)(37) + \$592.68(35)} \approx 0.1195, \text{ or } 11.95\%$$

Exercise Set 8.7

1. The present value of the Pennsylvania State Lottery

   $$= \$1,770,778 + \frac{\$1,770,778[1 - (1 + i)^{-N}]}{i}$$

   $$= \$1,770,778 + \frac{\$1,770,778[1 - (1 + 0.06)^{-25}]}{0.06}$$

   $$= \$24,407,263.85$$

3. The present value of the Pennsylvania State Lottery

   $$= \$1,770,778 + \frac{\$1,770,778[1 - (1 + i)^{-N}]}{i}$$

   $$= \$1,770,778 + \frac{\$1,770,778[1 - (1 + 0.07)^{-25}]}{0.07}$$

   $$= \$22,406,686.71$$

5. The present value of the Pennsylvania State Lottery

   $$= \$1,770,778 + \frac{\$1,770,778[1 - (1 + i)^{-N}]}{i}$$

   $$= \$1,770,778 + \frac{\$1,770,778[1 - (1 + 0.085)^{-25}]}{0.085}$$

   $$= \$19,893,257.88$$

7. The present value of the Pennsylvania State Lottery

   $$= \$1,770,778 + \frac{\$1,770,778[1 - (1 + i)^{-N}]}{i}$$

   $$= \$1,770,778 + \frac{\$1,770,778[1 - (1 + 0.09)^{-25}]}{0.09}$$

   $$= \$19,164,385.87$$

9. Divide to find the amount of each payment.
   $\$46,040,228/51 \approx \$902,750$

11. Divide to find the amount of each payment.
    $\$46,040,228/101 \approx \$455,844$

13. The \$2 million payoff will be paid in 26 equal payments of $\$2,000,000/26 \approx \$76,923$. The first \$76,923 will be paid immediately with payments of \$76,923 once a year for 25 years thereafter. Then for a \$2 million payoff and an 8% interest rate we have:

    The present value of the Pennsylvania State Lottery

    $$= \$76,923 + \frac{\$76,923[1 - (1 + i)^{-N}]}{i}$$

    $$= \$76,923 + \frac{\$76,923[1 - (1 + 0.08)^{-25}]}{0.08}$$

    $$= \$898,058.81$$

Exercise Set 9.1

1. a) Let "calling Plumber I" be state 1 and "calling Plumber II" be state 2. Since 50% of those who call Plumber I will call Plumber II the next time, the other 50% will call Plumber I the next time. 40% of those who call Plumber II will call Plumber I next time, so the other 60% will call Plumber II the next time. The transition matrix is

$$T = \begin{array}{c} \\ \text{I} \\ \text{II} \end{array} \begin{array}{cc} \text{I} & \text{II} \\ \begin{bmatrix} 0.5 & 0.5 \\ 0.4 & 0.6 \end{bmatrix} \end{array}.$$

   b) If the system is in state 1 (calling Plumber I), the probability it is in state 2 (calling Plumber II) after 2 repetitions is the (1, 2)-element in T^2.

$$T^2 = \begin{bmatrix} 0.5 & 0.5 \\ 0.4 & 0.6 \end{bmatrix} \begin{bmatrix} 0.5 & 0.5 \\ 0.4 & 0.6 \end{bmatrix} = \begin{bmatrix} 0.45 & 0.55 \\ 0.44 & 0.56 \end{bmatrix}$$

   The (1, 2)-element is 0.55, so the probability is 0.55, or 55%.

3. a) Let "selecting easy listening" be state 1, "selecting jazz" be state 2, and "selecting rock" be state 3. The transition matrix is

$$T = \begin{array}{c} \\ \text{E} \\ \text{J} \\ \text{R} \end{array} \begin{array}{ccc} \text{E} & \text{J} & \text{R} \\ \begin{bmatrix} 0.5 & 0.2 & 0.3 \\ 0.4 & 0.4 & 0.2 \\ 0.3 & 0.1 & 0.6 \end{bmatrix} \end{array} \begin{array}{l} \text{(Each row} \\ \text{must add} \\ \text{up to 1.)} \end{array}$$

   b) If the system is in state 2 (selecting jazz), the probability that it will be in state 1 (selecting easy listening) after 3 repetitions is given by the (2, 1)-element in T^3.

$$T^2 = \begin{bmatrix} 0.5 & 0.2 & 0.3 \\ 0.4 & 0.4 & 0.2 \\ 0.3 & 0.1 & 0.6 \end{bmatrix} \begin{bmatrix} 0.5 & 0.2 & 0.3 \\ 0.4 & 0.4 & 0.2 \\ 0.3 & 0.1 & 0.6 \end{bmatrix}$$

$$= \begin{bmatrix} 0.42 & 0.21 & 0.37 \\ 0.42 & 0.26 & 0.32 \\ 0.37 & 0.16 & 0.47 \end{bmatrix}$$

$$T^3 = T^2 \cdot T = \begin{bmatrix} 0.42 & 0.21 & 0.37 \\ 0.42 & 0.26 & 0.32 \\ 0.37 & 0.16 & 0.47 \end{bmatrix} \begin{bmatrix} 0.5 & 0.2 & 0.3 \\ 0.4 & 0.4 & 0.2 \\ 0.3 & 0.1 & 0.6 \end{bmatrix}$$

$$= \begin{bmatrix} 0.405 & 0.205 & 0.39 \\ 0.41 & 0.22 & 0.37 \\ 0.39 & 0.185 & 0.425 \end{bmatrix}$$

   The (2, 1)-element is 0.41, so the probability is 0.41, or 41%.

5. Let "sending a funny card" be state 1 and "sending a sentimental card" be state 2. The transition matrix is

$$T = \begin{array}{c} \\ \text{F} \\ \text{S} \end{array} \begin{array}{cc} \text{F} & \text{S} \\ \begin{bmatrix} 0.2 & 0.8 \\ 0.6 & 0.4 \end{bmatrix} \end{array}.$$

5. (continued)

   If the system is in state 2 (sending a sentimental card), the probability that it will be in state 2 after 2 repetitions is given by the (2, 2)-element of T^2.

$$T^2 = \begin{bmatrix} 0.2 & 0.8 \\ 0.6 & 0.4 \end{bmatrix} \begin{bmatrix} 0.2 & 0.8 \\ 0.6 & 0.4 \end{bmatrix} = \begin{bmatrix} 0.52 & 0.48 \\ 0.36 & 0.64 \end{bmatrix}$$

   The (2, 2)-element is 0.64, so the probability is 0.64, or 64%.

7. Let state 1 be "delivering a passenger to the first zone," let state 2 be "delivering a passenger to the second zone," and let state 3 be "delivering a passenger to the third zone." A taxi picking up a passenger in the first zone has a probability of 50%, or 1/2, of delivering the passenger to that zone. It is twice as likely to deliver a passenger to the second zone as to the third, so the probability of delivering a passenger to the second zone is $\left(\frac{2}{3}\right) \cdot \left(\frac{1}{2}\right)$, or $\frac{1}{3}$, and the probability of delivering a passenger to the third zone is $\left(\frac{1}{3}\right) \cdot \left(\frac{1}{2}\right)$, or $\frac{1}{6}$. A passenger picked up in the second zone will be let off there with a probability equal to that for being delivered to either other zone. Then the probabilities are: first zone, 1/3; second zone, 1/3; third zone, 1/3. A passenger picked up in the third zone is twice as likely to go to the first zone as either to go to the second zone or stay in the third zone. The probabilities are: first zone, 2/4, or 1/2; second zone: $\left(\frac{1}{2}\right) \cdot \left(\frac{1}{2}\right)$, or $\frac{1}{4}$; third zone: 1/4. The transition matrix is

$$T = \begin{array}{c} \\ \text{I} \\ \text{II} \\ \text{III} \end{array} \begin{array}{ccc} \text{I} & \text{II} & \text{III} \\ \begin{bmatrix} 1/2 & 1/3 & 1/6 \\ 1/3 & 1/3 & 1/3 \\ 1/2 & 1/4 & 1/4 \end{bmatrix} \end{array}.$$

   If the system is in state 2, the probability it will be in state 3 after 2 repetitions is the (2, 3)-element of T^2.

$$T^2 = \begin{bmatrix} 1/2 & 1/3 & 1/6 \\ 1/3 & 1/3 & 1/3 \\ 1/2 & 1/4 & 1/4 \end{bmatrix} \begin{bmatrix} 1/2 & 1/3 & 1/6 \\ 1/3 & 1/3 & 1/3 \\ 1/2 & 1/4 & 1/4 \end{bmatrix}$$

$$= \begin{bmatrix} 4/9 & 23/72 & 17/72 \\ 4/9 & 11/36 & 1/4 \\ 11/24 & 5/16 & 11/48 \end{bmatrix}$$

The (2, 3)-element is 1/4, so the probability is 1/4, or 25%.

9. The probability that the first selection will be easy listening is 40% and 20% that it will be jazz. Therefore, the probability that it will be rock is 100% - (40% + 20%), or 40%. The initial state vector is

$$P_0 = \begin{matrix} E & J & R \\ [0.4 & 0.2 & 0.4] \end{matrix}.$$

a) The state vector after 2 repetitions is $P_2 = P_0T^2$. (We found T^2 in Exercise 3.) The probability that the second selection will be easy listening is the (1, 1)-element of P_2.

$$P_2 = [0.4 \quad 0.2 \quad 0.4]\begin{bmatrix} 0.42 & 0.21 & 0.37 \\ 0.42 & 0.26 & 0.32 \\ 0.37 & 0.16 & 0.47 \end{bmatrix}$$

$$= [0.4 \quad 0.2 \quad 0.4]$$

The probability is 0.4, or 40%.

b) The probability that the fourth selection will be jazz is the (1, 2)-element of $P_4 = P_0T^4$.

$$T^4 = T^2 \cdot T^2 = \begin{bmatrix} 0.42 & 0.21 & 0.37 \\ 0.42 & 0.26 & 0.32 \\ 0.37 & 0.16 & 0.47 \end{bmatrix}\begin{bmatrix} 0.42 & 0.21 & 0.37 \\ 0.42 & 0.26 & 0.32 \\ 0.37 & 0.16 & 0.47 \end{bmatrix}$$

$$= \begin{bmatrix} 0.4015 & 0.202 & 0.3965 \\ 0.404 & 0.207 & 0.389 \\ 0.3965 & 0.1945 & 0.409 \end{bmatrix}$$

$$P_0T^4 = [0.4 \quad 0.2 \quad 0.4]\begin{bmatrix} 0.4015 & 0.202 & 0.3965 \\ 0.404 & 0.207 & 0.389 \\ 0.3965 & 0.1945 & 0.409 \end{bmatrix}$$

$$= [0.4 \quad 0.2 \quad 0.4]$$

The probability is 0.2, or 20%.

11. There is a 90% probability that Harry will send a funny card, so there is a 10% probability that he will send a sentimental card. Then

$$P_0 = \begin{matrix} F & S \\ [0.9 & 0.1] \end{matrix}.$$

The probability that the second card (two birthdays after her fiftieth) will be sentimental is the (1, 2)-element of P_0T^2. (We found T^2 in Exercise 5.)

$$P_0T^2 = [0.9 \quad 0.1]\begin{bmatrix} 0.52 & 0.48 \\ 0.36 & 0.64 \end{bmatrix} = [0.504 \quad 0.496]$$

The probability is 0.496, or 49.6%.

13. a) $P_2 = P_1T = P_0T^2$

$$T^2 = \begin{bmatrix} 0.4 & 0.4 & 0.2 \\ 0 & 0 & 1 \\ 1 & 0 & 0 \end{bmatrix}\begin{bmatrix} 0.4 & 0.4 & 0.2 \\ 0 & 0 & 1 \\ 1 & 0 & 0 \end{bmatrix}$$

$$= \begin{bmatrix} 0.36 & 0.16 & 0.48 \\ 1 & 0 & 0 \\ 0.4 & 0.4 & 0.2 \end{bmatrix}$$

13. (continued)

$$P_2 = P_0T^2 = [1 \quad 0 \quad 0]\begin{bmatrix} 0.36 & 0.16 & 0.48 \\ 1 & 0 & 0 \\ 0.4 & 0.4 & 0.2 \end{bmatrix}$$

$$= [0.36 \quad 0.16 \quad 0.48]$$

b) $P_4 = P_0T^4$

$$T^4 = T^2 \cdot T^2 = \begin{bmatrix} 0.36 & 0.16 & 0.48 \\ 1 & 0 & 0 \\ 0.4 & 0.4 & 0.2 \end{bmatrix}\begin{bmatrix} 0.36 & 0.16 & 0.48 \\ 1 & 0 & 0 \\ 0.4 & 0.4 & 0.2 \end{bmatrix}$$

$$= \begin{bmatrix} 0.4816 & 0.2496 & 0.2688 \\ 0.36 & 0.16 & 0.48 \\ 0.624 & 0.144 & 0.232 \end{bmatrix}$$

$$P_4 = P_0T^4 = [1 \quad 0 \quad 0]\begin{bmatrix} 0.4816 & 0.2496 & 0.2688 \\ 0.36 & 0.16 & 0.48 \\ 0.624 & 0.144 & 0.232 \end{bmatrix}$$

$$= [0.4816 \quad 0.2496 \quad 0.2688]$$

Rounding gives $P_4 = [0.48 \quad 0.25 \quad 0.27]$.

15. $P_0 = \begin{matrix} I & II & III \\ [0.6 & 0.3 & 0.1] \end{matrix}$

The distribution after all have picked up and dropped off four passengers is given by $P_4 = P_0T^4$. (We found T^2 in Exercise 7. We will now express the elements in decimal notation. Each row may not add up to exactly one because of rounding.)

$$T^4 = T^2 \cdot T^2 =$$

$$\begin{bmatrix} 0.4444 & 0.3194 & 0.2361 \\ 0.4444 & 0.3056 & 0.25 \\ 0.4583 & 0.3125 & 0.2292 \end{bmatrix}\begin{bmatrix} 0.4444 & 0.3194 & 0.2361 \\ 0.4444 & 0.3056 & 0.25 \\ 0.4583 & 0.3125 & 0.2292 \end{bmatrix}$$

$$= \begin{bmatrix} 0.4476 & 0.3133 & 0.2389 \\ 0.4479 & 0.3135 & 0.2386 \\ 0.4476 & 0.3135 & 0.2389 \end{bmatrix}$$

$$P_4 = P_0T^4 = [0.6 \quad 0.3 \quad 0.1]\begin{bmatrix} 0.4476 & 0.3133 & 0.2389 \\ 0.4479 & 0.3135 & 0.2386 \\ 0.4476 & 0.3135 & 0.2389 \end{bmatrix}$$

$$= [0.4477 \quad 0.3134 \quad 0.2388]$$

The distribution is first zone: approximately 44.8%, second zone: approximately 31.3%, third zone: approximately 23.9%.

17. If each of the choices is equally likely for the first cone, then

$$P_0 = \begin{matrix} \text{Vanilla} & \text{Chocolate} & \text{Strawberry} \\ [1/3 & 1/3 & 1/3]. \end{matrix}$$

The state vector after the first cone is

$$P_1 = P_0T = [1/3 \quad 1/3 \quad 1/3]\begin{bmatrix} 0 & 1/2 & 1/2 \\ 1/5 & 2/5 & 2/5 \\ 1/3 & 0 & 2/3 \end{bmatrix}$$

$$= [8/45 \quad 3/10 \quad 47/90], \text{ or}$$
$$[16/90 \quad 27/90 \quad 47/90].$$

The state vector after the second cone is

$$P_2 = P_1T = [8/45 \quad 3/10 \quad 47/90]\begin{bmatrix} 0 & 1/2 & 1/2 \\ 1/5 & 2/5 & 2/5 \\ 1/3 & 0 & 2/3 \end{bmatrix}$$

$$= [158/675 \quad 47/225 \quad 376/675], \text{ or}$$
$$[158/675 \quad 141/675 \quad 376/675].$$

The state vector after the third cone is

$$P_3 = P_2T = \begin{bmatrix} \dfrac{158}{675} & \dfrac{141}{675} & \dfrac{376}{675} \end{bmatrix}\begin{bmatrix} 0 & \dfrac{1}{2} & \dfrac{1}{2} \\ \dfrac{1}{5} & \dfrac{2}{5} & \dfrac{2}{5} \\ \dfrac{1}{3} & 0 & \dfrac{2}{3} \end{bmatrix}$$

$$= \begin{bmatrix} \dfrac{2303}{10,125} & \dfrac{677}{3375} & \dfrac{5791}{10,125} \end{bmatrix}, \text{ or}$$

$$\begin{bmatrix} \dfrac{2303}{10,125} & \dfrac{2031}{10,125} & \dfrac{5791}{10,125} \end{bmatrix}$$

The probability that the fourth cone will be chocolate is the (1, 2)-element in P_3. (Recall that P_3 is the state vector <u>after</u> the third cone.) The probability is 2031/10,125, or approximately 20%.

19. The probability that all three will be easy listening is (the probability that the first will be easy listening) × (the probability that the second will be easy listening) × (the probability that the third will be easy listening). These probabilities are the (1, 1)-elements of T, T^2, and T^3, respectively. (We found T, T^2, and T^3 in Exercise 3.)

$$0.5(0.42)(0.405) = 0.08505$$

The probability is approximately 8.5%.

21. This cannot be a transition matrix for a Markov chain, because it contains a negative entry.

23. This could be a transition matrix for a Markov chain, because all of the entries are nonnegative and the entries of each row sum to 1.

25. The probabilities that much homework will be given on each of the first five nights are the (1, 3)-elements of P_0, P_1, P_2, P_3, and P_4, respectively. (We will express the elements of P_0 and T in decimal notation.)

$$P_0 = [0 \quad 0.25 \quad 0.75]$$

$$P_1 = P_0T = [0 \quad 0.25 \quad 0.75]\begin{bmatrix} 0 & 0.4 & 0.6 \\ 0.25 & 0.5 & 0.25 \\ 0.6 & 0.2 & 0.2 \end{bmatrix}$$

$$= [0.5125 \quad 0.275 \quad 0.2125]$$

$$P_2 = P_1T = [0.5125 \quad 0.275 \quad 0.2125]\begin{bmatrix} 0 & 0.4 & 0.6 \\ 0.25 & 0.5 & 0.25 \\ 0.6 & 0.2 & 0.2 \end{bmatrix}$$

$$= [0.1963 \quad 0.385 \quad 0.4188]$$

$$P_3 = P_2T = [0.1963 \quad 0.385 \quad 0.4188]\begin{bmatrix} 0 & 0.4 & 0.6 \\ 0.25 & 0.5 & 0.25 \\ 0.6 & 0.2 & 0.2 \end{bmatrix}$$

$$= [0.3475 \quad 0.3548 \quad 0.2978]$$

$$P_4 = P_3T = [0.3475 \quad 0.3548 \quad 0.2978]\begin{bmatrix} 0 & 0.4 & 0.6 \\ 0.25 & 0.5 & 0.25 \\ 0.6 & 0.2 & 0.2 \end{bmatrix}$$

$$= [0.2674 \quad 0.3760 \quad 0.3568]$$

The probabilities of much homework being given for the first five nights are 0.75, 0.21, 0.42, 0.30, and 0.36. The numbers alternately decrease and increase, but each fluctuation is less than the previous one. They appear to approach some middle value.

Exercise Set 9.2

1. $\begin{bmatrix} 0.5 & 0.2 & 0.3 \\ 0.2 & 0 & 0.8 \\ 0 & 0.7 & 0.3 \end{bmatrix}$

It is possible to go from state 1 to states 2 or 3, from state 2 to states 1 or 3, and from state 3 to state 2 in one repetition. There is a 0 probability that the system will go from state 3 to state 1. However, it is possible to go from state 3 to state 2 in one stage and then from state 2 to state 1 in the next, so states 1 and 3 communicate. The matrix is irreducible.

3. $\begin{bmatrix} 0.5 & 0 & 0.5 \\ 0 & 1 & 0 \\ 0.6 & 0.3 & 0.1 \end{bmatrix}$

It is not possible to go from state 2 to state 1 or to state 3. The matrix is <u>not</u> irreducible.

$$5. \begin{bmatrix} 0.6 & 0 & 0.3 & 0 & 0.1 \\ 0 & 0.7 & 0 & 0.3 & 0 \\ 0 & 0 & 0.2 & 0 & 0.8 \\ 0 & 0.5 & 0 & 0.5 & 0 \\ 0.1 & 0 & 0 & 0 & 0.9 \end{bmatrix}$$

The transition diagram will be helpful in determining whether states communicate.

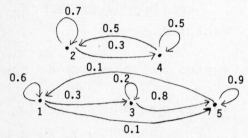

It is clear from the diagram that there is no communication between the following pairs of states: 1 and 2, 1 and 4, 2 and 3, 2 and 5, 3 and 4, 4 and 5. The matrix is not irreducible.

7. $T = \begin{bmatrix} 0 & 1 \\ 0.5 & 0.5 \end{bmatrix}$

a) The matrix is irreducible and at least one entry on the main diagonal is nonzero. Therefore, the matrix is regular.

b) To find the steady state vector $P = [p_1 \quad p_2]$ we solve the system of equations

$$P = PT$$
$$p_1 + p_2 = 1.$$

To find $P = PT$ we multiply and equate like components:

$$[p_1 \quad p_2] = [p_1 \quad p_2] \begin{bmatrix} 0 & 1 \\ 0.5 & 0.5 \end{bmatrix}$$

$$p_1 = 0.5p_2$$
$$p_2 = p_1 + 0.5p_2$$

Simplifying, we have

$$p_1 - 0.5p_2 = 0$$
$$p_1 - 0.5p_2 = 0.$$

Including the equation $p_1 + p_2 = 1$, we have

$$p_1 - 0.5p_2 = 0$$
$$p_1 - 0.5p_2 = 0$$
$$p_1 + p_2 = 1.$$

This system can be solved using the elimination method discussed in Chapter 2.

The initial echelon tableau is

p_1	p_2		1
1*	-0.5		0
1	-0.5		0
1	1		1

We proceed with the pivoting.

7. (continued)

p_1	p_2		1	
1	-0.5		0	
0	0		0	New Row 2 = -1(Row 1) + Row 2
0	1.5		1	New Row 3 = -1(Row 1) + Row 3

p_1	p_2		1	
1	-0.5		0	
0	1.5		1	Interchange Row 2 and Row 3
0	0		0	

p_1	p_2		1	
3	-1.5		0	New Row 1 = 3(Row 1)
0	1.5		1	
0	0		0	

p_1	p_2		1	
3	0		1	New Row 1 = Row 2 + Row 1
0	1.5		1	
0	0		0	

p_1	p_2		1	
1	0		$\frac{1}{3}$	New Row 1 = $\frac{1}{3}$(Row 1)
0	1		$\frac{2}{3}$	New Row 2 = $\frac{1}{1.5}$(Row 2)
0	0		0	

The solution is $p_1 = \frac{1}{3}$, $p_2 = \frac{2}{3}$. The steady state vector is $\left[\frac{1}{3} \quad \frac{2}{3}\right]$.

9. $T = \begin{bmatrix} 0 & 1 \\ 1 & 0 \end{bmatrix}$

a) T is irreducible, so it might be regular. All the entries on the main diagonal are zero, so we calculate successive powers of T. If some power of T contains all positive entries, the chain is regular.

$$T^2 = \begin{bmatrix} 0 & 1 \\ 1 & 0 \end{bmatrix}\begin{bmatrix} 0 & 1 \\ 1 & 0 \end{bmatrix} = \begin{bmatrix} 1 & 0 \\ 0 & 1 \end{bmatrix}$$

$$T^3 = T^2 \cdot T = \begin{bmatrix} 1 & 0 \\ 0 & 1 \end{bmatrix}\begin{bmatrix} 0 & 1 \\ 1 & 0 \end{bmatrix} = \begin{bmatrix} 0 & 1 \\ 1 & 0 \end{bmatrix}$$

Note that $T^3 = T$. Successive powers of T will alternate between T and T^2, both of which contain zero entries. Therefore, T is not regular.

11. $T = \begin{bmatrix} 0 & 1 & 0 \\ 2/3 & 0 & 1/3 \\ 1/3 & 2/3 & 0 \end{bmatrix}$

a) T is irreducible, so it might be regular. We calculate successive powers of T to determine if T is regular.

$$T^2 = \begin{bmatrix} 0 & 1 & 0 \\ 2/3 & 0 & 1/3 \\ 1/3 & 2/3 & 0 \end{bmatrix} \begin{bmatrix} 0 & 1 & 0 \\ 2/3 & 0 & 1/3 \\ 1/3 & 2/3 & 0 \end{bmatrix}$$

$$= \begin{bmatrix} 2/3 & 0 & 1/3 \\ 1/9 & 8/9 & 0 \\ 4/9 & 1/3 & 2/9 \end{bmatrix}$$

$$T^3 = T^2 \cdot T = \begin{bmatrix} 2/3 & 0 & 1/3 \\ 1/9 & 8/9 & 0 \\ 4/9 & 1/3 & 2/9 \end{bmatrix} \begin{bmatrix} 0 & 1 & 0 \\ 2/3 & 0 & 1/3 \\ 1/3 & 2/3 & 0 \end{bmatrix}$$

$$= \begin{bmatrix} 1/9 & 8/9 & 0 \\ 16/27 & 1/9 & 8/27 \\ 8/27 & 16/27 & 1/9 \end{bmatrix}$$

$$T^4 = T^3 \cdot T = \begin{bmatrix} 1/9 & 8/9 & 0 \\ 16/27 & 1/9 & 8/27 \\ 8/27 & 16/27 & 1/9 \end{bmatrix} \begin{bmatrix} 0 & 1 & 0 \\ 2/3 & 0 & 1/3 \\ 1/3 & 2/3 & 0 \end{bmatrix}$$

$$= \begin{bmatrix} 16/27 & 1/9 & 8/27 \\ 14/81 & 64/81 & 1/27 \\ 35/81 & 10/27 & 16/81 \end{bmatrix}$$

T^4 contains all positive entries, so T is regular.

b) Solve the system of equations

$$[p_1 \quad p_2 \quad p_3] = [p_1 \quad p_2 \quad p_3] \begin{bmatrix} 0 & 1 & 0 \\ 2/3 & 0 & 1/3 \\ 1/3 & 2/3 & 0 \end{bmatrix}$$

$$p_1 + p_2 + p_3 = 1.$$

Multiplying and equating like components, we have

$$p_1 = \frac{2}{3}p_2 + \frac{1}{3}p_3$$
$$p_2 = p_1 + \frac{2}{3}p_3$$
$$p_3 = \frac{1}{3}p_2$$
$$p_1 + p_2 + p_3 = 1.$$

Simplifying,

$$p_1 - \frac{2}{3}p_2 - \frac{1}{3}p_3 = 0$$
$$-p_1 + p_2 - \frac{2}{3}p_3 = 0$$
$$- \frac{1}{3}p_2 + p_3 = 0$$
$$p_1 + p_2 + p_3 = 1.$$

11. (continued)

Solve using the elimination method. The initial echelon tableau is

p_1	p_2	p_3		1
1	$-\frac{2}{3}$	$-\frac{1}{3}$		0
-1	1	$-\frac{2}{3}$		0
0	$-\frac{1}{3}$	1		0
1	1	1		1

The final tableau is

p_1	p_2	p_3		1
1	0	0		$\frac{7}{19}$
0	1	0		$\frac{9}{19}$
0	0	1		$\frac{3}{19}$
0	0	0		0

The solution is $p_1 = \frac{7}{19}$, $p_2 = \frac{9}{19}$, $p_3 = \frac{3}{19}$. The steady state vector is [7/19 9/19 3/19].

13. $T = \begin{bmatrix} 1 & 0 & 0 \\ 0 & 1 & 0 \\ 1/2 & 1/2 & 0 \end{bmatrix}$

a) It is not possible to leave state 1 or state 2. T is not irreducible and, therefore, not regular.

15. $T = \begin{bmatrix} 0 & 1 & 0 \\ 0 & 2/5 & 3/5 \\ 1 & 0 & 0 \end{bmatrix}$

a) T is irreducible, and at least one entry on the main diagonal is nonzero. Therefore, T is regular.

b) Solve the system of equations

$$[p_1 \quad p_2 \quad p_3] = [p_1 \quad p_2 \quad p_3] \begin{bmatrix} 0 & 1 & 0 \\ 0 & 2/5 & 3/5 \\ 1 & 0 & 0 \end{bmatrix}$$

$$p_1 + p_2 + p_3 = 1.$$

Multiplying and equating like components we have

$$p_1 = p_3$$
$$p_2 = p_1 + \frac{2}{5}p_2$$
$$p_3 = \frac{3}{5}p_2$$
$$p_1 + p_2 + p_3 = 1.$$

The solution is $p_1 = 3/11$, $p_2 = 5/11$, $p_3 = 3/11$. The steady state vector is [3/11 5/11 3/11].

235

17. $T = \begin{bmatrix} 0 & 1 & 0 \\ 0 & 0 & 1 \\ 0 & 0 & 1 \end{bmatrix}$

   a) It is not possible to go from state 2 to state 1 or from state 3 to either state 1 or state 2. T is not irreducible and, therefore, not regular.

19. $T = \begin{bmatrix} 0 & 1 & 0 & 0 \\ 1/3 & 1/3 & 1/3 & 0 \\ 0 & 1/3 & 1/3 & 1/3 \\ 0 & 0 & 1 & 0 \end{bmatrix}$

   a) Draw the transition diagram.

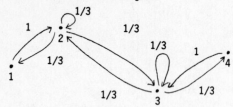

   The transition diagram shows that all the states communicate with each other, so T is irreducible. At least one entry on the main diagonal is nonzero, so T is regular.

   b) Solve the system of equations

$$[p_1 \quad p_2 \quad p_3 \quad p_4] = [p_1 \quad p_2 \quad p_3 \quad p_4] \begin{bmatrix} 0 & 1 & 0 & 0 \\ 1/3 & 1/3 & 1/3 & 0 \\ 0 & 1/3 & 1/3 & 1/3 \\ 0 & 0 & 1 & 0 \end{bmatrix}$$

   $p_1 + p_2 + p_3 + p_4 = 1$.
   Multiplying and equating like components we have

$$p_1 = \frac{1}{3}p_2$$
$$p_2 = p_1 + \frac{1}{3}p_2 + \frac{1}{3}p_3$$
$$p_3 = \frac{1}{3}p_2 + \frac{1}{3}p_3 + p_4$$
$$p_4 = \frac{1}{3}p_3$$
$$p_1 + p_2 + p_3 + p_4 = 1.$$

   The solution is $p_1 = 1/8$, $p_2 = 3/8$, $p_3 = 3/8$, $p_4 = 1/8$. The steady state vector is $[1/8 \quad 3/8 \quad 3/8 \quad 1/8]$.

21. $T = \begin{bmatrix} 1/2 & 1/2 & 0 & 0 \\ 0 & 1/2 & 1/2 & 0 \\ 0 & 0 & 1/2 & 1/2 \\ 0 & 0 & 1/2 & 1/2 \end{bmatrix}$

   a) Draw the transition diagram.

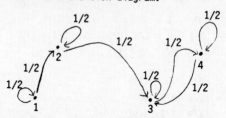

   The diagram shows it is not possible to go from state 2 to state 1 and from state 3 or state 4 to state 1 or state 2. Then T is not irreducible and, therefore, not regular.

23. $T = \begin{bmatrix} 0 & 2/5 & 0 & 3/5 \\ 0 & 0 & 0 & 1 \\ 1 & 0 & 0 & 0 \\ 0 & 0 & 1 & 0 \end{bmatrix}$

   a) Draw the transition diagram.

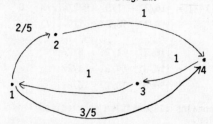

   The diagram shows that all the states communicate with each other, so T is irreducible and, therefore, might be regular. All entries on the main diagonal are zero, so we calculate successive powers of T to determine if T is regular.

$$T^2 = \begin{bmatrix} 0 & 2/5 & 0 & 3/5 \\ 0 & 0 & 0 & 1 \\ 1 & 0 & 0 & 0 \\ 0 & 0 & 1 & 0 \end{bmatrix} \begin{bmatrix} 0 & 2/5 & 0 & 3/5 \\ 0 & 0 & 0 & 1 \\ 1 & 0 & 0 & 0 \\ 0 & 0 & 1 & 0 \end{bmatrix}$$

$$= \begin{bmatrix} 0 & 0 & 3/5 & 2/5 \\ 0 & 0 & 1 & 0 \\ 0 & 2/5 & 0 & 3/5 \\ 1 & 0 & 0 & 0 \end{bmatrix}$$

23. (continued)

$$T^3 = T^2 \cdot T = \begin{bmatrix} 0 & 0 & 3/5 & 2/5 \\ 0 & 0 & 1 & 0 \\ 0 & 2/5 & 0 & 3/5 \\ 1 & 0 & 0 & 0 \end{bmatrix} \begin{bmatrix} 0 & 2/5 & 0 & 3/5 \\ 0 & 0 & 0 & 1 \\ 1 & 0 & 0 & 0 \\ 0 & 0 & 1 & 0 \end{bmatrix}$$

$$= \begin{bmatrix} 3/5 & 0 & 2/5 & 0 \\ 1 & 0 & 0 & 0 \\ 0 & 0 & 3/5 & 2/5 \\ 0 & 2/5 & 0 & 3/5 \end{bmatrix}$$

$$T^4 = T^3 \cdot T = \begin{bmatrix} 3/5 & 0 & 2/5 & 0 \\ 1 & 0 & 0 & 0 \\ 0 & 0 & 3/5 & 2/5 \\ 0 & 2/5 & 0 & 3/5 \end{bmatrix} \begin{bmatrix} 0 & 2/5 & 0 & 3/5 \\ 0 & 0 & 0 & 1 \\ 1 & 0 & 0 & 0 \\ 0 & 0 & 1 & 0 \end{bmatrix}$$

$$= \begin{bmatrix} 2/5 & 6/25 & 0 & 9/25 \\ 0 & 2/5 & 0 & 3/5 \\ 3/5 & 0 & 2/5 & 0 \\ 0 & 0 & 3/5 & 2/5 \end{bmatrix}$$

$T^5 = T^4 \cdot T$

$$= \begin{bmatrix} 2/5 & 6/25 & 0 & 9/25 \\ 0 & 2/5 & 0 & 3/5 \\ 3/5 & 0 & 2/5 & 0 \\ 0 & 0 & 3/5 & 2/5 \end{bmatrix} \begin{bmatrix} 0 & 2/5 & 0 & 3/5 \\ 0 & 0 & 0 & 1 \\ 1 & 0 & 0 & 0 \\ 0 & 0 & 1 & 0 \end{bmatrix}$$

$$= \begin{bmatrix} 0 & 4/25 & 9/25 & 12/25 \\ 0 & 0 & 3/5 & 2/5 \\ 2/5 & 6/25 & 0 & 9/25 \\ 3/5 & 0 & 2/5 & 0 \end{bmatrix}$$

$T^6 = T^5 \cdot T$

$$= \begin{bmatrix} 0 & 4/25 & 9/25 & 12/25 \\ 0 & 0 & 3/5 & 2/5 \\ 2/5 & 6/25 & 0 & 9/25 \\ 3/5 & 0 & 2/5 & 0 \end{bmatrix} \begin{bmatrix} 0 & 2/5 & 0 & 3/5 \\ 0 & 0 & 0 & 1 \\ 1 & 0 & 0 & 0 \\ 0 & 0 & 1 & 0 \end{bmatrix}$$

$$= \begin{bmatrix} 9/25 & 0 & 12/25 & 4/25 \\ 3/5 & 0 & 2/5 & 0 \\ 0 & 4/25 & 9/25 & 12/25 \\ 2/5 & 6/25 & 0 & 9/25 \end{bmatrix}$$

$T^7 = T^6 \cdot T$

$$= \begin{bmatrix} 9/25 & 0 & 12/25 & 4/25 \\ 3/5 & 0 & 2/5 & 0 \\ 0 & 4/25 & 9/25 & 12/25 \\ 2/5 & 6/25 & 0 & 9/25 \end{bmatrix} \begin{bmatrix} 0 & 2/5 & 0 & 3/5 \\ 0 & 0 & 0 & 1 \\ 1 & 0 & 0 & 0 \\ 0 & 0 & 1 & 0 \end{bmatrix}$$

$$= \begin{bmatrix} 12/25 & 18/125 & 4/25 & 27/125 \\ 2/5 & 6/25 & 0 & 9/25 \\ 9/25 & 0 & 12/25 & 4/25 \\ 0 & 4/25 & 9/25 & 12/25 \end{bmatrix}$$

23. (continued)

$T^8 = T^7 \cdot T$

$$= \begin{bmatrix} 12/25 & 18/125 & 4/25 & 27/125 \\ 2/5 & 6/25 & 0 & 9/25 \\ 9/25 & 0 & 12/25 & 4/25 \\ 0 & 4/25 & 9/25 & 12/25 \end{bmatrix} \begin{bmatrix} 0 & 2/5 & 0 & 3/5 \\ 0 & 0 & 0 & 1 \\ 1 & 0 & 0 & 0 \\ 0 & 0 & 1 & 0 \end{bmatrix}$$

$$= \begin{bmatrix} 4/25 & 24/125 & 27/125 & 54/125 \\ 0 & 4/25 & 9/25 & 12/25 \\ 12/25 & 18/125 & 4/25 & 27/125 \\ 9/25 & 0 & 12/25 & 4/25 \end{bmatrix}$$

$T^9 = T^8 \cdot T$

$$= \begin{bmatrix} 4/25 & 24/125 & 27/125 & 54/125 \\ 0 & 4/25 & 9/25 & 12/25 \\ 12/25 & 18/125 & 4/25 & 27/125 \\ 9/25 & 0 & 12/25 & 4/25 \end{bmatrix} \begin{bmatrix} 0 & 2/5 & 0 & 3/5 \\ 0 & 0 & 0 & 1 \\ 1 & 0 & 0 & 0 \\ 0 & 0 & 1 & 0 \end{bmatrix}$$

$$= \begin{bmatrix} 27/125 & 8/125 & 54/125 & 36/125 \\ 9/25 & 0 & 12/25 & 4/25 \\ 4/25 & 24/125 & 27/125 & 54/125 \\ 12/25 & 18/125 & 4/25 & 27/125 \end{bmatrix}$$

$T^{10} = T^9 \cdot T$

$$= \begin{bmatrix} 27/125 & 8/125 & 54/125 & 36/125 \\ 9/25 & 0 & 12/25 & 4/25 \\ 4/25 & 24/125 & 27/125 & 54/125 \\ 12/25 & 18/125 & 4/25 & 27/125 \end{bmatrix} \begin{bmatrix} 0 & 2/5 & 0 & 3/5 \\ 0 & 0 & 0 & 1 \\ 1 & 0 & 0 & 0 \\ 0 & 0 & 1 & 0 \end{bmatrix}$$

$$= \begin{bmatrix} 54/125 & 54/625 & 36/125 & 121/625 \\ 12/25 & 18/125 & 4/25 & 27/125 \\ 27/125 & 8/125 & 54/125 & 36/125 \\ 4/25 & 24/125 & 27/125 & 54/125 \end{bmatrix}$$

T^{10} contains all positive entries, so T is regular.

b) Solve the system of equations

$$[p_1 \quad p_2 \quad p_3 \quad p_4] = [p_1 \quad p_2 \quad p_3 \quad p_4] \begin{bmatrix} 0 & 2/5 & 0 & 3/5 \\ 0 & 0 & 0 & 1 \\ 1 & 0 & 0 & 0 \\ 0 & 0 & 1 & 0 \end{bmatrix}$$

$p_1 + p_2 + p_3 + p_4 = 1.$

Multiplying and equating like components we have

$$p_1 = p_3$$
$$p_2 = \frac{2}{5}p_1$$
$$p_3 = p_4$$
$$p_4 = \frac{3}{5}p_1 + p_2$$
$$p_1 + p_2 + p_3 + p_4 = 1.$$

23. (continued)

The solution is $p_1 = 5/17$, $p_2 = 2/17$, $p_3 = 5/17$, $p_4 = 5/17$. The steady state vector is [5/17 2/17 5/17 5/17].

25. $T = \begin{bmatrix} 0 & 0 & 0 & 1 \\ 1/2 & 0 & 1/2 & 0 \\ 0 & 1/2 & 0 & 1/2 \\ 0 & 0 & 1 & 0 \end{bmatrix}$

a) Draw the transition diagram.

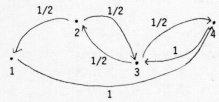

All the states communicate with each other, so T might be irreducible. Since all entries on the main diagonal are zero, we calculate successive powers of T to determine if T is regular.

$T^2 = \begin{bmatrix} 0 & 0 & 0 & 1 \\ 1/2 & 0 & 1/2 & 0 \\ 0 & 1/2 & 0 & 1/2 \\ 0 & 0 & 1 & 0 \end{bmatrix} \begin{bmatrix} 0 & 0 & 0 & 1 \\ 1/2 & 0 & 1/2 & 0 \\ 0 & 1/2 & 0 & 1/2 \\ 0 & 0 & 1 & 0 \end{bmatrix}$

$= \begin{bmatrix} 0 & 0 & 1 & 0 \\ 0 & 1/4 & 0 & 3/4 \\ 1/4 & 0 & 3/4 & 0 \\ 0 & 1/2 & 0 & 1/2 \end{bmatrix}$

$T^3 = T^2 \cdot T$

$= \begin{bmatrix} 0 & 0 & 1 & 0 \\ 0 & 1/4 & 0 & 3/4 \\ 1/4 & 0 & 3/4 & 0 \\ 0 & 1/2 & 0 & 1/2 \end{bmatrix} \begin{bmatrix} 0 & 0 & 0 & 1 \\ 1/2 & 0 & 1/2 & 0 \\ 0 & 1/2 & 0 & 1/2 \\ 0 & 0 & 1 & 0 \end{bmatrix}$

$= \begin{bmatrix} 0 & 1/2 & 0 & 1/2 \\ 1/8 & 0 & 7/8 & 0 \\ 0 & 3/8 & 0 & 5/8 \\ 1/4 & 0 & 3/4 & 0 \end{bmatrix}$

$T^4 = T^3 \cdot T$

$= \begin{bmatrix} 0 & 1/2 & 0 & 1/2 \\ 1/8 & 0 & 7/8 & 0 \\ 0 & 3/8 & 0 & 5/8 \\ 1/4 & 0 & 3/4 & 0 \end{bmatrix} \begin{bmatrix} 0 & 0 & 0 & 1 \\ 1/2 & 0 & 1/2 & 0 \\ 0 & 1/2 & 0 & 1/2 \\ 0 & 0 & 1 & 0 \end{bmatrix}$

$= \begin{bmatrix} 1/4 & 0 & 3/4 & 0 \\ 0 & 7/16 & 0 & 9/16 \\ 3/16 & 0 & 13/16 & 0 \\ 0 & 3/8 & 0 & 5/8 \end{bmatrix}$

25. (continued)

The pattern of zeros in successive powers of T will continue to alternate between the patterns in T^3 and T^4. Therefore, there is no power of T that will contain all positive entries, and T is not regular.

27. The transition matrix is

$$T = \begin{array}{cc} & \begin{array}{cc} I & II \end{array} \\ & \begin{bmatrix} 0.5 & 0.5 \\ 0.4 & 0.6 \end{bmatrix} \end{array}.$$

(See Exercise Set 9.1, Exercise 1.)

T is regular. (It is irreducible and at least one entry on the main diagonal is nonzero.) "In the long run" the state vector approaches a steady state vector P. The percent of the town's business that Plumber I will have is the (1, 1)-element of P. To find P we solve the following system of equations:

$$[p_1 \quad p_2] = [p_1 \quad p_2] \begin{bmatrix} 0.5 & 0.5 \\ 0.4 & 0.6 \end{bmatrix}$$

$p_1 + p_2 = 1$

Multiplying and equating like components we have

$p_1 = 0.5p_1 + 0.4p_2$

$p_2 = 0.5p_1 + 0.6p_2$

$p_1 + p_2 = 1.$

Simplifying, we obtain

$0.5p_1 - 0.4p_2 = 0$

$-0.5p_1 + 0.4p_2 = 0$

$p_1 + p_2 = 1.$

The solution of the system is $p_1 = 4/9$, $p_2 = 5/9$. In the long run Plumber I will have 4/9, or approximately 44%, of the town's plumbing business.

29. The transition matrix is

$$T = \begin{array}{ccc} & \begin{array}{ccc} E & J & R \end{array} \\ & \begin{bmatrix} 0.5 & 0.2 & 0.3 \\ 0.4 & 0.4 & 0.2 \\ 0.3 & 0.1 & 0.6 \end{bmatrix} \end{array}.$$

(See Exercise Set 9.1, Exercise 3.)

T is regular. (It is irreducible and at least one entry on the main diagonal is nonzero.) To find the percent of the songs played in a week that will be jazz we find the (1, 2)-element of the steady state vector. We solve the following system of equations:

$$[p_1 \quad p_2 \quad p_3] = [p_1 \quad p_2 \quad p_3] \begin{bmatrix} 0.5 & 0.2 & 0.3 \\ 0.4 & 0.4 & 0.2 \\ 0.3 & 0.1 & 0.6 \end{bmatrix}$$

$p_1 + p_2 + p_3 = 1$

29. (continued)

Multiply and equate like components.

$$p_1 = 0.5p_1 + 0.4p_2 + 0.3p_3$$
$$p_2 = 0.2p_1 + 0.4p_2 + 0.1p_3$$
$$p_3 = 0.3p_1 + 0.2p_2 + 0.6p_3$$
$$p_1 + p_2 + p_3 = 1$$

Simplify.

$$-0.5p_1 + 0.4p_2 + 0.3p_3 = 0$$
$$0.2p_1 - 0.6p_2 + 0.1p_3 = 0$$
$$0.3p_1 + 0.2p_2 - 0.4p_3 = 0$$
$$p_1 + p_2 + p_3 = 1$$

The solution is $p_1 = 0.4$, $p_2 = 0.2$, $p_3 = 0.4$. In a week 0.2, or 20%, of the songs played will be jazz.

31. T is regular. To find the distribution at the end of the day, find the steady state vector by solving the following system of equations:

$$[p_1 \quad p_2 \quad p_3] = [p_1 \quad p_2 \quad p_3]\begin{bmatrix} 1/2 & 1/3 & 1/6 \\ 1/3 & 1/3 & 1/3 \\ 1/2 & 1/4 & 1/4 \end{bmatrix}$$

$$p_1 + p_2 + p_3 = 1$$

Multiply and equate like components.

$$p_1 = \frac{1}{2}p_1 + \frac{1}{3}p_2 + \frac{1}{2}p_3$$
$$p_2 = \frac{1}{3}p_1 + \frac{1}{3}p_2 + \frac{1}{4}p_3$$
$$p_3 = \frac{1}{6}p_1 + \frac{1}{3}p_2 + \frac{1}{4}p_3$$

$$p_1 + p_2 + p_3 = 1$$

Simplify.

$$-\frac{1}{2}p_1 + \frac{1}{3}p_2 + \frac{1}{2}p_3 = 0$$
$$\frac{1}{3}p_1 - \frac{2}{3}p_2 + \frac{1}{4}p_3 = 0$$
$$\frac{1}{6}p_1 + \frac{1}{3}p_2 - \frac{3}{4}p_3 = 0$$

$$p_1 + p_2 + p_3 = 1$$

The solution is $p_1 = 30/67$, $p_2 = 21/67$, $p_3 = 16/67$. By the end of the day 30/67 of the taxis will be in zone I, 21/67 will be in zone II, and 16/67 will be in zone III.

33. The transition matrix is regular. In the long run, the percentage of sales that will be strawberry cones is the (1, 3)-element of the steady state vector P. Find P by solving the following system of equations:

$$[p_1 \quad p_2 \quad p_3] = [p_1 \quad p_2 \quad p_3]\begin{bmatrix} 0 & 1/2 & 1/2 \\ 1/5 & 2/5 & 2/5 \\ 1/3 & 0 & 2/3 \end{bmatrix}$$

$$p_1 + p_2 + p_3 = 1$$

Multiply and equate like components.

33. (continued)

$$p_1 = \frac{1}{5}p_2 + \frac{1}{3}p_3$$
$$p_2 = \frac{1}{2}p_1 + \frac{2}{5}p_2$$
$$p_3 = \frac{1}{2}p_1 + \frac{2}{5}p_2 + \frac{2}{3}p_3$$

$$p_1 + p_2 + p_3 = 1$$

Simplify.

$$-p_1 + \frac{1}{5}p_2 + \frac{1}{3}p_3 = 0$$
$$\frac{1}{2}p_1 - \frac{3}{5}p_2 = 0$$
$$\frac{1}{2}p_1 + \frac{2}{5}p_2 - \frac{1}{3}p_3 = 0$$

$$p_1 + p_2 + p_3 = 1$$

The solution is $p_1 = 3/13$, $p_2 = 5/26$, $p_3 = 15/26$. In the long run 15/26, or approximately 58%, of sales will be strawberry cones.

35. The transition matrix is

$$T = \begin{array}{c} FL \\ ME \\ AZ \end{array} \begin{array}{c} \begin{array}{ccc} FL & ME & AZ \end{array} \\ \begin{bmatrix} 0.4 & 0 & 0.6 \\ 0.3 & 0.5 & 0.2 \\ 0.2 & 0.6 & 0.2 \end{bmatrix} \end{array}$$

T is regular. In the long run, the probability they will vacation the next year in Maine is the (1, 2)-element of the steady state vector P. Find P by solving the following system of equations:

$$[p_1 \quad p_2 \quad p_3] = [p_1 \quad p_2 \quad p_3]\begin{bmatrix} 0.4 & 0 & 0.6 \\ 0.3 & 0.5 & 0.2 \\ 0.2 & 0.6 & 0.2 \end{bmatrix}$$

$$p_1 + p_2 + p_3 = 1$$

Multiply and equate like components.

$$p_1 = 0.4p_1 + 0.3p_2 + 0.2p_3$$
$$p_2 = 0.5p_2 + 0.6p_3$$
$$p_3 = 0.6p_1 + 0.2p_2 + 0.2p_3$$

$$p_1 + p_2 + p_3 = 1$$

Simplify.

$$-0.6p_1 + 0.3p_2 + 0.2p_3 = 0$$
$$-0.5p_2 + 0.6p_3 = 0$$
$$0.6p_1 + 0.2p_2 - 0.8p_3 = 0$$
$$p_1 + p_2 + p_3 = 1$$

The solution is $p_1 = 14/47$, $p_2 = 18/47$, $p_3 = 15/47$. In the long run, the probability they will vacation next year in Maine is 18/47, or approximately 38%.

37. The transition matrix is regular. The proportion of the population that will be in the middle class in the long run is the (1, 2)-element of the steady state vector P. Find P by solving the following system of equations:

$$[p_1 \quad p_2 \quad p_3] = [p_1 \quad p_2 \quad p_3] \begin{bmatrix} 0.55 & 0.40 & 0.05 \\ 0.20 & 0.60 & 0.20 \\ 0.20 & 0.30 & 0.50 \end{bmatrix}$$

$$p_1 + p_2 + p_3 = 1$$

Multiply and equate like components.

$$p_1 = 0.55p_1 + 0.20p_2 + 0.20p_3$$
$$p_2 = 0.40p_1 + 0.60p_2 + 0.30p_3$$
$$p_3 = 0.05p_1 + 0.20p_2 + 0.50p_3$$
$$p_1 + p_2 + p_3 = 1$$

Simplify.

$$-0.45p_1 + 0.20p_2 + 0.20p_3 = 0$$
$$0.40p_1 - 0.40p_2 + 0.30p_3 = 0$$
$$0.05p_1 + 0.20p_2 - 0.50p_3 = 0$$
$$p_1 + p_2 + p_3 = 1$$

The solution is $p_1 = 4/13$, $p_2 = 43/91$, $p_3 = 20/91$.

In the long run 43/91, or approximately 0.47, of the population will be in the middle class.

39. The two states must communicate with each other. Since all the elements of the transition matrix must be 1's and 0's the only possible transition diagram is the following:

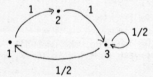

This gives the transition matrix $\begin{bmatrix} 0 & 1 \\ 1 & 0 \end{bmatrix}$.

41. False. All regular transition matrices are irreducible, but not all irreducible transition matrices are regular.

43. True. If the (i, i)-element of a transition matrix is 1, then it is not possible to move out of state i and the matrix is not regular.

45. False. Consider the following transition diagram:

This chain is irreducible. (All the states can communicate with each other.) The transition matrix is

$$T = \begin{bmatrix} 0 & 1 & 0 \\ 0 & 0 & 1 \\ 1/2 & 0 & 1/2 \end{bmatrix}.$$

There are more 0's than nonzero elements. Nevertheless, T is regular since at least one entry on the main diagonal is nonzero.

47. This is the matrix in Exercise 14 of this Exercise Set. T is irreducible and T^5 contains all positive entries, so T is regular. After 100 repetitions the rows of T will approach the steady state vector. We find this vector by solving the following system of equations:

$$[p_1 \quad p_2 \quad p_3] = [p_1 \quad p_2 \quad p_3] \begin{bmatrix} 0 & 1 & 0 \\ 0 & 0 & 1 \\ 2/5 & 3/5 & 0 \end{bmatrix}$$

$$p_1 + p_2 + p_3 = 1$$

Multiply and equate like components.

$$p_1 = \frac{2}{5}p_3$$
$$p_2 = p_1 + \frac{3}{5}p_3$$
$$p_3 = p_2$$
$$p_1 + p_2 + p_3 = 1$$

Simplify.

$$p_1 - \frac{2}{5}p_3 = 0$$
$$p_1 - p_2 + \frac{3}{5}p_3 = 0$$
$$p_2 - p_3 = 0$$
$$p_1 + p_2 + p_3 = 1$$

The solution is $p_1 = 1/6 \approx 0.167$, $p_2 = 5/12 \approx 0.417$, $p_3 = 5/12 \approx 0.417$. Thus, the rows of T^{100} to the nearest thousandths are each [0.167 0.417 0.417].

49. a) Let p(i, j) = the probability of going from state i to state j in a one-minute interval.

p(0, 0) = p(no one will join the queue) = 1 - 1/4 = 3/4

p(0, 1) = p(a person will join the queue) = 1/4

p(0, 2) = p(0, 3) = 0 (No more than one person joins the queue in every one-minute interval.)

p(1, 0) = p(one person is served and no one joins the queue) = $\frac{1}{3} \cdot \frac{3}{4} = \frac{1}{4}$

p(1, 1) = p(one person is served and a person joins the queue or no one is served and no one joins the queue) = $\frac{1}{3} \cdot \frac{1}{4} + \frac{2}{3} \cdot \frac{3}{4} = \frac{7}{12}$

p(1, 2) = p(no one is served and a person joins the queue) = $\frac{2}{3} \cdot \frac{1}{4} = \frac{1}{6}$

p(1, 3) = 0 (No more than one person is served or joins the queue in every one-minute interval.)

p(2, 0) = 0 (No more than one person is served in every one-minute interval.)

p(2, 1), p(2, 2), and p(2, 3) are computed exactly as p(1, 0), p(1, 1), and p(1, 2), respectively. Then p(2, 1) = 1/4, p(2, 2) = 7/12, and p(2, 3) = 1/6.

p(3, 0) = p(3, 1) = 0 (No more than one person is served in any one-minute interval.)

49. (continued)

p(3, 2) = p(a person is served) = 1/3 (Recall that if there are already 3 customers there, other customers will not stop.)

p(3, 3) = p(no one is served) = 1 - 1/3 = 2/3.

We can now write the transition matrix.

$$T = \begin{bmatrix} 3/4 & 1/4 & 0 & 0 \\ 1/4 & 7/12 & 1/6 & 0 \\ 0 & 1/4 & 7/12 & 1/6 \\ 0 & 0 & 1/3 & 2/3 \end{bmatrix}$$

b) It can be shown that T is regular. Therefore, the long-run probabilities for each state will be the elements of the steady state vector. Solve the following system of equations to find the steady state vector:

$$[p_1 \quad p_2 \quad p_3 \quad p_4]$$

$$= [p_1 \quad p_2 \quad p_3 \quad p_4] \begin{bmatrix} 3/4 & 1/4 & 0 & 0 \\ 1/4 & 7/12 & 1/6 & 0 \\ 0 & 1/4 & 7/12 & 1/6 \\ 0 & 0 & 1/3 & 2/3 \end{bmatrix}$$

$$p_1 + p_2 + p_3 + p_4 = 1$$

Multiply and equate like components.

$$p_1 = \frac{3}{4}p_1 + \frac{1}{4}p_2$$

$$p_2 = \frac{1}{4}p_1 + \frac{7}{12}p_2 + \frac{1}{4}p_3$$

$$p_3 = \frac{1}{6}p_2 + \frac{7}{12}p_3 + \frac{1}{3}p_4$$

$$p_4 = \frac{1}{6}p_3 + \frac{2}{3}p_4$$

$$p_1 + p_2 + p_3 + p_4 = 1$$

Simplify.

$$-\frac{1}{4}p_1 + \frac{1}{4}p_2 \qquad\qquad = 0$$

$$\frac{1}{4}p_1 - \frac{5}{12}p_2 + \frac{1}{4}p_3 \qquad = 0$$

$$\frac{1}{6}p_2 - \frac{5}{12}p_3 + \frac{1}{3}p_4 = 0$$

$$\frac{1}{6}p_3 - \frac{1}{3}p_4 = 0$$

$$p_1 + p_2 + p_3 + p_4 = 1$$

The solution is $p_1 = \frac{1}{3}$, $p_2 = \frac{1}{3}$, $p_3 = \frac{2}{9}$, $p_4 = \frac{1}{9}$. These are the long-run probabilities for states 0, 1, 2, and 3, respectively.

c) The probability distribution is

x_i	0	1	2	3
p_i	$\frac{1}{3}$	$\frac{1}{3}$	$\frac{2}{9}$	$\frac{1}{9}$

$$E(X) = 0 \cdot \frac{1}{3} + 1 \cdot \frac{1}{3} + 2 \cdot \frac{2}{9} + 3 \cdot \frac{1}{9}$$

$$= 0 + \frac{1}{3} + \frac{4}{9} + \frac{1}{3} = \frac{10}{9}, \text{ or } 1\frac{1}{9}$$

Exercise Set 9.3

1. $\begin{bmatrix} 1 & 0 \\ 0 & 1 \end{bmatrix}$

a) States 1 and 2 are both absorbing states since the system can move from state 1 only to state 1 and from state 2 only to state 2.

b) Since there are no nonabsorbing states, this matrix must represent an absorbing chain.

3. $\begin{bmatrix} 0 & 1 & 0 \\ 0 & 0 & 1 \\ 1 & 0 & 0 \end{bmatrix}$

a) There are no absorbing states, since there is not a 1 on the main diagonal.

b) The matrix does not represent an absorbing chain, because there are no absorbing states.

5. $\begin{bmatrix} 0 & 1 & 0 \\ 0 & 0 & 1 \\ 0 & 0 & 1 \end{bmatrix}$

a) The system can move from state 3 only to state 3, so this is an absorbing state.

b) The probability is 1 that the system will move from state 2 to state 3, the absorbing state. State 1 does not lead directly to state 3, but the system always moves from state 1 to state 2 and then from state 2 to state 3. Thus, the matrix represents an absorbing chain.

7. $\begin{bmatrix} 1/2 & 1/2 & 0 & 0 \\ 0 & 1 & 0 & 0 \\ 0 & 0 & 1/2 & 1/2 \\ 0 & 0 & 0 & 1 \end{bmatrix}$

a) The system can move from state 2 only to state 2 and from state 4 only to state 4, so states 2 and 4 are absorbing states.

b) There is a probability of 1/2 that the system will move from state 1 to state 2, an absorbing state. There is also a probability of 1/2 that the system will move from state 3 to state 4, another absorbing state. Since it is possible to go from any nonabsorbing state to some absorbing state, the matrix represents an absorbing chain.

9. $\begin{bmatrix} 0 & 1 & 0 & 0 \\ 1 & 0 & 0 & 0 \\ 0 & 0 & 1 & 0 \\ 0 & 0 & 0 & 1 \end{bmatrix}$

a) States 3 and 4 are absorbing states, since the system can move from state 3 only to state 3 and from state 4 only to state 4.

<u>9</u>. (continued)

b) States 1 and 2 communicate only with each other, so the matrix does not represent an absorbing chain.

<u>11</u>. There are four states of the system: (0, 3), (1, 2), (2, 1), and (3, 0). The first element in each ordered pair represents the amount Gambler A has and the second element represents Gambler B's holdings.

Since the total is always 3 we can look at the problem from Gambler A's perspective and consider the states to be 0, 1, 2, and 3. The transition matrix is

$$T = \begin{array}{c} \\ \$0 \\ \$1 \\ \$2 \\ \$3 \end{array} \begin{array}{cccc} \$0 & \$1 & \$2 & \$3 \\ \left[\begin{array}{cccc} 1 & 0 & 0 & 0 \\ 2/5 & 0 & 3/5 & 0 \\ 0 & 2/5 & 0 & 3/5 \\ 0 & 0 & 0 & 1 \end{array}\right] \end{array}.$$

Until one gambler is broke, there is a probability of 2/5 that A will lose the toss and move to the state below and a probability of 3/5 that he will win the toss and move to the state above. There are two absorbing states, $0 and $3. It is possible to move from state 2 ($1) to state 1 ($0) and from state 3 ($2) to state 4 ($3), so the chain is absorbing.

The probability that B will go broke is the same as the probability that A will go from state 2 ($1) to state 4 ($3). To find this probability we square T successively. (We will express the elements of T in decimal notation.)

$$T^2 = \left[\begin{array}{cccc} 1 & 0 & 0 & 0 \\ 0.4 & 0.24 & 0 & 0.36 \\ 0.16 & 0 & 0.24 & 0.6 \\ 0 & 0 & 0 & 1 \end{array}\right]$$

$$T^4 = \left[\begin{array}{cccc} 1 & 0 & 0 & 0 \\ 0.496 & 0.0576 & 0 & 0.4464 \\ 0.1984 & 0 & 0.0576 & 0.744 \\ 0 & 0 & 0 & 1 \end{array}\right]$$

$$T^8 = \left[\begin{array}{cccc} 1 & 0 & 0 & 0 \\ 0.5246 & 0.0033 & 0 & 0.4721 \\ 0.2098 & 0 & 0.0033 & 0.7869 \\ 0 & 0 & 0 & 1 \end{array}\right]$$

$$T^{16} = \left[\begin{array}{cccc} 1 & 0 & 0 & 0 \\ 0.5263 & 0.0000+ & 0 & 0.4737 \\ 0.2105 & 0 & 0.0000+ & 0.7895 \\ 0 & 0 & 0 & 1 \end{array}\right]$$

After 16 transitions we see that the probability that B will go broke is approximately 0.4737.

<u>13</u>. This system has four states which can be described as follows:

State 1: The engine is being used

State 2: The engine is undergoing the first test

State 3: The engine is undergoing the second test

State 4: The engine is scrapped

The transition matrix is

$$T = \begin{array}{c} \\ U \\ T1 \\ T2 \\ S \end{array} \begin{array}{cccc} U & T1 & T2 & S \\ \left[\begin{array}{cccc} 1 & 0 & 0 & 0 \\ 0 & 0 & 0.95 & 0.05 \\ 0.8 & 0.2 & 0 & 0 \\ 0 & 0 & 0 & 1 \end{array}\right] \end{array}$$

States 1 and 4 are absorbing states, and it is possible to go from states 2 and 3 to state 4, so this is an absorbing Markov chain. To examine the long-run probabilities, we square T successively.

$$T^2 = \left[\begin{array}{cccc} 1 & 0 & 0 & 0 \\ 0.76 & 0.19 & 0 & 0.05 \\ 0.8 & 0 & 0.19 & 0.1 \\ 0 & 0 & 0 & 1 \end{array}\right]$$

$$T^4 = \left[\begin{array}{cccc} 1 & 0 & 0 & 0 \\ 0.9044 & 0.0361 & 0 & 0.0595 \\ 0.952 & 0 & 0.0361 & 0.0119 \\ 0 & 0 & 0 & 1 \end{array}\right]$$

$$T^8 = \left[\begin{array}{cccc} 1 & 0 & 0 & 0 \\ 0.9370 & 0.0013 & 0 & 0.0616 \\ 0.9864 & 0 & 0.0013 & 0.0123 \\ 0 & 0 & 0 & 1 \end{array}\right]$$

$$T^{16} = \left[\begin{array}{cccc} 1 & 0 & 0 & 0 \\ 0.9383 & 0.0000+ & 0 & 0.0617 \\ 0.9877 & 0 & 0.0000+ & 0.0123 \\ 0 & 0 & 0 & 1 \end{array}\right]$$

The probability that an engine that is taking the first test will eventually be scrapped is the long-run probability of going from state 2 to state 4, or approximately 0.0617. The probability that an engine currently taking the second test will eventually be used is the long-run probability of going from state 3 to state 1, or approximately 0.9877.

15. The system has four states which can be described as follows:

State 1: Being a freshman

State 2: Being a sophomore

State 3: Dropping out

State 4: Graduating

The transition matrix is

$$T = \begin{array}{c}\\F\\S\\D\\G\end{array}\begin{array}{cccc}F & S & D & G\end{array}\\\left[\begin{array}{cccc}0.1 & 0.7 & 0.2 & 0 \\ 0 & 0.3 & 0.1 & 0.6 \\ 0 & 0 & 1 & 0 \\ 0 & 0 & 0 & 1\end{array}\right]$$

States 3 and 4 are absorbing states, and it is possible to go from states 1 and 2 to both state 3 and state 4, so this is an absorbing Markov chain. The probability that someone who is currently a freshman will graduate (that is, will go from state 1 to state 4) is found by successively squaring T.

$$T^2 = \left[\begin{array}{cccc}0.01 & 0.28 & 0.29 & 0.42 \\ 0 & 0.09 & 0.13 & 0.78 \\ 0 & 0 & 1 & 0 \\ 0 & 0 & 0 & 1\end{array}\right]$$

$$T^4 = \left[\begin{array}{cccc}0.0001 & 0.028 & 0.3293 & 0.6426 \\ 0 & 0.0081 & 0.1417 & 0.8502 \\ 0 & 0 & 1 & 0 \\ 0 & 0 & 0 & 1\end{array}\right]$$

$$T^8 = \left[\begin{array}{cccc}0.0000+ & 0.0002 & 0.3333 & 0.6665 \\ 0 & 0.0001 & 0.1428 & 0.8571 \\ 0 & 0 & 1 & 0 \\ 0 & 0 & 0 & 1\end{array}\right]$$

$$T^{16} = \left[\begin{array}{cccc}0.0000+ & 0.0000+ & 0.3333 & 0.6667 \\ 0 & 0.0000+ & 0.1429 & 0.8751 \\ 0 & 0 & 1 & 0 \\ 0 & 0 & 0 & 1\end{array}\right]$$

The probability of going from state 1 to state 4 is approximately 0.6667.

17. This system has four states which can be described as follows:

State 1: The mouse is in room I

State 2: The mouse is in room II

State 3: The mouse is in room III

State 4: The mouse is in the trap

17. (continued)

The transition matrix is

$$T = \begin{array}{c}\\I\\II\\III\\T\end{array}\begin{array}{cccc}I & II & III & T\end{array}\\\left[\begin{array}{cccc}0 & 1 & 0 & 0 \\ 2/3 & 0 & 1/3 & 0 \\ 0 & 1/2 & 0 & 1/2 \\ 0 & 0 & 0 & 1\end{array}\right]$$

State 4 is an absorbing state, and it is possible to go from each of the other states to state 4, so this is an absorbing Markov chain. In order to determine the expected number of visits to room I before the mouse enters the "absorbing" state we must compute the matrix $M = (I - NN)^{-1}$ used in the optional method described in this section for finding the limiting matrix for a chain. We must first reorder the states so the absorbing state is the first state of the transition matrix. Do this as follows:

State 1: The mouse is in the trap

State 2: The mouse is in room I

State 3: The mouse is in room II

State 4: The mouse is in room III

Write the transition matrix and divide it into submatrices with the absorbing state forming the 1×1 identity matrix in the upper left-hand corner.

$$T = \begin{array}{c}\\T\\I\\II\\III\end{array}\begin{array}{cccc}T & I & II & III\end{array}\\\left[\begin{array}{c|ccc}1 & 0 & 0 & 0 \\ \hline 0 & 0 & 1 & 0 \\ 0 & 2/3 & 0 & 1/3 \\ 1/2 & 0 & 1/2 & 0\end{array}\right]$$

NN is the 3×3 submatrix in the lower right-hand corner.

$$M = (I - NN)^{-1} = \left(\left[\begin{array}{ccc}1 & 0 & 0 \\ 0 & 1 & 0 \\ 0 & 0 & 1\end{array}\right] - \left[\begin{array}{ccc}0 & 1 & 0 \\ 2/3 & 0 & 1/3 \\ 0 & 1/2 & 0\end{array}\right]\right)^{-1}$$

$$= \left[\begin{array}{ccc}1 & -1 & 0 \\ -2/3 & 1 & -1/3 \\ 0 & -1/2 & 1\end{array}\right]^{-1} = \left[\begin{array}{ccc}5 & 6 & 2 \\ 4 & 6 & 2 \\ 2 & 3 & 2\end{array}\right]$$

Since the mouse begins in room II, the (2, 1)-element of M gives the expected number of visits to room I before the mouse enters the "absorbing" state. The expected number of visits is 4.

<u>19</u>. This system has 6 states: (0, 5), (1, 4), (2, 3), (3, 2), (4, 1), (5, 0). The first element of each ordered pair represent the amount that the first gambler has and the second element represents the second gambler's holdings. Since the total is always five we can look at the problem from the first gambler's perspective and consider the states to be 0, 1, 2, 3, 4, and 5. The transition matrix is

$$T = \begin{array}{c} \\ 0 \\ 1 \\ 2 \\ 3 \\ 4 \\ 5 \end{array} \begin{array}{cccccc} 0 & 1 & 2 & 3 & 4 & 5 \\ \left[\begin{array}{cccccc} 1 & 0 & 0 & 0 & 0 & 0 \\ 0.5 & 0 & 0.5 & 0 & 0 & 0 \\ 0 & 0.5 & 0 & 0.5 & 0 & 0 \\ 0 & 0 & 0.5 & 0 & 0.5 & 0 \\ 0 & 0 & 0 & 0.5 & 0 & 0.5 \\ 0 & 0 & 0 & 0 & 0 & 1 \end{array}\right] \end{array}$$

This is an absorbing Markov chain. The probability that the first gambler will lose (go broke) is the long-run probability of going from state 3 ($2) to state 1 ($0). We determine this probability by squaring T successively. We find that

$$T^{16} = \begin{bmatrix} 1 & 0 & 0 & 0 & 0 & 0 \\ 0.7865 & 0.0093 & 0 & 0.0151 & 0 & 0.1891 \\ 0.5824 & 0 & 0.0244 & 0 & 0.0151 & 0.3782 \\ 0.3782 & 0.0151 & 0 & 0.0244 & 0 & 0.5824 \\ 0.1891 & 0 & 0.0151 & 0 & 0.0093 & 0.7865 \\ 0 & 0 & 0 & 0 & 0 & 1 \end{bmatrix}$$

The probability that the first gambler will lose is 0.5824, or approximately 60%.

Exercise Set 10.1

<u>1</u>. a) Both players have the same two choices:

To hold out one finger, denoted 1, or

To hold out two fingers, denoted 2

There are four possible outcomes in this game. They can be represented by the ordered pairs (1, 1), (1, 2), (2, 1), and (2, 2) where Player I's strategy is the first element in each pair. Outcomes (1, 1) and (2, 2) have a payoff of 5¢ for Player I and -5¢ for Player II represented by (5, -5). (1, 2) and (2, 1) have a payoff of -5¢ for Player I and 5¢ for Player II represented by (-5, 5). We can represent this with a game tree. Since both players make their choices independently, we can let either one move first. We will choose Player I to move first.

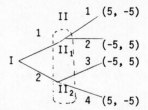

Note that at II_1 and II_2 Player II has the same set of choices, {1, 2}.

b) Each player can hold out one finger or two fingers. The set of pure strategies for each player is {1, 2}.

c) The rows of the payoff matrix correspond to Player I's strategies, and the columns correspond to Player II's strategies. Since this is a zero-sum game, the entries of the matrix can be written as the payoff to Player I.

$$I \begin{array}{c} \\ 1 \\ 2 \end{array} \begin{array}{cc} \overset{II}{\overset{1 \qquad 2}{\begin{bmatrix} 5 & -5 \\ -5 & 5 \end{bmatrix}}} \end{array}$$

<u>3</u>. a) Both players have the same two choices:

A "safe" strategy, denoted S, or

A "double cross" strategy, denoted D

There are four possible outcomes represented by the ordered pairs (S, S), (S, D), (D, S), and (D, D) where contestant I's strategy is the first element in each pair. The pairs have payoffs of (10, 10), (10, 1,000,000), (1,000,000, 10), and (-1, -1), respectively.

We will draw the game tree, choosing contestant I to move first.

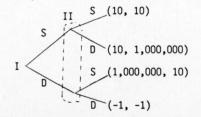

<u>3</u>. (continued)

b) Each player can choose a safe strategy or a double cross strategy. The set of pure strategies for each player is {S, D}.

c) The rows of the payoff matrix correspond to Player I's strategies, and the columns correspond to Player II's strategies. The entries are the payoffs for each pair of strategies.

$$I \begin{array}{c} \\ S \\ D \end{array} \begin{array}{cc} \overset{II}{\overset{S \qquad\qquad D}{\begin{bmatrix} (10, 10) & (10, 1{,}000{,}000) \\ (1{,}000{,}000, 10) & (-1, -1) \end{bmatrix}}} \end{array}$$

<u>5</u>. a) Both stations have the same two choices:

To switch programming, denoted S, or

Not to switch programming, denoted N

There are four possible outcomes represented by the ordered pairs (S, S), (S, N), (N, S), and (N, N) where ROWE's strategy is the first element in each pair. The pairs have payoffs of (0.7, 0.3), (0.65, 0.35), (0.6, 0.4), and (0.4, 0.6), respectively. We will draw the game tree, choosing ROWE to move first.

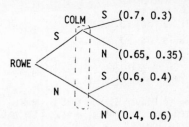

b) Each station can choose to switch its programming or not to switch its programming. The set of pure strategies for each station is {S, N}.

c) The rows of the payoff matrix correspond to ROWE's strategies, and the columns correspond to COLM's strategies. Since this is a constant-sum game, the entries of the matrix can be written as the payoffs to ROWE.

$$ROWE \begin{array}{c} \\ S \\ N \end{array} \begin{array}{cc} \overset{COLM}{\overset{S \qquad N}{\begin{bmatrix} 0.7 & 0.65 \\ 0.6 & 0.4 \end{bmatrix}}} \end{array}$$

Exercise Set 10.2

1.

$$
\begin{array}{c}
 & \text{II} \\
 & \begin{array}{ccc} \beta_1 & \beta_2 & \beta_3 \end{array} \\
\text{I}\begin{array}{c} \alpha_1 \\ \alpha_2 \\ \alpha_3 \end{array} \left[\begin{array}{ccc} 3 & 4 & 7 \\ 2 & 1 & 3 \\ 2 & 5 & 3 \end{array}\right]
\end{array}
$$

Comparing α_1 and α_2 we see that, regardless of what strategy Player II chooses, Player I can always gain more by playing α_1. Thus, strategy α_1 dominates α_2, or $\alpha_1 > \alpha_2$, and α_2 can be eliminated from the matrix. Neither α_1 nor α_3 dominates the other. We rewrite the matrix to show the elimination of α_2.

$$
\begin{array}{c}
 & \text{II} \\
 & \begin{array}{ccc} \beta_1 & \beta_2 & \beta_3 \end{array} \\
\text{I}\begin{array}{c} \alpha_1 \\ \alpha_2 \\ \alpha_3 \end{array} \left[\begin{array}{ccc} 3 & 4 & 7 \\ 2 & 1 & 3 \\ 2 & 5 & 3 \end{array}\right]
\end{array}
$$

Now we turn our attention to Player II's strategies. Recall that the dominant column strategy is the one with the smaller payoffs. Thus, now that α_2 is eliminated, $\beta_1 > \beta_2$ and $\beta_1 > \beta_3$. The recessive strategies can be eliminated, giving us the following:

$$
\begin{array}{c}
 & \text{II} \\
 & \begin{array}{ccc} \beta_1 & \beta_2 & \beta_3 \end{array} \\
\text{I}\begin{array}{c} \alpha_1 \\ \alpha_2 \\ \alpha_3 \end{array} \left[\begin{array}{ccc} 3 & 4 & 7 \\ 2 & 1 & 3 \\ 2 & 5 & 3 \end{array}\right]
\end{array}
$$

Returning to Player I, now that β_2 and β_3 are also eliminated, we see that $\alpha_1 > \alpha_3$. Eliminating α_3 we have the following:

$$
\begin{array}{c}
 & \text{II} \\
 & \begin{array}{ccc} \beta_1 & \beta_2 & \beta_3 \end{array} \\
\text{I}\begin{array}{c} \alpha_1 \\ \alpha_2 \\ \alpha_3 \end{array} \left[\begin{array}{ccc} 3 & 4 & 7 \\ 2 & 1 & 3 \\ 2 & 5 & 3 \end{array}\right]
\end{array}
$$

Player I will always play α_1, and Player II will always play β_1. Player I will gain 3, and Player II will lose 3. The optimum pure strategies are (α_1, β_1) with $v = 3$.

3.

$$
\begin{array}{c}
 & \text{II} \\
 & \begin{array}{cccc} \beta_1 & \beta_2 & \beta_3 & \beta_4 \end{array} \\
\text{I}\begin{array}{c} \alpha_1 \\ \alpha_2 \\ \alpha_3 \end{array} \left[\begin{array}{cccc} 7 & -8 & 1 & 2 \\ 2 & 5 & 5 & 4 \\ 9 & 7 & 8 & 5 \end{array}\right]
\end{array}
$$

Comparing rows we see that $\alpha_3 > \alpha_1$ and $\alpha_3 > \alpha_2$. We eliminate α_1 and α_2.

$$
\begin{array}{c}
 & \text{II} \\
 & \begin{array}{cccc} \beta_1 & \beta_2 & \beta_3 & \beta_4 \end{array} \\
\text{I}\begin{array}{c} \alpha_1 \\ \alpha_2 \\ \alpha_3 \end{array} \left[\begin{array}{cccc} 7 & -8 & 1 & 2 \\ 2 & 5 & 5 & 4 \\ 9 & 7 & 8 & 5 \end{array}\right]
\end{array}
$$

3. (continued)

Comparing columns, we see that $\beta_4 > \beta_1$, $\beta_4 > \beta_2$, and $\beta_4 > \beta_3$. We eliminate β_1, β_2 and β_3.

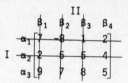

The optimum pure strategies are (α_3, β_4) with $v = 5$.

5.

$$
\begin{array}{c}
 & \text{II} \\
 & \begin{array}{ccc} \beta_1 & \beta_2 & \beta_3 \end{array} \\
\text{I}\begin{array}{c} \alpha_1 \\ \alpha_2 \\ \alpha_3 \\ \alpha_4 \end{array} \left[\begin{array}{ccc} 7 & 2 & 3 \\ 5 & 4 & 6 \\ 2 & 4 & 5 \\ 6 & 1 & 4 \end{array}\right]
\end{array}
$$

Comparing rows, we see that $\alpha_2 > \alpha_3$. Eliminate α_3.

$$
\begin{array}{c}
 & \text{II} \\
 & \begin{array}{ccc} \beta_1 & \beta_2 & \beta_3 \end{array} \\
\text{I}\begin{array}{c} \alpha_1 \\ \alpha_2 \\ \alpha_3 \\ \alpha_4 \end{array} \left[\begin{array}{ccc} 7 & 2 & 3 \\ 5 & 4 & 6 \\ 2 & 4 & 5 \\ 6 & 1 & 4 \end{array}\right]
\end{array}
$$

Comparing columns, we see that $\beta_2 > \beta_1$ and $\beta_2 > \beta_3$. Eliminate β_1 and β_3.

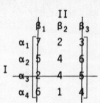

Comparing rows again, we see that $\alpha_2 > \alpha_1$ and $\alpha_2 > \alpha_4$. Eliminate α_1 and α_4.

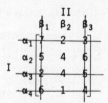

The optimum pure strategies are (α_2, β_2) with $v = 4$.

7.

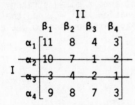

Comparing rows, we see that $\alpha_1 > \alpha_2$ and $\alpha_1 > \alpha_3$. Eliminate α_2 and α_3.

$$
\begin{array}{c}
 & & \text{II} \\
 & & \beta_1 \quad \beta_2 \quad \beta_3 \quad \beta_4 \\
 & \alpha_1 & \begin{bmatrix} 11 & 8 & 4 & 3 \end{bmatrix} \\
 & \alpha_2 & \begin{bmatrix} 10 & 7 & 1 & 2 \end{bmatrix} \\
\text{I} & \alpha_3 & \begin{bmatrix} 3 & 4 & 2 & 1 \end{bmatrix} \\
 & \alpha_4 & \begin{bmatrix} 9 & 8 & 7 & 3 \end{bmatrix}
\end{array}
$$

Comparing columns, we see that $\beta_4 > \beta_1$, $\beta_4 > \beta_2$, and $\beta_4 > \beta_3$. Eliminate β_1, β_2, and β_3.

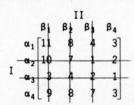

Comparing rows again, we see that α_1 and α_4 have the same value. Thus, the optimum pure strategies are (α_1, β_4) and (α_4, β_4) with $v = 3$.

9. First, write the payoff matrix. Let N = spending 2 days in the northern district and S = spending 2 days in the southern district. Let the rows correspond to Roy's strategies and the columns to Cal's strategies. Since this is a constant-sum game (the sum of the payoffs is always 1), we can write the entries of the matrix as the payoffs to Roy.

$$
\begin{array}{c}
 & & \text{Cal} \\
 & & \text{N} \quad\quad \text{S} \\
\text{Roy} & \text{N} & \begin{bmatrix} 0.65 & 0.55 \end{bmatrix} \\
 & \text{S} & \begin{bmatrix} 0.4 & 0.3 \end{bmatrix}
\end{array}
$$

We see that the first row dominates the second row, so we can eliminate row 2.

$$
\begin{array}{c}
 & & \text{Cal} \\
 & & \text{N} \quad\quad \text{S} \\
\text{Roy} & \text{N} & \begin{bmatrix} 0.65 & 0.55 \end{bmatrix} \\
 & \text{S} & \begin{bmatrix} 0.4 & 0.3 \end{bmatrix}
\end{array}
$$

Now the second column dominates the first. Eliminate column 1.

$$
\begin{array}{c}
 & & \text{Cal} \\
 & & \text{N} \quad\quad \text{S} \\
\text{Roy} & \text{N} & \begin{bmatrix} 0.65 & 0.55 \end{bmatrix} \\
 & \text{S} & \begin{bmatrix} 0.4 & 0.3 \end{bmatrix}
\end{array}
$$

The optimum pure strategies are (N, S). Roy should campaign two days in the northern district, and Cal should campaign two days in the southern district. Roy will get 55% of the vote.

11. Comparing rows we see that row 3 dominates both row 1 and row 2. Eliminate rows 1 and 2.

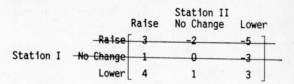

Now, column 2 dominates both columns 1 and 3. Eliminate the recessive columns.

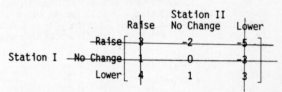

The optimum pure strategies are (Lower, No Change). Station I should lower its prices and station II should make no change. 1% of the business will go from station II to station I.

13. We found the payoff matrix in Exercise Set 10.1, Exercise 3.

$$
\begin{array}{c}
 & & \text{Player II} \\
 & & \text{S} \quad\quad\quad\quad \text{D} \\
 & \text{S} & \begin{bmatrix} (10,\ 10) & (10,\ 1{,}000{,}000) \end{bmatrix} \\
\text{Player I} & & \\
 & \text{D} & \begin{bmatrix} (1{,}000{,}000,\ 10) & (-1,\ -1) \end{bmatrix}
\end{array}
$$

If we start with the strategy pair SS with payoff (10, 10), then

a) Player I can gain by switching his strategy from S to D, so that the strategy pair is now DS with payoff (1,000,000, 10), or

b) Player II can gain by switching his strategy from S to D, so that the strategy pair is now SD with payoff (10, 1,000,000).

Similarly, if we start with the strategy pair DD with payoff (-1, -1), then

a) Player I can gain by switching his strategy from D to S, so that the strategy pair is now SD with payoff (10, 1,000,000), or

b) Player II can gain by switching his strategy from D to S, so that the strategy pair is now DS with payoff (1,000,000, 10).

If we start with the strategy pair SD with payoff (10, 1,000,000) or DS with payoff (1,000,000, 10), then neither player can gain by a unilateral change of strategy. Then SD and DS are the equilibrium pairs.

Exercise Set 10.3

1.

	II β₁	β₂	Row minimum	Maximum minimum
α₁	1	3	1	
α₂	4	2	2	2

I

Column maximum   4   3

Minimum maximum   3

There are no recessive strategies. Maximin = 2 and minimax = 3. Since the maximin and minimax values are not equal, we must seek a solution in terms of mixed strategies. Using the equation $XPY^t = v$ and the fact that X is a probability distribution, we get

i) $\quad x_1 + 4x_2 = v$

ii) $\quad 3x_1 + 2x_2 = v$

iii) $\quad x_1 + x_2 = 1, x_1 \geqslant 0, x_2 \geqslant 0.$

Subtracting ii) from i) we have

iv) $\quad -2x_1 + 2x_2 = 0.$

Solving the system of equations iii) and iv), we obtain the solution $x_1 = 1/2$, $x_2 = 1/2$. Substituting in either i) or ii) we find $v = 5/2$. Similarly,

i') $\quad y_1 + 3y_2 = v$

ii') $\quad 4y_1 + 2y_2 = v$

iii') $\quad y_1 + y_2 = 1, y_1 \geqslant 0, y_2 \geqslant 0.$

The solution is $y_1 = 1/4$, $y_2 = 3/4$, $v = 5/2$. The solution of the game is $x_1 = 1/2$, $x_2 = 1/2$, $y_1 = 1/4$, $y_2 = 3/4$, $v = 5/2$.

3.

	II β₁	β₂
α₁	3	5
α₂	2	6

I

Neither row is dominant. However, $\beta_1 > \beta_2$ so we can eliminate β_2.

	II β₁	β₂
α₁	3	5
α₂	2	6

I

Now $\alpha_1 > \alpha_2$. Eliminate α_2.

	II β₁	β₂
α₁	3	5
α₂	2	6

I

We have a pure strategy solution. Player I should always choose α_1 and Player II should always choose β_1. Thus, $x_1 = 1$, $x_2 = 0$, $y_1 = 1$, $y_2 = 0$, $v = 3$.

5.

	II β₁	β₂	Row minimum	Maximum minimum
α₁	1.3	-2.5	-2.5	
α₂	0.6	1.9	0.6	0.6

I

Column maximum   1.3   1.9

Minimum maximum   1.3

Since maximin = 0.6 and minimax = 1.3 (the values are not equal) we must seek a mixed strategy solution. Use the equation $XPY^t = v$ and the fact that X is a probability distribution to get a system of equations.

$\quad 1.3x_1 + 0.6x_2 = v$

$\quad -2.5x_1 + 1.9x_2 = v$

$\quad x_1 + x_2 = 1, x_1 \geqslant 0, x_2 \geqslant 0$

The solution is $x_1 = 13/51$, $x_2 = 38/51$, $v = 397/510$.

Similarly, we have

$\quad 1.3y_1 - 2.5y_2 = v$

$\quad 0.6y_1 + 1.9y_2 = v$

$\quad y_1 + y_2 = 1, y_1 \geqslant 0, y_2 \geqslant 0.$

The solution is $y_1 = 44/51$, $y_2 = 7/51$, $v = 397/510$.

The solution of the game is $x_1 = 13/51$, $x_2 = 38/51$, $y_1 = 44/51$, $y_2 = 7/51$, $v = 397/510$.

7.

	β₁,y₁	β₂,y₂	β₃,y₃	β₄,y₄
α₁, x₁	2	4	-1	5
α₂, x₂	3	7	2	4
α₃, x₃	2	3	8	9

No row dominates any other. However, $\beta_1 > \beta_2$ and $\beta_1 > \beta_4$, so we can eliminate β_2 and β_4.

	β₁,y₁	β₂,y₂	β₃,y₃	β₄,y₄
α₁, x₁	2	4	-1	5
α₂, x₂	3	7	2	4
α₃, x₃	2	3	8	9

Now $\alpha_2 > \alpha_1$, so we can eliminate α_1.

	β₁,y₁	β₂,y₂	β₃,y₃	β₄,y₄
α₁, x₁	2	4	-1	5
α₂, x₂	3	7	2	4
α₃, x₃	2	3	8	9

Neither remaining column dominates the other. We can continue the solution as for a 2 × 2 zero-sum or constant-sum game, assigning the probability of each recessive strategy the value 0.

We first get the following system of equations:

$\quad 3x_2 + 2x_3 = v$

$\quad 2x_2 + 8x_3 = v$

$\quad x_2 + x_3 = 1, x_2 \geqslant 0, x_3 \geqslant 0$

$\quad x_1 = 0$

The solution is $x_1 = 0$, $x_2 = 6/7$, $x_3 = 1/7$, $v = 20/7$.

<u>7</u>. (continued)

Similarly, we get
$$3y_1 + 2y_3 = v$$
$$2y_1 + 8y_3 = v$$
$$y_1 + y_3 = 1, \ y_1 \geqslant 0, \ y_3 \geqslant 0$$
$$y_2, y_4 = 0.$$

The solution is $y_1 = 6/7$, $y_2 = 0$, $y_3 = 1/7$, $y_4 = 0$, $v = 20/7$. The solution of the game is $x_1 = 0$, $x_2 = 6/7$, $x_3 = 1/7$, $y_1 = 6/7$, $y_2 = 0$, $y_3 = 1/7$, $y_4 = 0$, $v = 20/7$.

<u>9</u>.

	β_1, y_1	β_2, y_2	β_3, y_3	β_4, y_4
α_1, x_1	5	1	3	5
α_2, x_2	3	2	7	1
α_3, x_3	4	4	8	0
α_4, x_4	6	-1	0	5

No row dominates any other row. However, $\beta_2 > \beta_1$ and $\beta_2 > \beta_3$, so we can eliminate β_1 and β_3.

	β_1, y_1	β_2, y_2	β_3, y_3	β_4, y_4
α_1, x_1	5	1	3	5
α_2, x_2	3	2	7	1
α_3, x_3	4	4	8	0
α_4, x_4	6	-1	0	5

Now, $\alpha_1 > \alpha_4$, so we can eliminate α_4.

	β_1, y_1	β_2, y_2	β_3, y_3	β_4, y_4
α_1, x_1	5	1	3	5
α_2, x_2	3	2	7	1
α_3, x_3	4	4	8	0
α_4, x_4	6	-1	0	5

We can find no more recessive strategies. We now proceed to solve as for a 3×2 zero-sum or constant-sum game, assigning the probability of each recessive strategy the value 0. The mixed strategies for Player I can be obtained from
$$x_1 + 2x_2 + 4x_3 \geqslant v,$$
$$5x_1 + x_2 + 0x_3 \geqslant v,$$
$$x_1 + x_2 + x_3 = 1, \ x_1 \geqslant 0, \ x_2 \geqslant 0, \ x_3 \geqslant 0,$$
Max v.

Divide the constraints by v and make the substitutions $Z = \dfrac{X}{v}$, $z_0 = \dfrac{1}{v}$. We now have
$$z_1 + 2z_2 + 4z_3 \geqslant 1,$$
$$5z_1 + z_2 + 0z_3 \geqslant 1,$$
Min $z_0 = z_1 + z_2 + z_3$; $z_1, z_2, z_3 \geqslant 0.$

<u>9</u>. (continued)

The mixed strategies for Player II can be obtained from
$$y_2 + 5y_4 \leqslant v,$$
$$2y_2 + y_4 \leqslant v,$$
$$4y_2 + 0y_4 \leqslant v,$$
$$y_2 + y_4 = 1, \ y_2 \geqslant 0, \ y_4 \geqslant 0,$$
Min v.

Divide the constraints by v and make the substitutions $U = \dfrac{Y}{v}$, $u_0 = \dfrac{1}{v}$. We now have
$$u_2 + 5u_4 \leqslant 1,$$
$$2u_2 + u_4 \leqslant 1,$$
$$4u_2 + 0u_4 \leqslant 1,$$
Max $u_0 = u_2 + u_4$; $u_2, u_4 \geqslant 0.$

The linear programs in z and u are dual to each other. We can find the solution using the simplex method of Chapter 4.

u_2	u_4	z_1	z_2	z_3	u_0	1	q
1	5	1	0	0	0	1	1
2	1	0	1	0	0	1	1/2
4*	0	0	0	1	0	1	1/4 ← Min
①	-1	0	0	0	1	0	
0	20*	4	0	-1	0	3	3/20 ← Min
0	2	0	2	-1	0	1	1/2
4	0	0	0	1	0	1	-
0	④	0	0	1	4	1	
0	20	4	0	-1	0	3	
0	0	-4	20	-9	0	7	
4	0	0	0	1	0	1	
0	0	4	0	4	20	8	

Reading off the solutions for the primal U from the right-hand column and for the dual Z from the bottom row, we have
$$u_2 = 1/4, \ u_4 = 3/20, \ u_0 = 8/20 = 2/5,$$
$$z_1 = 4/20 = 1/5, \ z_2 = 0/20 = 0,$$
$$z_3 = 4/20 = 1/5, \ z_0 = 8/20 = 2/5.$$

Converting from Z and U to X and Y we have
$$x_1 = \frac{z_1}{z_0} = \frac{\frac{1}{5}}{\frac{2}{5}} = \frac{1}{2}, \quad x_2 = \frac{z_2}{z_0} = \frac{0}{\frac{2}{5}} = 0,$$
$$x_3 = \frac{z_3}{z_0} = \frac{\frac{1}{5}}{\frac{2}{5}} = \frac{1}{2}, \quad v = \frac{1}{z_0} = \frac{1}{\frac{2}{5}} = \frac{5}{2} \text{ and}$$
$$y_2 = \frac{u_2}{u_0} = \frac{\frac{1}{4}}{\frac{2}{5}} = \frac{5}{8}, \quad y_4 = \frac{u_4}{u_0} = \frac{\frac{3}{20}}{\frac{2}{5}} = \frac{3}{8},$$
$$v = \frac{1}{u_0} = \frac{1}{\frac{2}{5}} = \frac{5}{2}.$$

9. (continued)

The solution of the game is $x_1 = 1/2$, $x_2 = 0$, $x_3 = 1/2$, $x_4 = 0$, $y_1 = 0$, $y_2 = 5/8$, $y_3 = 0$, $y_4 = 3/8$, $v = 5/2$.

11.

		β_1, y_1	β_2, y_2	β_3, y_3	β_4, y_4
α_1,	x_1	0.5	-0.2	0.2	-0.1
α_2,	x_2	0.4	2.5	3.4	2.6
α_3,	x_3	-2.7	-1.5	0.1	-1.7
α_4,	x_4	0.5	-3.0	0.5	1.5

$\alpha_1 > \alpha_3$, so we can eliminate α_3.

		β_1, y_1	β_2, y_2	β_3, y_3	β_4, y_4
α_1,	x_1	0.5	-0.2	0.2	-0.1
α_2,	x_2	0.4	2.5	3.4	2.6
~~α_3~~,	x_3	~~-2.7~~	~~-1.5~~	~~0.1~~	~~-1.7~~
α_4,	x_4	0.5	-3.0	0.5	1.5

Now $\beta_2 > \beta_3$ and $\beta_2 > \beta_4$, so we can eliminate β_3 and β_4.

		β_1, y_1	β_2, y_2	β_3, y_3	β_4, y_4
α_1,	x_1	0.5	-0.2	0.2	-0.1
α_2,	x_2	0.4	2.5	3.4	2.6
~~α_3~~,	x_3	~~-2.7~~	~~-1.5~~	~~0.1~~	~~-1.7~~
α_4,	x_4	0.5	-3.0	0.5	1.5

$\alpha_1 > \alpha_4$, so we can eliminate α_4.

		β_1, y_1	β_2, y_2	β_3, y_3	β_4, y_4
α_1,	x_1	0.5	-0.2	0.2	-0.1
α_2,	x_2	0.4	2.5	3.4	2.6
α_3,	x_3	-2.7	-1.5	0.1	-1.7
α_4,	x_4	0.5	-3.0	0.5	1.5

We can find no more recessive strategies. We proceed as for a 2 × 2 zero-sum or constant-sum game, assigning the probability of each recessive strategy the value 0. We obtain a system of equations:

$$0.5x_1 + 0.4x_2 = v$$
$$-0.2x_1 + 2.5x_2 = v$$
$$x_1 + x_2 = 1, \; x_1 \geqslant 0, \; x_2 \geqslant 0$$

The solution is $x_1 = 3/4$, $x_2 = 1/4$, $v = 19/40$.

Similarly, we obtain another system of equations:

$$0.5y_1 - 0.2y_2 = v$$
$$0.4y_1 + 2.5y_2 = v$$
$$y_1 + y_2 = 1, \; y_1 \geqslant 0, \; y_2 \geqslant 0$$

The solution is $y_1 = 27/28$, $y_2 = 1/28$, $v = 19/40$.

The solution of the game is $x_1 = 3/4$, $x_2 = 1/4$, $x_3 = 0$, $x_4 = 0$, $y_1 = 27/28$, $y_2 = 1/28$, $y_3 = 0$, $y_4 = 0$, $v = 19/40$.

13.

		β_1, y_1	β_2, y_2	β_3, y_3	β_4, y_4
α_1,	x_1	1	4	7	5
α_2,	x_2	2	1	-2	-1

There are no recessive strategies. The mixed strategies for Player I can be obtained from

$$x_1 + 2x_2 \geqslant v,$$
$$4x_1 + x_2 \geqslant v,$$
$$7x_1 - 2x_2 \geqslant v,$$
$$5x_1 - x_2 \geqslant v,$$
$$x_1 + x_2 = 1, \; x_1 \geqslant 0, \; x_2 \geqslant 0,$$
Max v.

Divide the constraints by v and make the substitutions $Z = \dfrac{X}{v}$, $z_0 = \dfrac{1}{v}$. We obtain

$$z_1 + 2z_2 \geqslant 1,$$
$$4z_1 + z_2 \geqslant 1,$$
$$7z_1 - 2z_2 \geqslant 1,$$
$$5z_1 - z_2 \geqslant 1,$$
Min $z_0 = z_1 + z_2$; $z_1, z_2 \geqslant 0$.

The mixed strategies for Player II can be obtained from

$$y_1 + 4y_2 + 7y_3 + 5y_4 \leqslant v,$$
$$2y_1 + y_2 - 2y_3 - y_4 \leqslant v,$$
$$y_1 + y_2 + y_3 + y_4 = 1,$$
$$y_1 \geqslant 0, \; y_2 \geqslant 0, \; y_3 \geqslant 0, \; y_4 \geqslant 0,$$
Min v.

Divide the constraints by v and make the substitutions $U = \dfrac{Y}{v}$, $u_0 = \dfrac{1}{v}$. We obtain

$$u_1 + 4u_2 + 7u_3 + 5u_4 \leqslant 1,$$
$$2u_1 + u_2 - 2u_3 - u_4 \leqslant 1,$$
Max $u_0 = u_1 + u_2 + u_3 + u_4$;
$$u_1, u_2, u_3, u_4 \geqslant 0.$$

These linear programs in Z and U are dual to each other and can be solved using the simplex method. The solution is $u_1 = 6/11$, $u_2 = 0$, $u_3 = 0$, $u_4 = 1/11$, $u_0 = 7/11$ and $z_1 = 3/11$, $z_2 = 4/11$, $z_0 = 7/11$.

Converting from Z and U to X and Y, we obtain

$$x_1 = \frac{z_1}{z_0} = \frac{\frac{3}{11}}{\frac{7}{11}} = \frac{3}{7}, \; x_2 = \frac{z_2}{z_0} = \frac{\frac{4}{11}}{\frac{7}{11}} = \frac{4}{7},$$

$$v = \frac{1}{z_0} = \frac{11}{7} \text{ and } y_1 = \frac{u_1}{u_0} = \frac{\frac{6}{11}}{\frac{7}{11}} = \frac{6}{7},$$

$$y_2 = \frac{u_2}{u_0} = \frac{0}{\frac{7}{11}} = 0, \; y_3 = \frac{u_3}{u_0} = \frac{0}{\frac{7}{11}} = 0,$$

$$y_4 = \frac{u_4}{u_0} = \frac{\frac{1}{11}}{\frac{7}{11}} = \frac{1}{7}, \; v = \frac{1}{u_0} = \frac{11}{7}.$$

15. We found the payoff matrix in Exercise Set 10.1, Exercise 1.

$$\begin{array}{cc} & \text{II} \\ & \begin{array}{cc} 1 & 2 \end{array} \\ \text{I} \begin{array}{c} 1 \\ 2 \end{array} & \begin{bmatrix} 5 & -5 \\ -5 & 5 \end{bmatrix} \end{array}$$

This is a 2 × 2 zero-sum game. The first step in the solution is to solve the following system of equations:

$$5x_1 - 5x_2 = v$$
$$-5x_1 + 5x_2 = v$$
$$x_1 + x_2 = 1, \ x_1 \geqslant 0, \ x_2 \geqslant 0$$

The solution is $x_1 = 1/2$, $x_2 = 1/2$, $v = 0$.

Next, solve the following system of equations:

$$5y_1 - 5y_2 = v$$
$$-5y_1 + 5y_2 = v$$
$$y_1 + y_2 = 1, \ y_1 \geqslant 0, \ y_2 \geqslant 0$$

The solution is $y_1 = 1/2$, $y_2 = 1/2$, $v = 0$.

Each player should show one finger one-half of the time. The value of the game is 0.

17. Let C = going to the country and S = going to the stream. The payoff matrix is

$$\begin{array}{cc} & \text{Inspector} \\ & \begin{array}{cc} C & \ \ S \end{array} \\ \text{Plant} \begin{array}{c} C \\ S \end{array} & \begin{bmatrix} -100 & 200 \\ 200 & -2500 \end{bmatrix} \end{array}$$

We first solve the following system of equations:

$$-100x_1 + 200x_2 = v$$
$$200x_1 - 2500x_2 = v$$
$$x_1 + x_2 = 1, \ x_1 \geqslant 0, \ x_2 \geqslant 0$$

The solution is $x_1 = 9/10$, $x_2 = 1/10$, $v = -70$.

Next, solve this system of equations:

$$-100y_1 + 200y_2 = v$$
$$200y_1 - 2500y_2 = v$$
$$y_1 + y_2 = 1, \ y_1 \geqslant 0, \ y_2 \geqslant 0$$

The solution is $y_1 = 9/10$, $y_2 = 1/10$, $v = -70$.

The first system of equations tells us that the plant should dump in the country 9/10 of the time and in the stream 1/10 of the time. The second system of equations tells us that the inspector should go to the country 9/10 of the time and to the stream 1/10 of the time. The value of the game is -$70.

19. The first row of the matrix dominates the second, so we can eliminate the second row.

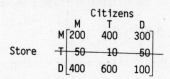

Column M dominates column T, so we can eliminate column T.

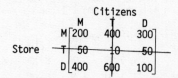

There are no more recessive strategies. We continue as for a 2 × 2 zero-sum or constant-sum game. First, solve the following system of equations:

$$200x_1 + 400x_3 = v$$
$$300x_1 + 100x_3 = v$$
$$x_1 + x_3 = 1, \ x_1 \geqslant 0, \ x_2 \geqslant 0$$

The solution is $x_1 = 3/4$, $x_3 = 1/4$, $v = 250$.

Next solve this system of equations:

$$200y_1 + 300y_3 = v$$
$$400y_1 + 100y_3 = v$$
$$y_1 + y_3 = 1, \ y_1 \geqslant 0, \ y_2 \geqslant 0$$

The solution is $y_1 = 1/2$, $y_3 = 1/2$, $v = 250$.

The first system of equations tells us that the store should use 3/4 mail and 1/4 door-to-door. The second system tells us that the citizen's coalition should use 1/2 mail and 1/2 door-to-door. The store can expect to get 250 signatures ($v = 250$). This is not enough to open.

21. This is a game against nature. The spring is not a volitional player nor is the game zero-sum. (There is no payoff to the spring, in fact.)

We determine the mixed strategies for the man.

$$3x_1 + 10x_2 \geqslant v,$$
$$24x_1 + 10x_2 \geqslant v,$$
$$x_1 + x_2 = 1, \ x_1 \geqslant 0, \ x_2 \geqslant 0,$$
$$\text{Max } v.$$

Changing variables from X to Z, we obtain

$$3z_1 + 10z_2 \geqslant 1,$$
$$24z_1 + 10z_2 \geqslant 1,$$
$$\text{Min } z_0 = z_1 + z_2; \ z_1, z_2 \geqslant 0.$$

The dual to this is

$$3u_1 + 24u_2 \leqslant 1,$$
$$10u_1 + 10u_2 \leqslant 1,$$
$$\text{Max } u_0 = u_1 + u_2; \ u_1, u_2 \geqslant 0.$$

The solution is $z_1 = 0$, $z_2 = 1/10$, $z_0 = 1/10$.

21. (continued)

Changing back to the original variables we have

$$x_1 = \frac{z_1}{z_0} = \frac{0}{\frac{1}{10}} = 0, \quad x_2 = \frac{z_2}{z_0} = \frac{\frac{1}{10}}{\frac{1}{10}} = 1,$$

$$v = \frac{1}{z_0} = 10.$$ Thus, the man should not drink.

His expected time of survival is 10 hours ($v = 10$).

23. No row dominates another. However column II dominates column III, so column III can be eliminated.

$$\begin{array}{cc} & \text{Champion} \\ & \begin{array}{ccc} \text{I} & \text{II} & \text{III} \end{array} \\ \text{Challenger} \begin{array}{c} \text{I} \\ \text{II} \\ \text{III} \end{array} & \left[\begin{array}{ccc} 0.85 & 0.20 & 0.30 \\ 0.10 & 0.45 & 0.45 \\ 0.20 & 0.25 & 0.30 \end{array}\right] \end{array}$$

There are no more recessive strategies. We proceed as for a 3×2 zero-sum or constant-sum game.

The mixed strategies for the challenger can be obtained from

$$0.85x_1 + 0.10x_2 + 0.20x_3 \geqslant v,$$
$$0.20x_1 + 0.45x_2 + 0.25x_3 \geqslant v,$$
$$x_1 + x_2 + x_3 = 1,$$
$$x_1 \geqslant 0, \ x_2 \geqslant 0, \ x_3 \geqslant 0,$$
Max v.

Divide the constraints by v and make the substitutions $Z = \frac{X}{v}$, $z_0 = \frac{1}{v}$.

$$0.85z_1 + 0.10z_2 + 0.20z_3 \geqslant 1,$$
$$0.20z_1 + 0.45z_2 + 0.25z_3 \geqslant 1,$$
Min $z_0 = z_1 + z_2 + z_3$;
$$z_1, z_2, z_3 \geqslant 0$$

The mixed strategies for the champion can be obtained from

$$0.85y_1 + 0.20y_2 \leqslant v,$$
$$0.10y_1 + 0.45y_2 \leqslant v,$$
$$0.20y_1 + 0.25y_2 \leqslant v,$$
$$y_1 + y_2 = 1, \ y_1 \geqslant 0, \ y_2 \geqslant 0,$$
Min v.

Divide the constraints by v and make the substitution $U = \frac{Y}{v}$, $u_0 = \frac{1}{v}$.

$$0.85u_1 + 0.20u_2 \leqslant 1,$$
$$0.10u_1 + 0.45u_2 \leqslant 1,$$
$$0.20u_1 + 0.25u_2 \leqslant 1,$$
Max $u_0 = u_1 + u_2$; $u_1, u_2 \geqslant 0$

The linear programs in Z and U are dual to each other and can be solved using the simplex method. The solution is $u_1 = \frac{0.25}{0.3625}$, $u_2 = \frac{0.75}{0.3625}$,

$u_0 = \frac{1}{0.3625}$ and $z_1 = \frac{0.35}{0.3625}$, $z_2 = \frac{0.65}{0.3625}$,

$z_3 = 0$, $z_0 = \frac{1}{0.3625}$.

23. (continued)

Converting from Z and U to X and Y, we obtain $x_1 = 0.35$, or $7/20$; $x_2 = 0.65$, or $13/25$, $x_3 = 0$, $v = 0.3625$, or $29/80$ and $y_1 = 0.25$, or $1/4$, $y_2 = 0.75$, or $3/4$; $v = 0.3625$, or $29/80$. Thus, the challenger should prepare for serve I $7/20$ of the time and for serve II $13/20$ of the time. The champion should use serve I $1/4$ of the time and serve II $3/4$ of the time.

25. Let p = the payoff for not getting caught that makes the game fair. The payoff matrix is

$$\begin{array}{cc} & \text{Inspector} \\ & \begin{array}{cc} \text{C} & \quad \text{S} \end{array} \\ \text{Plant} \begin{array}{c} \text{C} \\ \text{S} \end{array} & \left[\begin{array}{cc} -100 & p \\ p & -2500 \end{array}\right] \end{array}$$

For payoff p, $v = 0$, so we get the following system of equations:

i) $-100x_1 + px_2 = 0$

ii) $px_1 - 2500x_2 = 0$

iii) $x_1 + x_2 = 1$

Solve i) for x_1: $-100x_1 = -px_2$

$$x_1 = \frac{px_2}{100}$$

Substitute $\frac{px_2}{100}$ for x_1 in ii) and solve for p:

$$\frac{p^2 x_2}{100} - 2500x_2 = 0$$
$$\frac{p^2 x_2}{100} = 2500x_2$$
$$\frac{p^2}{100} = 2500 \quad (x_2 \neq 0)$$
$$p^2 = 250,000$$
$$p = 500$$

(For $p = 500$, $x_1 = 5/6$ and $x_2 = 1/6 \neq 0$, so our solution is valid.)

For the game to be fair, the payoff for not getting caught should be $500.